Nature's Geography

Nature's Geography

New Lessons for Conservation in Developing Countries

Edited by

Karl S. Zimmerer and Kenneth R. Young

THE UNIVERSITY OF WISCONSIN PRESS

The University of Wisconsin Press
2537 Daniels Street
Madison, Wisconsin 53718

3 Henrietta Street
London WC2E 8LU, England

5   4   3   2   1

Printed in the United States of America

Library of Congress Cataloging-in-Publication Data
Nature's geography: new lessons for conservation in developing
    countries / edited by Karl S. Zimmerer and Kenneth R. Young.
        368 p.   cm.
    Includes bibliographical references and index.
    ISBN 0-299-15910-8 (cloth : alk. paper).
    ISBN 0-299-15914-0 (pbk. : alk. paper)
    1. Landscape changes—Developing countries.   2. Biological
    diversity conservation—Developing countries.   3. Conservation of
    natural resources—Developing countries.   4. Economic development—
    Environmental aspects.   I. Zimmerer, Karl S. II. Young, Kenneth R.
    GF900.N37   1998
    333.7′2′091724—dc21   98-25743

Dedicated to those who live and work in the landscapes of developing countries, including the many who have helped and guided us over the years.

Contents

Figures

Maps

Tables

Contributors

Mark Blumler is assistant professor of geography at the State University of New York at Binghamton. He studies succession, ecology of wild cereal grasses, Mediterranean environments, plant species conservation, and environmental history. His articles have appeared in such journals as *Current Anthropology, Economic Botany,* and *Fremontia: Journal of the California Native Plant Society.*

Barbara Brower is associate professor of geography at Portland State University. She studies mountain peoples, resource policy, and environmental history in the Himalaya and the American West, and edits *Himalayan Research Bulletin.* Her publications include *The Sherpa of Khumbu: People, Livestock, and Landscape* (Oxford University Press, 1991) and articles in *Mountain Research and Development* and *Society and Natural Resources.*

Ann Dennis is research plant ecologist at the USDA Forest Service Pacific Southwest Research Station. Her research concerns human impacts on plant biodiversity, especially the effects of logging and grazing on flora of the Sierra Nevada, and on the development of electronic information resources appropriate at the landscape scale. Her most recent work appears in *Conservation in Highly Fragmented Landscapes* (Chapman and Hall, 1997).

Fernando R. Echavarria is currently an American Association for the Advancement of Science Diplomacy Fellow in the U.S. Department of State's Office of Ecology. As assistant professor of geography at the University of Nebraska–Lincoln, his teaching and research focuses on the use of remote sensing for mapping forests, land use, and land cover change in the Neotropics. His recent publications have appeared in *Revista Geográfica, Geocarto International,* and in specialized journals on remote sensing and GIS.

Sally P. Horn is associate professor of geography, and an adjunct professor in the departments of Ecology and Evolutionary Biology, and Botany, at the University of Tennessee–Knoxville. She studies fire ecology and vegetation, fire, and climate history in Latin America and the Caribbean. Her publications have appeared in *Quarternary Research, The Holocene, Biotropica,* and *Climate Research.*

Kimberly E. Medley is associate professor of geography at Miami University, Oxford, Ohio. She studies the human and environmental factors affecting forest biogeography, ecology, and conservation in the United States, Kenya, and Madagascar. She has recently published in such journals as *Conservation Biology, Economic Botany,* and *Professional Geographer.*

John Metz is associate professor of geography at Northern Kentucky University. He has studied farming systems, forest product use, and forest ecology in western and central Nepal. His recent publications have appeared in *Geographical Review, Environmental Management, Human Ecology,* and *World Development.*

Francisco L. Pérez is associate professor of geography at the University of Texas–Austin. His research focuses on the geoecology of high-mountain areas in the Venezuelan Andes, the Cascade Mountains, Haleakala in Hawaii, and the Hill Country of Central Texas. His research has appeared in such journals as *Geoderma, Geografiska Annaler, Journal of Biogeography, Physical Geography, Plant Ecology,* and *Zeitschrift für Geomorphologie.*

Alan H. Taylor is associate professor of geography at Pennsylvania State University. His interests include the vegetation dynamics, disturbance ecology, and biological conservation of montane forests in southwestern China, Rwanda, and western North America. His work has appeared in numerous journals, including *Nature, Journal of Biogeography, Journal of Ecology,* and *Canadian Journal of Forest Research.*

Matthew D. Turner is assistant professor of geography at the University of Wisconsin–Madison. He studies rangeland dynamics, land-use change, and human and political ecology in semiarid West Africa. His recent publications have appeared in *Economic Geography, Journal of Biogeography, Journal of Applied Ecology,* and *International Journal of Remote Sensing.*

Robert Voeks is professor of geography at California State University–Fullerton. His main research interests are tropical forest ecology and ethnobotany. In addition to recent articles in *Economic Botany, Journal of Ethnobiology,* and *Geographical Review,* he is the author of *Sacred Leaves of Candomblé: African Magic, Medicine, and Religion in Brazil* (University of Texas Press, 1997).

Kenneth R. Young is associate professor of geography at the University of Maryland, Baltimore County. He studies the biogeography and biological conservation of tropical environments in Central and South America. He is the coeditor of *Biogeografía, ecología, y conservación del bosque montano en el Perú* (Museo de Historia Natural, Lima, 1992). His articles have appeared in such journals as *Mountain Research and Development, Conservation Biology, Physical Geography,* and *Landscape and Urban Planning.*

Karl S. Zimmerer is professor of geography, environmental studies, and development studies at the University of Wisconsin–Madison. He studies the biogeography, soil-plant ecology, and human and political ecology of landscape change. His publications include *Changing Fortunes: Biodiversity and Peasant Livelihood in the Peruvian Andes* (University of California Press, 1996) and articles in *Annals of the Association of American Geographers, Journal of Biogeography, Nature,* and *BioScience.*

Qin Zisheng is associate professor of biology at Sichuan Teacher's College in Nanchong, China. Her primary research interests are plant taxonomy, vegetation of montane forests in Sichuan, and biological conservation. She has published articles with Alan H. Taylor in such journals as *Environmental Conservation, Bulletin of the Torrey Botanical Club,* and *Vegetatio.*

Nature's Geography

Introduction
The Geographical Nature of Landscape Change

Karl S. Zimmerer and Kenneth R. Young

Places in the developing countries of Asia, Africa, and Latin America are made up of diverse environments and varied landscapes. Many people in these countries appreciate their environments and desire the sustainability of landscapes that range from the heavily utilized to mostly wild. Yet their landscapes are marked by environmental changes that are widespread and multifaceted. These environmental changes include both human-induced alterations and natural processes. Understanding such changes and responding to them must be based on the awareness that environmental change is remarkably more complex than was previously thought (Botkin 1990; Wu and Loucks 1995). This realization can be used to rethink the present and future strategies for conservation and sound resource use in developing countries. Our volume is spurred by this realization and by its call for the multiscalar and multitemporal analysis of environmental change.

The volume's analysis of the complex character of environmental change is intended to expand and, to an extent, even reformulate the current state of knowledge about nature and its relation to human societies. We approached this goal by combining a series of theoretical overviews with related groups of case studies. Our case studies cover both the landscapes that are vegetated predominantly by forests and grasslands *and* those utilized landscapes in settled places where changes affect the resource base of primary economic activities such as agriculture and livestock-raising. In the conclusion we derive comparisons among landscapes and regional settings in order to further examine the complex nature of dynamic changes.

The geographical nature of change in this diverse array of landscapes is the sustained focus of our volume. Recent thinking in geography and related fields is used to provide a new perspective on environmental change and conservation strategies that is applied to the management of protected areas and the assessment of sound resource use. Thus our volume formulates a theoretical framework that is based on landscape and regional analysis. After this initial discussion about the general aspects of our project, each subsequent part of the introduction is devoted to a theoretical overview.

A firmer knowledge of changes in the diverse landscapes of poorer countries must aid the urgent need to join environmental management that is sound with economic development that is viable in the long term. The imperative for conservation-with-development has been labeled "sustainable development." Much-publicized mandates for sustainable development echo forth as a *sine qua non* of conservation in the developing countries, a seeming panacea for the world's environmental problems. Nonetheless, sustainable development has remained a general concept and one that is subject to much debate (Meffe and Carroll 1994; Murdoch and Clark 1994; Redclift 1987; Wilbanks 1994). Indeed the more exact meanings of sustainability are typically lacking. It is clear, however, that our capacity not only for definition but also for appraising and usefully debating the prospects of sustainable development must depend on the rigorous analysis of environmental change. Such work—whether research, applied interventions, or both—will help us to better grasp the role of nature's own capacities and constraints in the possible scenarios for sustainability.

The title of this volume draws attention to its combined focus on the biogeography and human ecology of environmental changes. Geography offers much to such understanding—with its core of human-environment relations, geography integrates various areas of research that are highly relevant (figure I.1). Biogeography, which addresses the distributions and dynamics of organisms, is one subfield that contributes a much-needed perspective on the nature of environmental change. The generally similar approaches of cultural, human, and political ecology offer other key perspectives (Zimmerer 1996a). We rely on all these perspectives in our series of case studies. A comparative framework helps to explore what these findings on the geography of environmental change imply for conservation strategies. The tone of the studies reflects an overarching concern for sus-

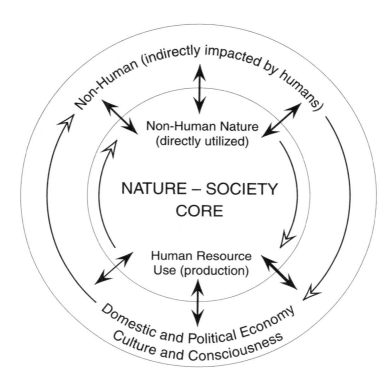

Fig. I.1. A schematic model of human-environment relations. Biogeography focuses on environmental changes that involve the distributions and dynamics of organisms. Human, political, and cultural ecology offer a perspective on the causes and consequences of anthropogenic environmental change.

tainable human relations with nature, a timely and abiding preoccupation for geography and geographers.

Our selection of case studies addresses the geographical nature of changes that are at the center of pressing problems of worldwide importance. The studies are grouped into three parts: (1) Part 1 addresses the conservation challenges presented by forest fragmentation as these change processes impact on communities of forest users and such wildlife species as endangered primates in East Africa, rare giant pandas in China, and forest ecosystems in the South American Andes; (2) Part 2 covers the conservation challenges of long- and short-term change processes due to livestock grazing, human-set fires, and other disturbances that occur in the major mountain watersheds of South Asia and Central and South America; and (3) Part 3 focuses on the conservation challenges of utilized environments such as agriculture and rangeland in the settled areas of the Near East, the Sahel of West Africa, the South American Andes, and the inhabited forests of mountains and tropical lowlands in South and Southeast Asia.

The claim to newness in our volume is based on twin foundations. In the first place the claim reflects the recentness of the geographical studies themselves. Each chapter presents an original analysis of extensive field research that has been conducted within the past ten years. As importantly, the volume offers a newness of overall perspective. Our fresh vantage point arises from a geographical integration of the new views of environmental change that are developing in the field of geography and in its cognate fields such as ecology, environmental studies, environmental history, and conservation biology.

Three interrelated views in particular combine to inform our new perspective on the geography of environmental change. These are: (1) the salience of nonequilibrium conditions in environments and change processes (even in those environments that are perceived as relatively unaltered by human activities); (2) the widespread and, in some cases, pervasive nature of human impacts on natural environments (even in the environments that may appear unutilized at first glance); and (3) the heightened awareness that environmental change analysis must also include basic resource activities such as agriculture and grazing, and thus the nature of settled areas.

Quickening awareness of these views is creating a perspective on nature and its relations to humans that is distinct, at least in degree, from those that predominated earlier. Distinct tendencies of the growing views are evidenced in a general way by a pair of recent books: *Discordant Harmonies: A New Ecology for the Twenty-first Century* by ecologist Daniel Botkin (1990) and *Uncommon Ground: Rethinking the Human Place in Nature* edited by historian William Cronon (1996a). The two books are intended to summarize and to advance the recent views, as well as to emphasize their relevance to the philosophy and practice of United States conservation. *Discordant Harmonies* and *Uncommon Ground* extend a great deal further the perspectives that were established earlier in scholarly and scientific literature (Egerton 1973; Glacken 1967; Sauer 1956, 1958 [1967]; Worster 1990; Wu and Loucks 1995). Other aspects of their arguments existed in western Europe and especially in the traditions of "geoecology" and "geobotany" (Lauer 1984, 1993; Troll 1968, 1971). Still other elements of the new views and what they mean for the future of conservation and resource use are attaining special significance in the developing countries (Gómez-

Pompa and Klaus 1992; Guha 1997). Our volume draws much inspiration from the crucial importance of such considerations as well as from the debates that surround them.

Since the triad of views mentioned above is so basic to current thought on environmental change and conservation, each view merits a brief introduction. As we raise the perspectives we attempt to show how this volume takes a particular position on each view. (Subsequent sections within the introduction describe each perspective in more detail.) We begin with the widespread recognition that nonequilibrium conditions and change processes are prevalent tendencies even in environments that may be impacted little if at all by human activities (Botkin 1990; Wu and Loucks 1995). Nonequilibrial change is to be expected in most environmental systems, which tend to be unstable to varying degrees, thus expressing what Botkin has popularized as "discordant harmonies," a metaphor adopted from the ancient Greek philosopher Plotinus. This term reflects the role of environmental stability as partial at best, leading to the current ecological view that equilibrium-tending conditions may be uncommon. The import of this view has led Botkin to dub it "new ecology" while others claim proof of a paradigm shift (Wu and Loucks 1995).

Our volume furnishes much new evidence of the common forms and the variation of nonequilibrium conditions at several spatial scales. The most important corollary of this advance is that much environmental change is not even generally similar. Our studies adopt the growing insight that it is the frequency, rate, kind, and degree of change that must matter in the analysis of human environmental change (Botkin 1990; Vale 1982; Worster 1995). To the parameters identified by such studies we append the key parameter of spatial scale and we emphasize the environments of developing countries. Most of our cases address tropical settings. The premise that not all change is the same must serve as a cornerstone for the conservation debates and the political contention that surrounds them. Botkin stresses this point, which nonetheless is still unheeded at least in much public discussion: As he advises, "To accept certain kinds of change is not to accept all kinds of change," this crucial caveat being essential to avoiding a "Pandora's box of problems for environmentalists" (Botkin 1990: 11, 156). Such emphasis on the differences among types of environmental change urgently needs to be placed at the center of conservation analysis and strategies.

The second view—new in that it has reached unprecedented levels of awareness—is that human activities of various sorts are creating widespread modifications of the global biosphere (Glacken 1967; Sauer 1956, 1958; Turner et al. 1990). While some human-induced alterations are ancient, a number of current environmental changes are unique or accelerating. Mounting awareness of a pervasive human impact on the biosphere has renewed a debunking of the once widespread "pristine myth" of untouched virgin wilderness that was held as the state of nature throughout much of the world prior to European colonization and settlement (Butzer 1993; Cronon 1996b; Denevan 1992). This second view has led to a highly publicized challenge of the idea of purely natural wilderness.

Awareness of wide-ranging human impacts has also unleashed a critique that argues for the end of the conventional wilderness concept as a cornerstone for nature protection. This aspect of the second view is equally central to our concerns. According to Cronon "wilderness . . . is quite profoundly a human creation" and one that as an idea cannot help in making much progress for the sake of conservation at least in terms of the United States scene (Cronon 1996b: 69, 73). Interestingly, western Europeans have been less susceptible to the myth of pristine wilderness, due in large part to the undeniable human imprint on the state of Europe's nature. Still it is a fact that the protected areas of many poorer countries derive some of their justification and design from the wilderness idea and so our studies cannot avoid engaging the critique offered by Cronon and others. Like them, we show how the protected areas should not be seen as free of human influence—in the context of both environmental reality and human-constructed ideas.

Our position offers a notable addition to Cronon's criticisms of the wilderness idea. A definable amount of progress in protecting environments and lessening degradation can in fact arise from the establishment of protected areas. To do so our view must transcend the dichotomy of having to choose between either parks or people. In certain cases, the protected areas and peoples of developing countries have thus far managed to benefit from the reality of an "intact-if-not-untouched" nature. We suggest that since the actual degrees of human modification are all-important, the practical reality of an intact-if-not-untouched nature can serve as a sort of working ideal for nature protection. In these cases the ideal may serve not only as a mental model but also as part of a political strategy. In various cases the nature ideal has been helpful in thwarting the onslaught of outside development efforts while local residents in and near the protected areas have benefited from improvements in resource management.

Third of the newly influential views is that the analysis of environmental conservation must also address the nature of habitats that are established for human use. Until recently most works on conservation have focused solely on national parks and other protected areas. Their focus may be difficult to defend since only six percent of the world's land area is designated as "protected," and since only 19 of 131 countries in Asia, Africa, and Latin America have marked substantial areas as protected (Schelhas and Greenberg 1996). Still, the ecological nature of landscapes that are utilized for economic purposes such as farming and livestock-raising has tended to escape the recent overviews such as *Discordant Harmonies*. It is not a little ironic that a "new ecology for the twenty-first century" is so largely devoted to nature protection in parks and reserve areas. The apparent aversion to utilized landscapes by this recent manifesto on the human place in nature is not a peculiar prejudice. Indeed the progress of ecological ideas is still apprehended chiefly as an historical march of scientific advances and humanistic interpretations across landscapes made up principally of designated parks and perceived wilderness.

Our volume helps to redress the omission of utilized environments such as farm and rangelands from the purview of new ecological interpretations. Taken together, the studies here advance a series of insights that these perspectives, with their emphasis on nonequilibrium dynamics, can offer much to research on the biogeography and human ecology of heavily utilized environments such as farm and rangeland (Scoones 1994; Vandermeer 1995; Zimmerer 1994, 1996b). It is striking that the ecological interpretation of land use in settled areas has until quite recently remained among the most wed to a view of nature as equilibrium-tending or even static. Our studies integrate the geographical analysis of the nature that is inhabited and directly worked by humans, on the one hand, with the increasingly shared views on ecological theory and environmental history that have been summarized by Botkin (1990) and advanced by Cronon (1996a).

By integrating the new views, our volume makes use of a series of theoretical concepts and ideas that are presented in the following sections of the introduction. The first section introduces a pair of geographical concepts, the landscape and the region. The twin concepts serve as foundations in the analytical models and explanatory narratives of our case studies. Our use of the landscape and region as core concepts underscores the volume's central concern with complex change processes—including human alterations—and its special emphasis on the role of spatial scale in biogeographical and human ecological relationships. The following section discusses natural disturbances and biogeographical landscapes and related ideas such as patch dynamics and nonequilibrium processes. It is prompted by the expectation of change in most environmental systems, which are nonequilibrial and variable at many scales. Next we discuss anthropogenic disturbances and biogeographical landscapes, introducing concepts such as cultural landscape, anthropogenic fire ecology, and environmental degradation. The final two sections of the introduction describe the particular role of regional analysis for conservation strategies and

present a brief overview of the book's contents, including its central findings and the reasons for its regional coverage.

Concepts of Landscape and Region

Landscape, as used in this volume, refers to the concept of a "heterogeneous land area composed of a cluster of interacting ecosystems that is
repeated in similar form" (Forman 1995: 13; Forman and Godron 1986:
11; Troll 1971: 44; Turner 1989; for the common geographical meanings
of landscape see Meinig 1979). The particular definition of the landscape
concept that we adopt here stems from earlier research in geography. It
derives from the approach that Troll had defined as "landscape ecology,"
an approach that he saw as joining the geographical traditions of spatial
and biophysical analysis (Troll 1971). In the 1980s and the early '90s the
approach was expanded further by incorporating into it a group of ecological concepts, especially those of community ecology, and by infusing additional spatial concepts from geography, landscape architecture, and
planning (Forman 1995; Forman and Godron 1986; Turner 1989; Urban
et al. 1987; Zimmerer 1994).

Our use of the term "biogeographical landscape" is meant to specify a
spatially distinctive system of interacting biological processes and physical
features in the environment. These processes and features are in many
cases impacted upon by human activities. Biogeographical landscapes are
thus landscapes interpreted through the joint lens of biogeography and
human ecology. Such landscapes are conceptualized, measured, and evaluated using the theories, models, and methods of persons interested in the
human-altered distributions of plants, animals, and ecosystems. Each case
study in this volume presents the analysis of one or more biogeographical
landscapes. The human-impacted riverine forest landscape, for example,
sets the scene for the analysis of forest and primate conservation (chapter
1), while the landscape of panda conservation is a bamboo-dominated
montane forest that is being altered by human activities (chapter 2). Some
studies assess the nature of more than one biogeographical landscape, as
in those chapters on the montane forest and treeless landscapes located at
the high elevations of inhabited mountain environments (chapters 3–7).
In subsequent chapters, it becomes clear that biogeographical landscapes
include environments that bear the daily imprint of primary economic
activities, such as grazing and farming, as well as those forest landscapes
utilized for the extraction of tree products (chapters 8–12).

These landscapes, in turn, make up regions of a more inclusive spatial
scale that are now seen as crucial for the analysis of environmental problems and the planning of sustainable development interventions including
protected areas (Blaikie and Brookfield 1987; Forman 1995; Ives and Mes-

serli 1989; Zimmerer 1993, 1996b; an example of biologically defined
"ecoregions" is Dinerstein et al. 1995). In spatial terms, the concept of
region used in this volume adopts the conventional meaning of a subdivi-
sion of something larger (Schmidt 1954; Smith 1996). Regions often occupy subnational territories formed by processes that are national and
global in scale. In reality, our concept of regions is based on a regional-
national-global framework. Several of our studies use the regional concept
to indicate the surroundings or geographical context of smaller-sized
areas. In this latter sense, the concept of region is taken to mean the surrounding contexts of conservation reserves, protected areas, and the utilized landscapes of forest, grassland, and farmland.

The concept of regions within national and global scales makes explicit
the spatial interconnections, both nonhuman and human, that exist
among key biogeographical landscapes (Eden 1990; Noss 1983). Inadequate specificity in the analysis of biophysical and human conditions, as
well as its converse of poor contextualization, can easily weaken the analysis of landscape change and conservation. Such oversights create the risk
of naively short-sighted and impracticable conservation strategies. For example, the analyses of protected areas without reference to their broader
surroundings—both ecologic and socioeconomic—are likely to be so limited in use and insight that they produce inappropriate prescriptions for
management and conservation.

In summary, the concepts of landscape and region offer spatial scales
that are well-suited to the study of environmental change and to assessing
its importance to conservation and resource use in developing countries.
In the following sections we distinguish two sorts of landscape change:
that stemming from nonhuman or "natural" agents, and that due directly
to human-induced modifications, the latter referred to as anthropogenic
change. The twin themes are closely related and equally central to the
challenges being encountered by the people and institutions of the poorer
countries. Our point in drawing the distinction is a heuristic one intended
to aid in the presentation of complex, highly varied, and interrelated processes of landscape change. The distinction is not meant to infer that humans are not in a certain sense "natural," or that there are inevitable
differences between natural and anthropogenic forms of landscape
change. In fact, a chief goal throughout this volume is to enable the comparative analysis of all sorts of landscape change.

Natural Disturbances and Biogeographical Landscapes

The first theme woven throughout this volume is the role in conservation
and resource use of landscape changes that are not caused directly by
human action. The introduction and concepts discussed here thus concen-

trate on natural disturbances and their spatial dynamics, the role of habitat patches resulting from natural disturbances, the impact of long-term historical changes in environments, and the relevance of natural disturbances to human land use.

Ecological disturbances of nonhuman origin are initiated by a wide assortment of agents. Increasingly, studies have stressed that natural disturbances such as drought, floods, fire, windstorms, and disease outbreaks are prevalent in a variety of temperate and tropical landscapes, including coral reefs and other oceanic "seascapes" (Connell 1978; Forman and Godron 1986; Kellman 1997; Pickett 1980; Pickett and White 1985; Stoddart 1972; Vale 1982; Veblen 1992; Watt 1947; White 1979). Such disturbances may occur at frequent intervals and sometimes across extensive areas. Other disturbances, such as ice storms and landslides, can be regionally or locally common. The disturbances may be distinguished on the basis of initiation and as belonging to either a primarily biotic category (also known as endogenous or autogenic disturbance), or to a primarily physical group (also referred to as exogenous or allogenic). Both categories are considered to be natural disturbances.

Landscapes of developing countries, like those of developed ones, are seen as subject to a great array of natural disturbances. Examples include landslides and other geomorphic disturbances in mountain areas, windstorms, landslips, and lightning strikes documented in tropical rain forests worldwide (Connell 1978; Denslow 1987), and lightning-set fires and short-term climate change (such as drought) in subhumid and semiarid environments (Minnich 1983). Natural disturbances in many landscapes of developing countries may differ in frequency and extent from those of developed countries since biophysical environments themselves may differ (such as the tropical rain forests and tropical mountain environments that are concentrated in poorer countries). Notwithstanding such contrasts, in general the spatial and temporal characteristics of natural disturbance are key features for conservation analysis.

Contrary to many earlier assumptions, a burgeoning collection of studies shows that certain disturbances may help to maintain the diversity of plant and animal species in a large variety of landscapes (Connell 1978; Denslow 1987; Grubb 1977; Pickett 1980; Pimm 1984, 1991; Sauer 1988; Veblen 1992; Walker 1989). Various natural disturbances help to facilitate the coexistence of diverse species by creating habitat heterogeneity and by making available resources such as sunlight and nutrients that many plants need for regeneration. A large share of species may even depend on natural disturbances within the landscape in order to reproduce, and thereby to persist in the evolutionary sense. Such natural disturbances might also contribute to the prospects of coexistence for trees and other plant species by reducing the level of competition among ecologically similar taxa.

The now-predominant view of natural disturbance as common, highly varied, and biologically integral has helped to accent the prevalence of nonequilibria, instability, and even chaotic change in biophysical environments (Botkin 1990; Colwell 1985, 1992; Worster 1990; Wu and Loucks 1995). Botkin's term "new ecology" is not fully agreed upon and hence merits the quotation marks. Nonetheless his eye-grabbing term clearly signals the eclipse of an earlier biological ecology that rested on equilibrium-based systems theory. This assumed that environments tend strongly toward homeostasis and in the direction of an end point that invariably was the "climax community," the latter a highly influential concept deeded by the pioneer American ecologist Frederic Clements (Laszlo 1972; Odum 1969). Since not all systems ecology is anchored to equilibrium assumptions, it is useful to distinguish the earlier view as systems ecology qua systems (Zimmerer 1994). That earlier view conceived of species diversity, for instance, as maintained in natural systems with strong equilibrating tendencies and a minor or negligible role of disturbance. Our newer ecological view is finding that the role of natural disturbances is major and may be essential to maintain high species diversity in many environments.

The added weight given to the role of natural disturbances in landscapes has led to new interest in the spatial properties of such changes. Growing attention has turned to the analysis of the discrete subunits or "patches" that typically compose a landscape. The role of natural disturbances in the formation of these patches has ignited interest in what is now known as "patch dynamics." Patches initiated by disturbance are referred to as "disturbance patches" (Baker 1989; Belsky and Canham 1994; Levin 1992; Pickett 1980; Schelhas and Greenberg 1996; Veblen 1992; Pickett and White 1985). A high priority for conservation planning is to evaluate the size and dynamics of patches, a pair of attributes that prefigure the suitability of each disturbed area to patch-dependent species. The overall combinations of patches, as well as features such as their ages and their spatial relations to one another (or configuration), also matter greatly to conservation goals such as biodiversity and ecosystem protection. Renewed consideration of these latter features has helped to broaden the study of patches, making it more holistic and less focused solely on size and distance parameters (Schelhas and Greenberg 1996).

In many landscapes the full suite of disturbance-created patches may not, however, be easily incorporated into conservation designs. This difficulty is not for lack of interest or insight on the part of conservation science, for growing weight has been granted to the role of disturbance patches in the conservation of species and ecosystems in protected areas (Baker 1989; Botkin 1990; Chapman 1947; Forman and Godron 1986; Pickett and Thompson 1978; Soulé and Simberloff 1986; Soulé and Wilcox 1980). Rather the difficulty emerges from the findings inspired by the "new ecology" perspective. It was once believed that the widespread regularity of disturbances, when summed across a landscape, serves to maintain a "shifting-mosaic steady state" of diverse patches that tended toward an overall equilibrium. Yet the model of the shifting-mosaic steady state seems applicable only at certain large-area spatial scales, and even then only in certain landscapes or time periods. The lack of equilibrium among patches means that the relevance of natural disturbance for the design of conservation strategies must be considered in other terms. Given the commonness of nonequilibrium situations, one priority for reserve design is to evaluate the disturbance regimes that create existing patches and whether the resident populations of plants and animals in those patches are viable for conservation purposes.

Long-term irregular changes of physical conditions, especially those related to climate, are a major source of nonequilibrium conditions and commonly confound the search for steady-state mosaics of patch types. Take the case of climate changes during and following the last glacial advance, changes that notably altered the conditions of biota and soils. Associated environmental changes were pronounced in temperate biomes. Yet these long-term climate shifts also resulted in certain widespread alterations of the biota and soils of tropical and subtropical areas. For example, some paleoenvironmental studies suggest that during glacial periods a tropical savanna covered sizeable areas of the present-day tropical rain forests of the Amazon and Orinoco basins in South America (Absy 1982; Haffer 1969, 1982; Klammer 1982; Prance 1978). High-elevation landscapes near treeline in the neighboring Andes Mountains meanwhile also changed as a result of climate warming and glacial retreat, capped by a notable contraction of glaciers during the present century (Schubert 1992; Thompson et al. 1995). Examples such as these of long-term biophysical changes preclude the ready identification of single environmental states or natural conditions that otherwise could furnish the benchmarks for conservation efforts.

While long-term biophysical changes offer a major conceptual puzzle for conservation, it is the short-term changes that most often are involved in practical challenges. The prevalence of landscape change over the course of a few years or decades is seen as posing a special dilemma for parks and other protected areas (Davis 1976, 1984; Sprugel 1991; Vale 1982, 1987). The picture of environments in flux depicted by scientific findings frequently differs from the image of long-term environmental stability and balance held by many people, including conservationists. In the United States a much-discussed example of the seeming paradox (and practical difficulties) of conserving an ever-changing environment is derived from the experience of national parks such as Yosemite and Sequoia that occupy the western slopes of the Sierra Nevada in California (Dilsaver and Tweed 1990; Graber 1995; Vale 1987; Vankat and Major 1978). Here, ongoing landscape change entails such processes as the enhanced survival of young trees that most people, including the majority of park visitors, perceive to be a desired state of protected nature, although in this case landscape change is due instead to the parks' suppression of natural wildfires. A similar experience at Yellowstone National Park is discussed by Despain (1990). Meanwhile analogous conservation dilemmas are occurring in many developing countries as protected areas become established within their territories.

Finally, the study of land use in settled areas where environments are subject to economic activities such as agriculture and livestock-raising must be seen as lagging noticeably behind the advances discussed above. Oddly, the proliferating mass of land-use assessments has remained slow to account for the important role of such disturbances. This halting awareness may be due to the especially strong legacy of concepts from equilibrium ecology or systems ecology qua systems. Influential models of livestock-carrying capacity and explanations for the biodiversity of economic plants, for example, persist in overlooking the prevalence of natural ecological disturbances in the rangeland and agricultural land of poorer countries (Harlan 1992; Pratt and Gwynne 1977). Such oversights may be due to the fact that primary economic landscapes are so strongly marked by human modifications that they tend to be presumed as stable or free of natural disturbance. Recent case studies and critiques have begun to shift our view of the ecological dynamics in such landscapes toward the direction of a nonequilibrial perspective (Brookfield and Padoch 1994; Scoones 1994; Vandermeer 1995; Westoby et al. 1989; Zimmerer 1994, 1996b, 1998).

Anthropogenic Disturbances and Biogeographical Landscapes

The nature of human-caused landscape change is the second main theme of our volume. Analysis of anthropogenic disturbance is firmly rooted in the traditional study of biogeographical landscapes. In fact it was the human-altered landscapes of Europe and North Africa that furnished the study material for George Perkins Marsh, a statesman-scholar and founder of modern environmental studies and environmental geography. Marsh's pioneering treatise *Man and Nature,* published in 1864, chronicled the complex and far-reaching history of human influences on landscapes as diverse as forests, grasslands, rivers and riparian habitats, farmland, and rangeland (Marsh 1864 [1965]).

In this century geographer Carl Sauer established an approach that may be termed "cultural-historical ecology" (Zimmerer 1996a). Sauer followed the path that was opened by Marsh's theme of the human modification of nature, but enlarged the geographical compass of these studies to include a wide array of tropical and subtropical environments. Tropical rain forests, tropical savannas, subtropical deserts, and mixed mountain biomes were seen in the studies of Sauer and his students as forming biogeographical landscapes that were being shaped to varying degrees by human activities (Sauer 1956, 1958 [1967]). According to Sauer, cultural landscapes were expressed where the human impact was especially discernable.

Knowledge of the historical impact of human activities is adding a crucial dimension to our understanding of environmental change in developing countries. For example, a mounting number of studies reveal that many forests and shrublands in the humid tropics have typically burned at occasional intervals and that at least some vegetation-altering fires were ignited by humans (Horn 1993; Horn and Sanford 1992; Kellman 1975; Sanford et al. 1985). Anthropogenic fire now appears to have altered certain vegetation types in both the tropical lowlands and the highlands since the end of the last glacial period more than 10000 years ago. Fire and other causes of long-term vegetation change also led to the major modification of soils. In some cases human activities could be seen to alter the overall complex of the soil-plant system, a nexus of many environmental functions. For instance, the initial clearing of forests and other plant cover for the expansion of agriculture in the subtropical highlands of present-day Mexico, begun several thousand years ago, appears to have triggered landscape-transforming episodes of aggravated soil erosion (Butzer 1993; Metcalfe et al. 1989; Street-Perrott et al. 1989).

A biogeographical view of landscapes seeks to elucidate the ecological processes or dynamics *per se* of human-caused disturbance, including its spatial and temporal traits (figure I.2). Human activities are seen either to alter the natural disturbance processes of vegetation or to initiate novel types of disturbance (Christenson 1989; Kellman and Tackaberry 1997; Savage 1991; Vale 1982; Veblen and Lorenz 1988). In the former case, the anthropogenic disturbances can be compared to natural ones, although the human impacts may, and often do, differ substantially in degree. Regimes of anthropogenic disturbances must be distinguished on the basis of frequency, scale, and magnitude. One framework for the classification of disturbances identifies two basic types of events (Vale 1982: 16). First there are the low-to-moderate types of chronic disturbances that are noticeable over a few years to several decades or more, such as livestock grazing, trampling, off-road vehicle use, and air pollution. The second class of disturbances are more intense and of shorter duration, exemplified by fires, logging, construction activities, and farm abandonment or estab-

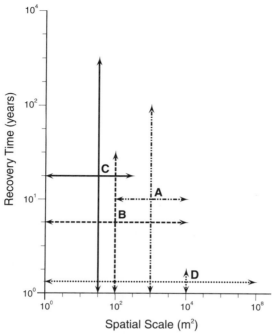

Fig. I.2. Environmental disturbances at the landscape scale. The diagram is designed to show the relationship between spatial scale and temporal scale that can characterize all disturbances. Approximations of a few disturbances taken from our case studies are presented as examples: (a) forest clearings in western China; (b) fire in Chirripó National Park, Costa Rica; (c) natural hazards including geomorphic disturbances in a farm landscape of the southern Peruvian Andes; (d) short-term climate and vegetation change in rangeland of the African Sahel. A cross is used to approximate the spatial scale and temporal scale of a major disturbance discussed in each case.

lishment. By focusing on the ecological properties of disturbances, the biogeographer's perspective offers an important antidote to "relativized" portrayals that would tend to gloss over the key contrasts that usually separate the different sorts of landscape change (Botkin 1990; Worster 1995; Wu and Loucks 1995).

Human alterations are seen to have shaped the majority of landscapes as new studies reveal more about the historical modification of these habitats. For example, the invasion of shrubs into grassland areas has often been driven by the overgrazing of livestock and the artificial suppression of fires (Blumler 1993; Johannessen 1963; Savage 1991; Veblen and Lorenz 1988). The consequence of human activities on landscape change in these examples is frequently complex. Shrub invasion in the above example may be thwarted by a moderate frequency of human-set fires. Such insights into the altered ecologies of anthropogenic landscapes also indicate that two or more human activities can be: 1) synergistic, compounding the effects of one another, or 2) antagonistic and operating contrary to one other, as in the case of heavy grazing that favors shrub invasion versus moderate grazing and fire that maintains grassland vegetation (Myers 1995).

Shrub, grass, and forest landscapes in many regions of developing countries reflect a whole host of past and present disturbances owing to human activities. Ecological analysis of these human impacts points to certain regimes of disturbance that characterize particular places (Brower 1991; Metz 1990, 1994; on the tropical rain forests of eastern Brazil, see Voeks 1988 and Voeks and Vinha 1988). In addition to the various practices of livestock grazing and fire, other human-induced disturbances that need to be accounted for include fuelwood and fodder collecting, logging, hunting, and the gathering of nontimber plant products. At least in some cases, the response of landscapes to anthropogenic changes such as these may be more resilient than was previously assumed (Blumler 1993). Our interpretation of the relations of human modifications to ecological outcomes is thus more nuanced and less certain than was often thought.

Much research on the dynamics of anthropogenic change details a clear and, in some cases, alarming deterioration of environments during the recent past. Connections between human-created landscape change and the decline of certain plant and wildlife populations furnish several examples of such impacts (Goudie 1990; Vale 1982; Westman 1978). In many cases the fragmentation of formerly contiguous forest into separate patches as a result of human activities like logging, agricultural clearing, and road-building is also a prime example (Burgess and Sharpe 1981; Forman and Godron 1986; Schelhas and Greenberg 1996; Young 1994). Forest fragmentation can result not only in the direct loss of habitats, but it also alters the ecological dynamics of remaining forest patches, including their disturbance regimes and regeneration modes. In fact there is growing awareness that these alterations of disturbance and regeneration dynamics are driving the highly publicized plight of several endangered wildlife species such as a variety of primates in Africa and the giant panda in China (Medley 1993; Taylor and Qin 1989).

Alteration of soil-plant dynamics by human influences govern the degree of environmental deterioration that is found in many landscapes. Many sorts of change alter the nature of aboveground plant cover and modify the nature of rooting, in either case imparting modifications to the soil environment (Thomas and Allison 1993; Vale 1982; Zimmerer 1993). Since tropical and subtropical vegetation plays an especially critical role in cycling nutrients and binding soil particles, the more severe forms of vegetation change in these environments can inflict considerable consequences on soils. Activities like large-scale logging and agricultural clearing, severe overgrazing, and road-building imperil the range of normal functioning characteristic of soil-plant systems. Overgrazing of the high-elevation flora of some tropical mountains, for instance, is resulting in acute damage due to soil erosion and nutrient loss (Pérez 1992, 1993), although this problem has to date drawn far less publicity than the soil-plant systems of the lowland rain forests.

Conservation, Resource Use, and Regions

The evident influence of change processes of both the natural and anthropogenic sorts is triggering a new flood of attention in conservation and resource thinking to the spatial attributes of landscape dynamics, environments, and related aspects of human activity and experience. Evidence of the nonequilibrial character of many change processes is redoubling this increased interest. Several existing concepts of spatial scale need to be considered more directly in order to better understand the actual practices and implementation of conservation and sound resource use. Patch, landscape, and region are prime examples of concepts that offer the potential for further use.

Areas of environmental interest or resource use often make up patches within the landscapes that in turn make up a region. Forest patches within landscapes and within regions are a building focus for conservation interests (Forman 1995; Forman and Godron 1986; Kellman and Tackaberry 1997; Schelhas and Greenberg 1996). The areas of forest cover may be woodland remnants that have been left uncleared by humans, natural forest plantations managed by people for agroforestry and other purposes, sacred forests preserved for cultural and religious reasons, and gallery or riparian forests. Ecotone areas may also be viewed as landscape patches that are situated within regions. While ecotones are often sensitive indicators and thus potentially serve as biotic monitors of environmental change, the influence of surrounding environments on ecotone habitats suggests the framing of these analyses at both the landscape and regional scales (Gosz and Sharpe 1989; Thomas and Allison 1993).

The success of protected and managed areas, and, more broadly, the environmental soundness of land and resource use also require us to consider more fully spatial scale. Regions, which form in relation to forces and conditions that are national and global in scale, are often key units for conservation analysis. Consideration of regions leads in turn to an awareness that the support and involvement of local inhabitants are among the most crucial conditions that ultimately determine the outcome of many conservation efforts. All-important arbiters of their support and involvement include the nature of their economic activities and livelihoods, cultural perceptions and educational awareness, sociopolitical organizations, and psychological desires. The approach of "regional political ecology," like the related human and cultural ecology, offers a regional-national-global framework for the analysis of human conditions that bear on environmental change and nature-society relations in general. Regional political ecology and related approaches are grounded in geography as well as the cognate subfields of anthropology, sociology, and environmental history (Blaikie and Brookfield 1987; Peluso 1992; Zimmerer 1996a, 1996b).

A pair of recent advances in political ecology and related fields warrants a special note when mentioning the regional-national-global framework for environmental analysis. First, a charge of "green imperialism" can be leveled against those sustainability and conservation projects in developing countries that fail to entail the fully democratic involvement of inhabitants and their local institutions (Alcorn 1994; Friedmann and Rangan 1993; Guha 1997; Peluso 1993; Redclift 1987). If imposed undemocratically, conservation projects violate the political rights of citizens and may infringe on their even more basic civil and human rights. Unfortunately these injustices have become commonplace. Such abuses will get worse as some governments of developing countries expand their environmental protection efforts in socially unjust ways. Ironically, environmental rights may be expanded in these cases even while political, civil, and human rights are curtailed.

Conservation programs that are imposed entirely by external interests are typically beset by social problems since they may lack local support and can easily generate the active opposition of the people and groups that are affected. Such projects are frequently spawned by international donors or activists with single-minded or naive outlooks. Their oversights may threaten projects, including elements that otherwise are feasible. For this reason, and for those of humanitarianism and social justice, the persons and institutions promoting environmental protection and sound resource use should make sure that the rights of residents are fully protected and empowered. Although not easily undertaken, these steps will help ensure that environments also are protected and kept viable as resources. Environmental interventions may thus require working in close cooperation with regional or other local political groups.

The second advance in political ecology is its renewed interest in environmental modification *per se.* Motivated largely by the grave conditions of many people in poorer countries due to environment-related problems, some research prioritized purely political and social analysis. Its social science centered on the close interconnections between various environmental problems, on the one hand, and vastly unequal relations of social and political power on the other. In at least a few of its philosophical forms this political and social analysis portrays nature as a blank slate that is inscribed solely by ideas and ideology. Although perhaps unfamiliar to biogeographers and other natural scientists, these forms of thinking appear to question whether nature itself exists outside human social life (Gandy 1996). In response, some recent work in political ecology has urged a bridging of social science analysis and the study of environmental modification or the "nature of nature" (Blaikie 1994; Peet and Watts 1996; Roberts 1995; Zimmerer 1996a). Philosophically, this trend acknowledges the objective existence of environments *per se*—not as blank slates but as parts of a highly differentiated natural world made up of beings, matter, and processes whose existence is at least partly independent of our own.

Historical accounts of environmental change in developing countries illustrate the merits of the regional-national-global framework. Historical geography, historical ecology, and environmental history—three closely related subfields—have commonly adopted this concept of regional space and its interconnections. Findings sensitive to these scales highlight the historical role of nature-society relations in environmental change, including the influences of social power, mental ideas or "constructions" of nature, and the state of nature itself (Dean 1983; Peluso 1992; Zimmerer 1996b). Historical studies cast in the regional-national-global framework have demonstrated an array of cases where the rights of particular resource users (often poor people and ethnic minorities) were unjustly usurped by more powerful individuals and groups. In these cases the control of regional environments and resources has been unfairly coerced by private interests, undemocratic governments, and even conservation proponents that claim to support indigenous environmental traditions.

In the specific case of established protected areas, a variety of diverse impacts may originate in the nearby areas that are more heavily utilized. Some modifications may resemble those that are already familiar problems in the conservation efforts of developed countries. Inadequate environmental planning and regulation can result in the construction of roads, bridges, and dams that impact on landscapes that are slated for protection. These regional-scale activities may directly damage the habitats of the protected area or they may act indirectly by modifying the types, magnitudes, frequencies, and locations of disturbances that are ecologically integral. The location of these activities, as well as their other properties, determine the degree of modification they impart. Changes like these can lead to a serious elimination of habitats and to the possible loss of species (Savage

1993; Young 1994, 1996). Even the implementation of nature tourism or so-called ecotourism can, if not properly designed and implemented, easily undermine the objectives of conservation.

Protected areas also commonly demonstrate the effect of human-caused disturbances that impinge on the expectations of visitors. A classic dilemma involves the region that borders a protected area and that on occasion gives rise to fires. Lightning-set fires may arise in these surrounding regions. Also many farmers and herders in developing countries tend to rely on the intentional burning of vegetation for range management and agricultural clearing (Denevan 1992; Sauer 1956, 1967; Young 1993; Uhl et al. 1990). Yet in a variety of cases the occasional burns ignited by rural inhabitants do not degrade present-day vegetation since their practices are of similar magnitude and frequency to long-term fire regimes. Still the reality of a burned landscape is often at odds with the expectations of nature tourists that choose to visit protected areas. Environmental education is a priority in these cases although the visitors' expectations still may encourage a policy of fire suppression. Similarly, policies that prohibit all hunting in a protected area—often enforced by the expectations of visitors and conservation administrators—may be at odds with local practices that do not actually threaten the viability of wildlife populations (Alcorn 1994; Guha 1997).

Beyond the case of protected areas, the region-state-global framework must be seen as an equally crucial guide for the sustainability efforts that involve primary economic activities. Analysis of economic development, the land use practices and related knowledge of local people, and the changing biogeographical landscapes of settled areas are topics at the core of current thinking on sustainable development. Research on these topics depends on a studied familiarity with spatial concepts. Analysis of economic prospects and pressures needs to reach beyond the scale of a single biogeographical landscape to that of the larger region and, in many cases, to processes that are national and global in scale (Blaikie and Brookfield 1987; Redclift 1987; Zimmerer 1993). Similarly, cultural ways of knowing and being that are related to the environment may be poorly grasped and easily misinterpreted when the scope of geographical space is reduced to a single landscape or over-generalized at the scale of an entire country (Alcorn 1989, 1994; Crumley 1994; Murdoch and Clark 1994; Peluso 1993; Posey 1992; Redford and Mansour 1996; Sponsel et al. 1996; Zimmerer 1996b).

Overview of the Book

The general themes discussed in the above sections must be advanced by way of specific case studies. The new views place greater emphasis on the character of ecological processes and conservation prospects that are regionally specific. We firmly believe that the reasoned extension of such regional-scale processes and prognoses over large areas is apt for certain general sorts of evaluation and insights about conservation and resource use. However, the unwarranted extrapolation of local findings into universal prescriptions for protecting nature cannot offer the insights needed for sustainable development. Mindful of this new awareness, our volume is designed to present the analyses of specific cases that are united by shared themes and that use these bases to discuss general concepts and theory. These generalizations are often applicable at intermediate scales that include the areas of regions and landscapes described in the preceding sections.

The twelve case studies chosen for our volume conform to a general style. The studies use an overview of recent research findings to develop the foundations for conservation and resource use. In constructing these foundations, the chapters critique a number of relevant concepts, theories, and ideas. Due to that emphasis, each chapter limits its discussion of research methodologies and data analysis, although they are detailed in the author's other publications. The majority of studies originated as presentations in special sessions organized at the Annual Meeting of the Association of American Geographers in San Francisco in 1994. Subsequently we invited manuscripts from the presenters and a few other geographers. (All chapters are authored or coauthored by geographers who are professionally interested and active in the nature of resource management and biological conservation in developing countries.) The chapters draw from the extensive field experience of the authors in developing countries that include regions of Asia, Africa, and Latin America (map I.1). As outlined below, the case studies cover a wide range of landscapes in mountain and lowland biomes, riverine and semiarid environments, and agricultural and range ecosystems.

It is significant that five of the studies in our collection are set in Latin America and that four are clustered in the Andes Mountains of South America. This aspect of the volume's coverage reflects a particular strength and dynamically evolving tradition in geographical research. The emphasis may be seen as instilled by the early studies of Andean environments and landscapes by the geographer Alexander von Humboldt (von Humboldt 1822). His Andean studies founded the modern biogeographical synthesis of plant morphology, climate, and elevation effects and helped to establish our modern concept of human-environment relations. In Europe those foundations continued to serve as a cornerstone of geographical research on the Andes (Gade 1996; Lauer 1984, 1993; Troll 1968, 1971). In the United States, a Latin American emphasis of biogeography and human ecology research also reflects the historical influ-

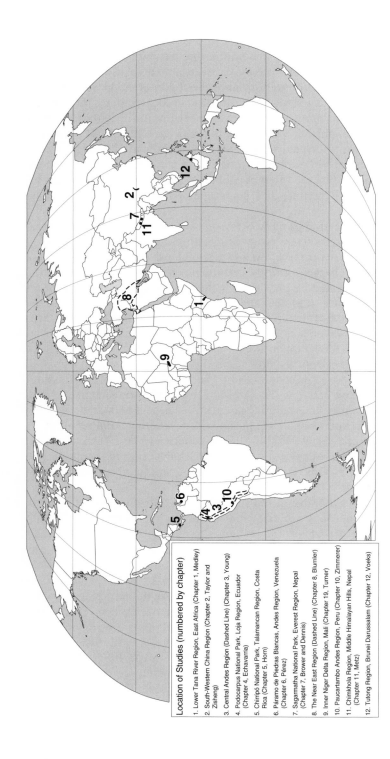

Location of Studies (numbered by chapter)

1. Lower Tana River Region, East Africa (Chapter 1, Medley)
2. South-Western China Region (Chapter 2, Taylor and Zisheng)
3. Central Andes Region (Dashed Line) (Chapter 3, Young)
4. Podocarpus National Park, Loja Region, Ecuador (Chapter 4, Echavarria)
5. Chirripó National Park, Talamancan Region, Costa Rica (Chapter 5, Horn)
6. Paramo de Piedras Blancas, Andes Region, Venezuela (Chapter 6, Pérez)
7. Sagarmatha National Park, Everest Region, Nepal (Chapter 7, Brower and Dennis)
8. The Near East Region (Dashed Line) (Chapter 8, Blumler)
9. Inner Niger Delta Region, Mali (Chapter 19, Turner)
10. Paucartambo Andes Region, Peru (Chapter 10, Zimmerer)
11. Chimkhola Region, Middle Himalayan Hills, Nepal (Chapter 11, Metz)
12. Tutong Region, Brunei Darussalam (Chapter 12, Voeks)

Map I.1. Locations of the study areas.

ence of Carl Sauer and his colleagues and students at the University of California at Berkeley (Sauer 1956, 1958). James Parsons in particular helped forge a distinctive style of research in the United States with his studies of the biogeographical-cultural-historical ecology of Latin America (Parsons 1971). In turn, some of Parson's students and colleagues have imprinted a noticeable influence on biogeographical studies that involve Latin America, as well as Latin American studies that involve biogeography (Denevan 1961; Johannessen 1963; Veblen and Lorenz 1988).

The first part of the book entitled "The Conservation Challenges of Forest Loss and Fragmentation" provides several examples of how both natural forest dynamics and forest degradation or deforestation affect forest cover and connectivity in developing countries. The section begins with an overview of the conservation challenges surrounding landscape-level changes taking place in the riverine forests of the Tana River in eastern Kenya (chapter 1, Medley). This is followed by a discussion of the relation between forest dynamics, forestry practices, and giant pandas in mountain areas of Southwest China (chapter 2, Taylor and Qin). Finally, the humid forests of the central Andes Mountains are described and analyzed in terms of past and present deforestation in a general overview (chapter 3, Young), and in terms of the changes that are occurring in the protected area of Podocarpus National Park located in the southern Ecuadorian Andes (chapter 4, Echavarria).

The second part of the book entitled "The Conservation Challenges of Disturbances in High Mountains" addresses the environmental changes of landscapes and regions in the highlands of Asia and Latin America. This section includes an analysis of the fire ecology and biogeography of a tropical alpine (or páramo) landscape of Chirripó National Park in Costa Rica and its implications for park management (chapter 5, Horn); an evaluation of disturbances caused by livestock and other human-related activities in timberline landscape of the Venezuelan Andes (chapter 6, Pérez), and an interpretation of livestock and related impacts in the Himalayas of Sagarmatha (Mt. Everest) National Park in Nepal (chapter 7, Brower & Dennis).

The third part of the book entitled "Land Use in Settled Areas and the Prospects for Conservation-with-Development" addresses the nature of landscape change associated with agriculture, livestock-raising, and the extractive use of forest products. This section addresses the role of long-term historical as well as short-term successional change in the diverse landscapes of the Near East (chapter 8, Blumler), the interannual dynamics of change in the drylands landscape and pastoral economies of the African Sahel (chapter 9, Turner), the ecological versatility of diverse food plants and the dynamics of short-term disturbance (intra-annual and interannual) in the agricultural landscape of the Andes Mountains of South America (chapter 10, Zimmerer), and the ecological dynamics of extractive use and natural disturbances in the forest landscapes of the Nepal Himalaya (chapter 11, Metz) and the Borneo rain forest of Southeast Asia (chapter 12, Voeks).

The twin themes of landscape disturbance and human-induced modification serve as common threads that weave through all the chapters. A third common thread spins out of the volume's steady focus on the implications of these studies for nature conservation and sustainable resource use. Our entire volume, and especially the third section, is intended to provide carefully derived insights for the suitable joining of conservation and development. This emphasis is especially necessary since the majority of land cover in the developing countries is already being put to economic purposes (Turner et al. 1990). Agriculture, livestock-raising, the extractive use of forest products, and other resource-related activities thus strongly shape the bulk of biogeographical landscapes in these world regions.

Case studies in the volume furnish a carefully chosen array of examples, analyses, and discussions that are addressed primarily to scholars, students, and practitioners in geography, ecology, environmental studies, conservation biology, and international and development studies (especially conservation-and-development). It will also interest persons in environmental history, ecological anthropology, development and environmental sociology, range management, forestry, and agronomy. We especially hope that our studies will aid and challenge those students interested in combining a commitment to fieldwork with the latest efforts at integrative analyses involving each of these individual areas of study. Insightful integration at the core of such studies, rather than merely multidisciplinary amalgamation, must be founded in a new appreciation for the complexities of nature itself.

Acknowledgments

For their helpful comments on draft versions of the introduction we would like to thank Mark Blumler, Sally Horn, Kimberly Medley, John Metz, Matthew Turner, and Thomas Vale. Karl Zimmerer's research on the theoretical overviews was funded in 1994–1995 by a grant from the Graduate School of the University of Wisconsin–Madison. The Cartography Laboratory of the University of Wisconsin–Madison, especially Onno Brouwer, its director, and Qingling Wang, one of its talented cartographers, produced many fine maps and illustrations for the volume. We are also grateful for the interest, cooperation, and guidance of Mary Elizabeth Braun, Mark Crawford, and Raphael Kadushin of the University of Wisconsin Press.

References

Absy, M. L. 1982. Quaternary Palynological Studies in the Amazon Basin. In *Biological Diversification in the Tropics,* ed. G. T. Prance, pp. 67–73. New York: Columbia University Press.

Alcorn, J. B. 1989. Process as Resource: The Traditional Agricultural Ideology of Bora and Huastec Resource Management and its Implications for Research. In *Advances in Economic Botany,* eds. D. A. Posey and W. Balée, pp. 63–77. New York: New York Botanical Garden.

Alcorn, J. B. 1994. Noble Savage or Noble State? Northern Myths and Southern Realities in Biodiversity Conservation. *Etnoecología* 11(3): 7–19.

Baker, W. L. 1989. Landscape Ecology and Nature Reserve Design in the Boundary Waters Canoe Area, Minnesota. *Ecology* 70(1): 23–35.

Belsky, A. J., and C. D. Canham. 1994. Forest Gaps and Isolated Savanna Trees: An Application of Patch Dynamics in Two Ecosystems. *BioScience* 44(2): 77–84.

Blaikie, P. 1994. *Political Ecology in the 1990s: An Evolving View of Nature and Society.* Distinguished Speaker Series Number 13, Center for Advanced Study of International Development (CASID). East Lansing: Michigan State University.

Blaikie, P., and H. Brookfield. 1987. *Land Degradation and Society.* London: Methuen.

Blumler, M. A. 1993. Successional Pattern and Landscape Sensitivity in the Mediterranean and Near East. In *Landscape Sensitivity,* eds. D. S. G. Thomas and R. J. Allison, pp. 287–305. New York: John Wiley and Sons.

Botkin, Daniel B. 1990. *Discordant Harmonies: A New Ecology for the Twenty-first Century.* New York: Oxford University Press.

Brookfield, H., and C. Padoch. 1994. Appreciating Agrodiversity: A Look at the Dynamism of Indigenous Farming Practices. *Environment* 36: 7–11, 37–45.

Brower, B. 1991. *Sherpa of Khumbu: People, Livestock, and Landscape.* Delhi: Oxford University Press.

Burgess, R. L., and D. M. Sharpe, eds. 1981. *Forest Island Dynamics in Man-Dominated Landscapes.* New York: Springer-Verlag.

Butzer, K. W. 1993. No Eden in the New World. *Nature* 362: 15–16.

Chapman, H. H. 1947. Natural Areas. *Ecology* 28(2): 193–194.

Christensen, N. L. 1989. Landscape History and Ecological Change. *Journal of Forest History* 33: 116–125.

Colwell, R. K. 1985. The Evolution of Ecology. *American Zoologist* 25: 771–777.

Colwell, R. K. 1992. Making Sense of Ecological Complexity. *Biotropica* 24(2b): 226–232.

Connell, J. H. 1978. Diversity in Tropical Rain Forests and Coral Reefs. *Science* 199: 1302–1309.

Cronon, William, ed. 1996a. *Uncommon Ground: Rethinking the Human Place in Nature.* New York: W. W. Norton.

Cronon, William. 1996b. The Trouble with Wilderness; Or, Getting Back to the Wrong Nature. In *Uncommon Ground: Rethinking the Human Place in Nature,* ed. William Cronon, pp. 69–90. New York: W. W. Norton.

Crumley, C. L. 1994. *Historical Ecology: Cultural Knowledge and Changing Land-scapes.* Seattle: University of Washington Press.

Davis, M. B. 1976. Pleistocene Plant Geography. *Geoscience and Man* 13: 13–26.

Davis, M. B. 1984. Climatic Instability, Time Lags, and Community Disequilib-rium. In *Community Ecology,* eds. J. Diamond and T. J. Case, pp. 269–284. New York: Harper and Row.

Dean, W. 1983. Deforestation in Southeastern Brazil. In *Global Deforestation and the 19th Century World Economy,* eds. R. Tucker and J. F. Richards, pp. 50–67. Durham: Duke University Press.

Denevan, W. M. 1961. The Upland Pine Forests of Nicaragua: A Study in Cultural Plant Geography. *University of California Publications in Geography* 12: 251–320.

Denevan, W. M. 1992. The Pristine Myth: The Landscape of the Americas in 1492. *Annals of the Association of American Geographers* 82(3): 369–385.

Denslow, J. S. 1987. Tropical Rainforest Gaps and Tree Species Diversity. *Annual Review of Ecology and Systematics* 18: 431–451.

Despain, D. G. 1990. *Yellowstone Vegetation.* Boulder: Roberts Rinehart.

Dilsaver, L. M., and W. C. Tweed. 1990. *Challenge of the Big Trees.* Three Rivers, California: Sequoia Natural History Association.

Dinerstein, E., D. M. Olson, D. J. Graham, A. L. Webster, S. A. Primm, M. P. Bookbinder, and G. Ledec. 1995. *A Conservation Assessment of the Terrestrial Ecoregions of Latin America and the Caribbean.* Washington, D. C.: The World Wildlife Fund and The World Bank.

Eden, M. J. 1990. *Ecology and Land Management in Amazonia.* London: Belha-ven Press.

Egerton, F. N. 1973. Changing Concepts of the Balance of Nature. *The Quarterly Review of Biology* 48: 322–350.

Forman, R. T. T. 1995. *Land Mosaics: The Ecology of Landscapes and Regions.* Cambridge: Cambridge University Press.

Forman, R. T. T., and M. Godron. 1986. *Landscape Ecology.* New York: John Wi-ley and Sons.

Friedmann, J., and H. Rangan, eds. 1993. *In Defense of Livelihood: Comparative Studies on Environmental Action.* West Hartford, Connecticut: Kumarian Press.

Gade, D. W. 1996. Carl Troll on Nature and Culture in the Andes. *Erdkunde* 50: 301–316.

Gandy, M. 1996. Crumbling Land: The Postmodernity Debate and the Analysis of Environmental Problems. *Progress in Human Geography* 20(1): 23–40.

Glacken, C. J. 1967. *Traces on the Rhodian Shore.* Los Angeles and Berkeley: Uni-versity of California Press.

Gómez-Pompa, A., and A. Klaus. 1992. Taming the Wilderness Myth. *BioScience* 42(4): 271–279.

Gosz, J. R., and P. J. H. Sharpe. 1989. Broad-Scale Concepts for Interactions of Climate, Topography, and Biota at Biome Transitions. *Landscape Ecology* 3: 229–243.

Goudie, A. 1990. *The Human Impact on the Natural Environment.* Cambridge: The Massachusetts Institute of Technology Press.

Graber, David M. 1995. Resolute Biocentrism: The Dilemma of Wilderness in National Parks. In *Reinventing Nature? Responses to Postmodern Deconstructionism,* eds. Michael E. Soulé and Gary Lease, pp. 123–136. Covelo: Island Press.

Grubb, P. J. 1977. The Maintenance of Species-Richness in Plant Communities: The Importance of the Regeneration Niche. *Biological Review* 52: 107–145.

Guha, R. 1997. The Authoritarian Biologist and the Arrogance of Anti-humanism. *The Ecologist* 21: 14–20.

Haffer, J. 1969. Speciation in Amazon Forest Birds. *Science* 165: 131–137.

Haffer, J. 1982. General Aspects of the Refuge Theory. In *Biological Diversification in the Tropics,* ed. G. T. Prance, pp. 6–24. New York: Columbia University.

Harlan, J. R. 1992. *Crops and Man.* 2d Edition. Madison: American Society of Agronomy and Crop Science Society of America.

Horn, S. P. 1993. Postglacial Vegetation and Fire History in the Chirripó Páramo of Costa Rica. *Quaternary Research* 40: 107–116.

Horn, S. P., and R. L. Sanford Jr. 1992. Holocene Fires in Costa Rica. *Biotropica* 24(3): 354–361.

Ives, J., and B. Messerli. 1989. *The Himalayan Dilemma: Reconciling Development and Conservation.* London: Routledge.

Johannessen, C. L. 1963. Savannas of Interior Honduras. *Ibero-Americana* 46: 1–173.

Kellman, M. 1975. Evidence for Late-Glacial Age Fire in a Tropical Montane Savanna. *Journal of Biogeography* 2: 57–63.

Kellman, M., and R. Tackaberry. 1997. *Tropical Environments: The Functioning and Management of Tropical Ecosystems.* London: Routledge.

Klammer, G. 1982. Die Paläowüste des Pantanal von Mato Grosso und die pleistozäne Klimageschichte der brasilianischen Randtropen. *Zeitschrift für Geomorphologie* 26: 393–416.

Lauer, W., ed. 1984. *Natural Environment and Man in Tropical Mountain Ecosystems.* Stuttgart: Franz Steiner.

Lauer, W. 1993. Human Development and Environment in the Andes: A Geoecological Overview. *Mountain Research and Development* 13(2): 157–166.

Laszlo, E. 1972. The Systems View of Nature. In *The Systems View of the World,* ed. E. Laszlo, pp. 19–46. New York: George Braziller.

Levin, S. A. 1992. The Problem of Pattern and Scale in Ecology. *Ecology* 73(6): 1943–1967.

Marsh, G. P. 1864 [1965]. *Man and Nature: Or, Physical Geography as Modified by Human Action,* ed. D. Lowenthal. Cambridge: Harvard University Press.

Medley, K. E. 1993. Primate Conservation along the Tana River, Kenya. An Examination of Forest Habitat. *Conservation Biology* 7: 109–121.

Meffe, G. K., and C. R. Carroll, eds. 1994. *Principles of Conservation Biology.* Sunderland, Massachusetts: Sinauer Associates.

Meinig, D. W., ed. 1979. *The Interpretation of Ordinary Landscapes: Geographical Essays.* New York: Oxford University Press.

Metcalfe, S. E., F. A. Street-Perrott, R. B. Brown, P. E. Hales, R. A. Perrott, and F. M. Steininger. 1989. Late Holocene Human Impact on Lake Basins in Central Mexico. *Geoarchaeology* 4(2): 119–141.

Metz, J. J. 1990. Forest Product Use in Upland Nepal. *Geographical Review* 80(3): 279–287.

Metz, J. J. 1994. Forest Product Use at an Upper Elevation Village in Nepal. *Environmental Management* 18(3): 371–390.

Minnich, R. A. 1983. Fire Mosaics in Southern California and Northern Baja California. *Science* 219: 1287–1294.

Murdoch, J., and J. Clark. 1994. Sustainable Knowledge. *Geoforum* 25(2): 115–132.

Myers, N. 1995. Environmental Unknowns. *Science* 269: 358–360.

Noss, R. F. 1983. A Regional Landscape Approach to Maintain Diversity. *BioScience* 33: 700–706.

Odum, E. P. 1969. The Strategy of Ecosystem Development. *Science* 164: 262–270.

Parsons, J. J. 1971. Ecological Problems and Approaches in Latin American Geography. In *Geographic Research on Latin America, Benchmark 1970*, eds. B. Lentnek, R. L. Carmin, and T. L. Martinson, pp. 13–32. Muncie, Indiana: Ball State University.

Peet, R., and M. Watts, eds. 1996. *Liberation Ecologies: Environment, Development, Social Movements.* London: Routledge.

Peluso, D. 1993. Conservation and Indigenismo. *Hemisphere* 5: 6–8.

Peluso, N. L. 1992. *Rich Forests, Poor People: Resource Control and Resistance in Java.* Los Angeles and Berkeley: University of California Press.

Peluso, N. L. 1993. Coercing Conservation?—The Politics of State Resource Control. *Global Environmental Change* 3(2): 199–216.

Pérez, F. 1992. The Ecological Impact of Cattle on Caulescent Andean Rosettes in a High Venezuelan Páramo. *Mountain Research and Development* 12: 29–46.

Pérez, F. 1993. Turf Destruction by Cattle in the High Equatorial Andes. *Mountain Research and Development* 13: 107–110.

Pickett, S. T. A. 1980. Non-equilibrium Coexistence of Plants. *Bulletin of the Torrey Botanical Club* 107(2): 238–248.

Pickett, S. T. A., and John N. Thompson. 1978. Patch Dynamics and the Design of Nature Reserves. *Biological Conservation* 13: 27–37.

Pickett, S. T. A., and P. S. White, eds. 1985. *The Ecology of Natural Disturbance and Patch Dynamics.* New York: Academic Press.

Pimm, S. L. 1984. The Complexity and Stability of Ecosystems. *Nature* 307(26): 321–326.

Pimm, S. L. 1991. *The Balance of Nature? Ecological Issues in the Conservation of Species and Communities.* Chicago: University of Chicago Press.

Posey, D. A. 1992. Interpreting and Applying the "Reality" of Indigenous Concepts: What is Necessary to Learn from the Natives? In *Conservation of Neotropical Forests: Working from Traditional Resource Use,* eds. K. H. Redford and C. Padoch, pp. 21–34. New York: Columbia University Press.

Prance, G. T. 1978. The Origin and Evolution of the Amazon Flora. *Interciencia* 3/4: 207–222.

Pratt, D. J., and M. D. Gwynne, eds. 1977. *Rangeland Management and Ecology in East Africa.* New York: R. E. Krieger.

Redclift, M. 1987. *Sustainable Development: Exploring the Contradictions.* London: Routledge.

Redford, K. H. 1990. The Ecologically Noble Savage. *Orion Nature Quarterly* 9: 25–29.

Redford, K. H., and J. A. Mansour. 1996. *Traditional Peoples and Biodiversity Conservation in Large Tropical Landscapes.* Washington, D. C.: The Nature Conservancy.

Roberts, R. 1995. Taking Nature-Culture Hybrids Seriously in Agricultural Geography. *Environment and Planning A* 27: 673–675.

Sanford, R. L., Jr., J. Saldarriaga, K. E. Clark, C. Uhl, and R. Herrera. 1985. Amazon Rain-Forest Fires. *Science* 227: 53–55.

Sauer, C. O. 1956. The Agency of Man on the Earth. In *Man's Role in Changing the Face of the Earth,* volume 1, ed. W. L. Thomas Jr., pp. 49–69. Chicago: University of Chicago Press.

Sauer, C. O. 1958 [1967]. Man in the Ecology of Tropical America. In *Land and Life: A Selection from the Writings of Carl Ortwin Sauer,* ed. J. Leighly, pp. 182–193. Los Angeles and Berkeley: University of California Press.

Sauer, J. D. 1988. *Plant Migration: The Dynamics of Geographic Patterning in Seed Plant Species.* Los Angeles and Berkeley: University of California Press.

Savage, M. 1991. Structural Dynamics of a Southwestern Pine Forest Under Chronic Human Influence. *Annals of the Association of American Geographers* 81(2): 271–289.

Savage, M. 1993. Ecological Disturbance and Nature Tourism. *Geographical Review* 83(3): 290–300.

Schelhas, J., and R. Greenberg, eds. 1996. *Forest Patches in Tropical Landscapes.* Covelo: Island Press.

Schmidt, K. P. 1954. Faunal Realms, Regions and Provinces. *Quarterly Review of Biology* 29: 322–336.

Schubert, C. 1992. The Glaciers of the Sierra Nevada de Merida (Venezuela): A Photographic Comparison of Recent Deglaciation. *Erdkunde* 46: 58–64.

Scoones, I. 1994. *Living with Uncertainty: New Directions in Pastoral Development in Africa.* London: Intermediate Technology Publications Ltd.

Smith, J. M. 1996. Ramifications of Region and Senses of Place. In *Concepts in Human Geography,* eds. C. Earle, K. Mathewson, and M. S. Kenzer, pp. 189–212. London: Rowman and Littlefield.

Soulé, Michael E., and Daniel Simberloff. 1986. What Do Genetics and Ecology Tell Us About the Design of Nature Reserves? *Biological Conservation* 35: 19–40.

Soulé, Michael E., and B. A. Wilcox. 1980. *Conservation Biology: An Evolutionary-Ecological Perspective.* Sunderland, Massachusetts: Sinauer Associates.

Sponsel, L. E., T. N. Headland, and R. C. Bailey. 1996. *Tropical Deforestation: The Human Dimension.* New York: Columbia University Press.

Sprugel, D. G. 1991. Disturbance, Equilibrium, and Environmental Variability: What Is "Natural" Vegetation in a Changing Environment? *Biological Conservation* 58: 1–18.

Street-Perrott, F. A. et al. 1989. Anthropogenic Soil Erosion around Lake Patzcuaro, Michoacan, Mexico, during the Preclassic and Late Postclassic-Hispanic Periods. *American Antiquity* 54: 759–765.

Stoddart, D. R. 1972. Catastrophic Damage to Coral Reef Communities by Earthquake. *Nature* 239: 51.

Taylor, A. H., and Qin Zisheng. 1989. Structure and Composition of Uncut and Selectively Cut *Abies-Tsuga* Forest in Wolong Natural Reserve and Implications for Panda Conservation in China. *Biological Conservation* 47: 83–108.

Thomas, D. S. G., and R. J. Allison. 1993. *Landscape Sensitivity.* New York: John Wiley and Sons.

Thompson, L. G., E. Mosley-Thompson, M. E. Davis, P.-N. Lin, K. A. Henderson, J. Cole-Dai, J. F. Bolzan, and K.-b. Liu. 1995. Late Glacial Stage and Holocene Tropical Ice Core Records from Huascarán, Peru. *Science* 269: 46–50.

Troll, C., ed. 1968. *Geo-ecology of the Mountainous Regions of the Tropical Americas.* Proceedings of the Mexico UNESCO Symposium. Bonn: Ferd Dummlers Verlag.

Troll, C. 1971. Landscape Ecology (Geoecology) and Biogeocenology—A Terminological Study. *Geoforum* 8: 43–46.

Turner, B. L. II, W. C. Clark, R. W. Kates, J. F. Richards, J. F. Mathews, and W. B. Meyers. 1990. *The Earth as Transformed by Human Action: Global and Regional Changes in the Biosphere over the Past 300 Years.* Cambridge: Cambridge University Press.

Turner, M. G. 1989. Landscape Ecology: the Effect of Pattern on Process. *Annual Review of Ecology and Systematics* 20: 171–197.

Uhl, C., D. Nepstad, R. Buschbacher, C. Clark, B. Kauffman, and S. Subler. 1990. Studies of Ecosystem Response to Natural and Anthropogenic Disturbances Provide Guidelines for Designing Sustainable Land-Use Systems in Amazonia. In *Alternatives to Deforestation: Steps Toward Sustainable Use of the Amazon Rainforest,* ed. A. Anderson, pp. 25–42. New York: Columbia University Press.

Urban, D. L., R. V. O'Neill, and H. H. Shugart, Jr. 1987. Landscape Ecology. *BioScience* 37(2): 119–127.

Vale, T. R. 1982. *Plants and People: Vegetation Change in North America.* Resource Publications in Geography. Washington, D. C.: Association of American Geographers.

Vale, T. R. 1987. Vegetation Change and Park Purposes in the High Elevations of Yosemite National Park, California. *Annals of the Association of American Geographers* 77(1): 1–18.

Vandermeer, J. 1995. The Ecological Bases of Alternative Agriculture. *Annual Review of Ecology and Systematics* 26: 201–224.

Vankat, J. L., and J. Major. 1978. Vegetation Changes in Sequoia National Park, California. *Journal of Biogeography* 5: 377–402.

Veblen, T. T. 1992. Regeneration Dynamics. In *Plant Succession in Theory and Prediction,* eds. D. C. Glenn-Lewin, R. K. Peet, and T. T. Veblen, pp. 152–187. New York: Chapman and Hall.

Veblen, T. T., and D. C. Lorenz. 1988. Recent Vegetation Changes along the Forest/Steppe Ecotone of Northern Patagonia. *Annals of the Association of American Geographers* 78: 93–111.

Voeks, R. A. 1988. The Brazilian Fiber Belt: Harvest and Management of the Piassava Fiber Palm (*Attalea funifera* Mart.). *Advances in Economic Botany* 6: 262–275.

Voeks, R. A., and S. G. Vinha. 1988. Fire Management of the Piassava Fiber Palm (*Attalea funifera*) in Eastern Brazil. *Yearbook of the Conference of Latin Americanist Geographers* 14: 7–13.

von Humboldt, Alexander. 1822. *Personal Narrative of Travels to the Equinoctial Regions of the New Continent, 1799–1804,* trans. Helen Maria Williams. London: Longman.

Walker, B. 1989. Diversity and Stability in Ecosystem Conservation. In *Conservation for the Twenty-First Century,* eds. D. Western and M. C. Pearl, pp. 121–130. New York: Oxford University Press.

Watt, A. S. 1947. Pattern and Process in the Plant Community. *Journal of Ecology* 35(1/2): 1–22.

Westman, W. E. 1978. Measuring the Inertia and Resilience of Ecosystems. *BioScience* 28(11): 705–710.

Westman, W. E. 1985. *Ecology, Impact Assessment, and Environmental Planning.* New York: John Wiley and Sons.

Westoby, M., B. Walker, and I. Noy-Meir. 1989. Opportunistic Management for Rangelands Not at Equilibrium. *Journal of Range Management* 42(4): 266–273.

White, P. S. 1979. Pattern, Process, and Natural Disturbance in Vegetation. *The Botanical Review* 45: 229–299.

Wilbanks, T. J. 1994. "Sustainable Development" in Geographic Perspective. *Annals of the Association of American Geographers* 84(4): 541–556.

Worster, D. 1990. The Ecology of Order and Chaos. *Environmental History Review* 14(1/2): 1–18.

Worster, D. 1995. Nature and the Disorder of History. In *Reinventing Nature? Responses to Postmodern Deconstructionism,* eds. Michael E. Soulé and Gary Lease, pp. 65–86. Covelo: Island Press.

Wu, J., and O. L. Loucks. 1995. From Balance of Nature to Hierarchical Patch Dynamics: A Paradigm Shift in Ecology. *The Quarterly Review of Biology* 70(4): 439–466.

Young, K. R. 1993. National Park Protection in Relation to the Ecological Zonation of a Neighboring Human Community: An Example from Northern Peru. *Mountain Research and Development* 13(3): 267–280.

Young, K. R. 1994. Roads and the Environmental Degradation of Tropical Montane Forests. *Conservation Biology* 8(4): 972–976.

Young, K. R. 1996. Threats to Biological Diversity Caused by Coca/Cocaine Deforestation in Peru. *Environmental Conservation* 23: 7–15.

Zimmerer, K. S. 1993. Soil Erosion and Labor Shortages in the Andes with Special Reference to Bolivia, 1953–91: Implications for "Conservation-with-Development." *World Development* 21(10): 1659–1675.

Zimmerer, K. S. 1994. Human Geography and the "New Ecology": The Prospect and Promise of Integration. *Annals of the Association of American Geographers* 84(1): 108–125.

Zimmerer, K. S. 1996a. Ecology as Cornerstone and Chimera in Human Geography. In *Concepts in Human Geography,* eds. C. Earle, K. Mathewson, and M. S. Kenzer, pp. 161–188. London: Rowman and Littlefield.
Zimmerer, K. S. 1996b. *Changing Fortunes: Biodiversity and Peasant Livelihood in the Peruvian Andes.* Los Angeles and Berkeley: University of California Press.
Zimmerer, K. S. 1998. The Ecogeography of Major Crops: Versatility and Conservation. *BioScience* 48(6): 445–454.

PART 1

THE CONSERVATION CHALLENGES OF FOREST LOSS AND FRAGMENTATION

Introduction

More than a third of the earth's land mass was originally covered by forest and woodlands. Natural changes, like long-term climatic shifts or catastrophic disturbances, have always acted to change the boundaries of forested regions and the continuity of forest within those regions. As is well known, many types of land-use practices also result in changes in forest extent and cover. What is difficult, yet important, for biological conservation in developing countries is to assess landscape forest patterns imposed by environmental gradients and by natural disturbances, and to compare such patterns to those created by human practices such as forest clearing, extraction, and burning. In addition, synergistic and complex interactions are likely and to be expected. Examples of these phenomena are discussed in the following case studies.

Kimberly Medley reviews landscape-level changes in the important riverine forests of the Tana River in eastern Kenya. This tropical forest landscape is long and narrow, centered on the river, and surrounded by a semiarid region dominated by thorn scrub. The forest's dynamics are notably affected by shifts in the river and by large herbivores such as elephants and buffalo. Additional disturbance is obviously anthropogenic in origin and results in extraction of forest products and some deforestation on substrates in the floodplain that are suitable for agriculture and extraction of forest products. As Medley shows, the external and internal processes that cause forest heterogeneity also affect wildlife populations, in particular the red colubus and crested mangabey monkeys that are species of concern. An integrated approach to conservation must include population analyses of the endangered species, nested within assessments of landscape dynamics, and interfaced with rural development programs that reduce pressure on the land while allowing the participation of local people.

At first, the forested landscape described by Alan Taylor and Qin Zisheng in the humid montane region of Southwest China appears to be quite different. However, there are similarities in that both natural disturbances and anthropogenically-caused changes in forest composition are critical to understand because they collectively affect the extent and quality of habitat for the endangered giant panda. Solutions in terms of conservation programs require an emphasis on the design of forestry practices that permit or even foster postharvest use of forest stands by pandas, in addition to ecological restoration practices that restore forest cover and increase connectivity of forest habitats. Taylor and Qin feel that control of agricultural expansion and poaching are also critical. An interesting constraint is the restriction of the giant panda's diet to understory bamboo. As a result, it is the interaction between forest dynamics and the growth and mortality of the bamboo species that is one of the critical features of this system.

Kenneth Young provides an overview of similar issues in the humid montane forest region of the central Andes. In his conceptualization, two general types of montane forest landscapes are found in humid sites above about 2000 meters in Ecuador, Peru, and Bolivia: landscapes originally covered by continuous cover of forest (except for elevations above timberline), and those that even without human impact would have had naturally fragmented forest because of the heterogeneity of environmental conditions in the rugged mountainous terrain. Each of these two types has been additionally modified by forest clearing and degradation, some undoubtedly dating back to the arrival of people at the start of the Holocene. Forests on the old deforestation fronts are small, quite isolated from each other, and often chronically disturbed by local people. The forested landscapes that only recently have been deforested tend to be the wettest montane forests and contain the most biological diversity, including plant and animal species that need large tracts of forest and interior forest habitats. These landscapes require the establishment and maintenance of large nature reserves, while the forest patches on old deforestation frontiers are instead best dealt with in rural development programs that include ecological restoration using native tree and shrub species.

Fernando Echavarria provides an example of a study in which the rate of landscape change is measured on an active deforestation front located in and near a national park in southern Ecuador. Remote sensing provides baseline and quantifiable measures of deforestation that are of fundamental importance to conservationists and development planners. However, one of the sources of uncertainty in using these data is the problem inherent in moving from spectral reflection captured by satellite-borne sensors to the next steps of spectral normalization, classification, and change detection. For some of these steps, in fact, Echavarria is able to quantify the uncertainty by providing accuracy assessments and amounts of variation for the deforestation rates. As he makes clear, the location and tempo of deforestation in any particular landscape are quite variable. Regional assessments only provide general guidance, and do not allow for planning of specific conservation-with-development programs in particular locations. Remote sensing potentially fills a valuable role in providing actualized assessments of forest cover and spatial patterns in a manner that also takes account of the limitations of the technique.

1

Landscape Change and Resource Conservation along the Tana River, Kenya

Kimberly E. Medley

Conservation Dilemmas for the Tana River Primate National Reserve

Tropical riverine forests in East Africa, some obtaining canopy heights 20 meters above an otherwise thorn-scrub environment, are small in total area, particularly sensitive to changes in the stream system and hydrologic regime, and a valuable land resource for human populations (Medley and Hughes 1996). Allogenic "natural" disturbances imposed by the river and human-landscape modification greatly influence the distribution, structure, and dynamics of forest patches. Riverine forests in semiarid regimes are among the more fragmented and consequently threatened ecosystems of developing countries.

The riverine forest landscape along the Tana River, Kenya, provides the only habitat for two endangered monkeys, the Tana River red colobus (*Colobus badius rufomitratus*) and crested mangabey (*Cercocebus galeritus galeritus*) (Marsh 1976). Early research by Marsh (1978) and Homewood (1976) quantified their populations and guided the establishment of the Tana River Primate National Reserve in 1976. Upon establishment, the reserve captured over 50% of their populations and some of the best riverine forest habitat.

Tana River Primate National Reserve is one of the smallest reserves in

39

East Africa (171 km²); the forest area, which occurs as 26 patches near the river, is about 9.5 square kilometers in area. A resurvey conducted ten years after its establishment revealed an 80% decline in the red colobus and a 25% decline in the crested mangabey populations (Marsh 1986). This crash in the populations of both primates may be attributable to a loss and/or decline in the condition of their forest habitat (Marsh 1986). By 1986 the National Museums of Kenya, Kenya Wildlife and Management Department (now the Kenya Wildlife Service), Wildlife Conservation International–New York Zoological Society (now the Wildlife Conservation Society), World Wildlife Fund, and African Wildlife Foundation began the collaborative "Tana River Primate Project." These agencies sponsored research on natural and human factors that contribute to forest fragmentation, the adaptive characteristics of the two endangered primates to their forest habitat, and management options for sustainable development.

This chapter reviews some major findings from my study of forest ecology and fragmentation in the reserve, and describes their implications for conservation in a riverine biogeographical landscape. The study defines the *landscape* by the composition, structure, and position of forest patches along the river; the *region* includes the semiarid floodplain corridor in East Kenya. The primates' dependence on mature forest cover, coupled with the reality of patch dynamics and landscape change, mean that both intensive restorative management and community participation in the protection of nonreserve lands are needed for conservation. Landscape fragmentation resulting from the interplay between natural environmental processes and human activities (see chapters 2 and 3) threatens the endangered primate populations and other resources in the protected area. The study is biogeographical in its attempt to: (1) couple ecology with a geographic perspective that examines spatial patterns at multiple scales (regionally along the river corridor, among forest patches in the landscape, and within a forest patch); (2) focus on human-environment interactions; and (3) emphasize conservation and compatible relationships between humans and resources as the management alternative for sustainable development.

Geographic Setting

The Tana River Primate National Reserve is located along the lower floodplain section of the Tana River, approximately 100 kilometers from the Indian Ocean (map 1.1). The Tana is the largest river in Kenya, beginning in the highlands near Mount Kenya and downcutting as far as Kora Rapids, then continuing as a meandering stream with an associated

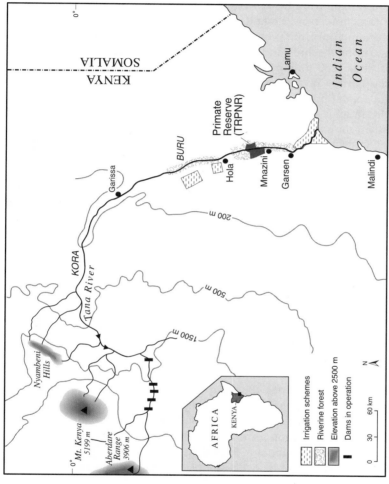

Map 1.1. The Tana River basin, Kenya. The new dam sites at Mutonga and Grand Falls are shown as triangles south of the Nyambeni Hills (Butynski 1995). The map is modified from Medley (1992).

41

floodplain until it meets the delta at Garsen. The watershed begins near the equator and extends to about four degrees south latitude at the Indian Ocean. Temperatures are consistently warm with minimum and maximum temperatures averaging between 21.4°C and 33°C at Hola, about 40 kilometers upstream from the reserve (Muchena 1987).

The Tana flows from a humid montane environment through a semiarid region. Annual precipitation along the river corridor is more than 1000 millimeters in the highlands, less than 300 millimeters at Garissa, and greater than 1000 millimeters at the coast. The stream loses volume as water evaporates or seeps into the ground along the floodplain corridor. Average annual precipitation at Hola is about 470 millimeters (Muchena 1987). Rainfall is seasonal, with long rains occurring between March and May, when the sun is north of the equator (southeast monsoon), and between October and December, when the sun is south of the equator (northeast monsoon).

The floodplain section of the Tana River is characterized by a gradient upstream toward lower precipitation and by unpredictability in the temporal and spatial distribution of rainfall events that is typical of a semiarid climate regime. According to the Holdridge (1967) classification scheme, the life zone is thorn woodland. Riverine forest is groundwater-dependent and its lateral extent, limited to about one kilometer, is determined by access to the water table, which declines with distance from the river. Consequently, the hydrological characteristics of the river system strongly influence forest composition, structure, and dynamics (Hughes 1988, 1990; Medley and Hughes 1996).

Pastoral and agricultural groups rely on the floodplain for their subsistence. The Orma and Somali-Wardei pastoral ethnic groups reside on the upland plains (all outside the reserve) and use the river only during the dry seasons. The Pokomo are an agricultural group that relies on low-lying depositional settings along the river, such as old oxbows and point bars for farming. When the reserve was established, approximately 550 Pokomo lived and/or maintained agricultural plots inside its boundary (Marsh 1976). Annual population growth between 1975 and 1991 was about 1%, increasing the total to about 630 residents (Seal et al. 1991). The clearing of land for farms and settlements, coupled with the environmental disturbances of a meandering stream, create a mosaic of forest patches along the floodplain corridor (map 1.2).

Research Approach

The Tana River Primate National Reserve became a focal region for research in the mid-1970s. Ecological surveys conducted by Andrews et al.

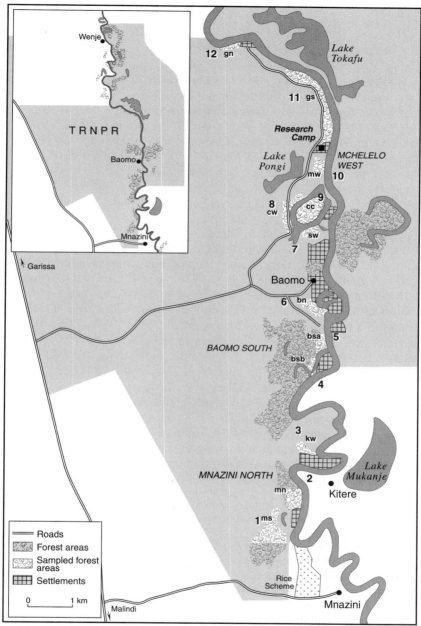

Map 1.2. The study area located in the south-central sector of the Tana River Primate National Reserve. The twelve forest areas from south to north are: (1) Mnazini South (ms); (2) Mnazini North (mn); (3) Kitere West (kw); Baomo South B (bsb); (5) Baomo South A (bsa); (6) Baomo North (bn); (7) Sifa West (sw); (8) Congolani West (cw); (9) Congolani Central (cc); (10) Mchelelo West (mw); (11) Guru South (gs); (12) Guru North (gn). The top inset shows the reserve boundary and riverine forest (shaded). Modified from Medley (1992).

(1975) documented the biological diversity of the region and provided preliminary checklists for the birds, mammals, and plants. Research by Marsh (1978) on the red colobus and by Homewood (1976) on the crested mangabey provided baseline information on the behavior and demography of the two primates and descriptive analyses of the forest community types. Allaway (1979) examined elephant populations during this same time period. Later, Hughes (1985) conducted a regional study on forest ecology, hydrology, and the environmental impacts of large-scale development projects along the Tana River. Her study included sample plots at Bura and in the reserve. Concurrent with my field research, Decker (1989) and Odhiambo (1990) conducted research on the red colobus, Kinnaird (1990) studied the crested mangabey, and Njue (1992) examined hydrologic and edaphic influences on forests in the reserve. Together, these projects helped direct my selection of sites, provided a basis for regional/temporal observations, and made possible the study of primate-habitat relationships (Medley 1990).

Field data were collected for my study in May–July 1986 and September 1987–October 1988. I selected 12 areas that were representative of the variations in forest types and for which population and habitat-use data were available on the endangered primates (map 1.2). In each study area, sample points were located at 50-meter distances along transects placed about 100 meters apart. The number of transects, transect length, and total points varied according to the size of the forest patch. This sampling design was used to compare the woody plant composition and structure of forests in the reserve (pooled for the 12 study areas) with those located outside at upstream (Hughes 1985) and downstream (Medley et al. 1989) localities, to examine community diversity within the reserve based on the relative densities of tree species in the study areas, and to look at spatial patterns within forest patches.

Plant Diversity and Geography

The riverine forest flora of the Tana River is a product of continental (Richards 1973) and regional (Hamilton 1974) influences. Compared with other tropical forests, the woody vegetation is floristically poor (<100 tree species), but taxonomic affinities offer important insights into the biogeographical history of East Africa. I documented 172 woody plant species in 127 genera and 49 plant families in the reserve (Medley 1992). Trees, shrubs, and lianas are the predominant growth forms, herbs are uncommon, and epiphytes, ferns, and mosses are absent. Floristic diversity and structure relies almost entirely on access to water from floods for early establishment, and groundwater for continued growth. Of particular importance are the occurrences of *Populus ilicifolia,* a riverbank endemic, *Cynometra lukei,* a newly described tree species that dominates low-lying backswamp areas (Beentje 1994), and *Pachystela msolo,* a tree that dominates low-levee settings and is only found outside West Central Africa in the East Usambara Mountains of Tanzania and along the lower Tana River. The geographic isolation of forests along the Tana River and their biogeographical history contribute to the presence of some rare, endemic, or disjunct plant species (Medley 1992).

Based on interviews in the region, 98 plant species in 88 genera and 38 families are of resource value to the Pokomo (Medley 1993a). Fifteen species are used for food, 34 for construction, 43 for technology, 23 for remedy, 2 for commerce, 9 for firewood, 4 for ritual, and 7 for other uses such as beehives, games, or poisons. Plants vary in their quality for particular uses and the relative impact of extraction on their growth structure (Medley 1993a). What was apparent from my discussions with the Pokomo is that only a few of them frequent the forests, local knowledge of forest-plant uses is dwindling, and the value of resource diversity is not widely recognized.

The diversity of forest ecosystems is best explained by the site conditions of different riverine landforms, laterally with distance from the river or longitudinally along the river course (Medley 1992; Medley and Hughes 1996). Forest types include flooded low-levee positions along the river dominated by *Pachystela msolo* and *Ficus sycomorus,* high-levee positions with a more open and diverse canopy (often with *Sorindeia madagascariensis* and *Diospyros mespiliformis*), clay-backwater swamps dominated by trees adapted to long periods of inundation (*Garcinia livingstonei* and *Mimusops obtusifolia*), and the plains edge with an open canopy of more drought-resistant species such as *Acacia rovumae* and *Salvadora persica* (Medley 1992). A shift in the river greatly influences the persistence of a forest type or the configuration of ecosystems (Medley and Hughes 1996). For instance, the meander cutoff at Congolani Central in 1961 reduced low-levee forest, encouraged pioneer vegetation in the filled oxbow, and shifted the position of point bar and cut banks along the river (figure 1.1). Upstream forests at Bura along the floodplain corridor have a low heterogeneity of depositional features; downstream forests near Garsen have a low occurrence of well-drained sandy levees and an abundance of tall grassland that is adapted to long periods of inundation by flooding (Hughes 1988; Medley and Hughes 1996). *Pachystela msolo,* the dominant tree in the reserve and frequent downstream, is absent from the Bura forests; *Barringtonia racemosa,* an abundant tree in the forests near Garsen, is absent from upstream areas; and *Acacia elatior,* a tree common in the Bura forests, is absent from forests downstream (Medley 1992). These

Fig. 1.1. Aerial view of the meander cutoff and oxbow at Congolani Central, along the Tana River. The photograph identifies some major community types that have been disturbed by a shift in the river course.

findings suggest that local forest-landscape diversity patterns are greatly affected by changes in the river course and that the regional diversity of forest community types is not adequately captured in the protected section of the Tana River.

Forest-Primate Relationships

Primate conservation depends on the protection of suitable forest habitat (Medley 1993b; see chapter 2). My objective was to identify the habitat attributes that are most related with their successful growth, and potentially important to their survival. Between 1985 and 1988 primate abundances did not change significantly in the reserve (Decker and Kinnaird 1992), suggesting that their populations were temporarily stable and a reflection of the relative carrying capacities of the existing forest patches.

Primate habitat was defined by attributes of forest structure, food-resource abundances, levels of disturbance, and the shapes and sizes of the forest study areas. A model of relative habitat quality was constructed by correlations between these attributes and primate abundances, and by the characteristics of high-use areas determined from annual ranging patterns within three forests (Decker and Kinnaird 1992; Medley 1993b).

Primate group abundances show positive relationships with the height of the forest canopy, forest-patch size, and the relative amount of interior forest (Medley 1993b). Although the primates occur in a dynamic riverine environment, disturbances that disrupt homogeneity in a forest patch, reduce its area, and increase edge effects show notable negative relationships with primate abundances. Primates prefer interior-forest conditions, but their adaptation to a low diversity and cover of food resources suggests that food requirements may be easily met in a forest of suitable structure (Medley 1993b).

A preference for forest-patch interiors is also shown by primate ranging patterns in three forest patches. High-use areas (greater than 3% of total site records within a forest) corresponded mostly with vegetation plots near the patch center in Mchelelo West, near the river in Mnazini North, and near the oxbow in Baomo South (map 1.2). Figs are a valuable primate resource and at least partially influence ranging patterns (Decker 1994; Kinnaird 1990). *Ficus sycomorus* occurs in most of the plots and the other figs *F. bussei, F. natalensis,* and *F. bubu,* despite their overall low occurrence in the reserve (4 out of 363 points), were present in six of seven 50 × 50 m high-use plots.

Forest-Landscape Disturbances

Forest area, fragmentation, and ecological structure are principal landscape characteristics that define the condition of riverine forest vegetation along the Tana River. Forests in the landscape are disturbed primarily by floodplain dynamics, human activities, and large mammals such as the elephant and buffalo. I evaluated these disturbances relative to the impact they have on habitat for the endangered primates and the potential for forest regeneration.

The primates prefer forest-patch interiors; therefore a loss in forest area and/or an increase in edge effects are negative impacts on their populations. Using airphotos, I measured a 56% loss in forest cover from 1960 to 1975 for the south-central portion of the reserve. Five forests were fragmented into 15 forest patches and the mean area-to-perimeter ratio dropped by 44% (Medley 1993b). Decker and Kinnaird (1992) suggest that the sharp decline in primate populations between 1975 and 1985 may be due to their compression into smaller forest patches. From the establishment of the reserve in 1976 to 1988 I measured an approximate 4% decline in forest cover in the 12 study areas: 7.5 hectares were lost from riverbank erosion and 8.8 hectares from clearing. Forest losses attributable to "natural" and "human" disturbances were approximately equal in the reserve.

River meanders and floods are allogenic factors that continually alter conditions for trees species establishment and growth (Medley and Hughes 1996). Mature forest depends on access to the groundwater table, but seedling establishment and early growth must rely on new sediment deposits such as point bars and/or surface soil replenishment with overbank flooding (Medley and Hughes 1996). Along a short section of the river, such as that flowing through the reserve, it is not clear whether lateral gains through deposition are equal to erosional losses. Furthermore, assuming a link between upstream and downstream processes, it is not clear to what degree floodplain building is impacted by the trapping of sediments in upstream dam reservoirs (Medley and Hughes 1996). One possible explanation for the low regeneration observed in *Pachystela msolo* is that successful establishment depends on a rare, big flood event and the deposition of much coarser-textured sediments. A decrease in flood levels, reduction in sediment loads, or change in sediment texture (from sands to clays) are all possible outcomes of dam construction in the upper basin (Hughes 1994; Medley and Hughes 1996).

The local Pokomo populations directly influence forest cover through the clearing of mature forest near the river for farm plots and the cultivation of young floodplain point bars. Their demand for land resources is in direct conflict with the conservation of riverine forest in the protected area. The extraction of forest products, however, does not directly result in a loss of forest area. In contrast, plants are selected from the forest with consequent effects on the structure and composition of the existing habitat, such as the formation of a canopy gap when canoes or beehives are constructed from large trees, coppiced trees by the cutting of stems for poles, and a reduction in mean palm height with the harvesting of fronds (Medley 1993a). High-extraction intensities occur near settlements, trails, or roads, or are associated with the acquisition of a large tree. The impacts of extraction on forest composition and regeneration potential are difficult to project and are complicated by the influences of a spatially heterogeneous and dynamic physical environment.

Large mammal disturbances of the understory forest vegetation include bark removal, browsing, and the breaking of stems. *Rinorea elliptica* had nearly 80% of its stem-basal area recorded as damaged (Medley1990). Particularly susceptible locations are along the forest-scrubland edge, near oxbows, or along dry-season tracks to the river. Cape buffalo remain in fair numbers and are still an important influence on the forests. A significant decline in elephant abundances in the Tana River District, mostly because of poaching, has greatly reduced overall animal disturbances in the reserve. Forest structure may be changing from the absence of this impact.

Discussion: Implications for Conservation

Most reserves are established for the long-term protection of biotic resources. A principal goal for the Tana River Primate National Reserve is the long-term preservation of the endangered Tana River red colobus and crested mangabey. *In situ* preservation of the primates depends on the maintenance of suitable forest habitat and viable population sizes. Primate habitat is negatively influenced by forest loss, fragmentation, and ecological decline in structure and composition; all are consequences of landscape disturbances along the Tana River. The underlying assumption is that species may be preserved through management, or the conservation of resources (Mangel et al. 1996). Three interrelated questions arise when my results are applied to conservation in this region: (1) Can the existing reserve effectively promote sustainability in primate populations? (2) Does primate conservation ensure protection of the biogeographical landscape? and (3) In what ways can local management remedy the degradation of forest resources?

From an ecological perspective, the findings from my research suggest that the existing reserve is too small. Historical trends for this region show a relaxation in primate numbers with significant forest decline (Decker and Kinnaird 1992), adding support to the conservation implications of island biogeography (Shafer 1990). Another major shift in the river or forest-clearing event could further threaten already endangered species populations. The new view that ecosystems are dynamic or nonequilibrial (Botkin 1990; Zimmerer 1994) can be applied to the Tana River forests—the placement of the reserve's boundaries, however, do not explicitly account for this complex disturbance regime (Pickett and Thompson 1978; White and Bratton 1980) and are subject to question.

Two approaches serve to reevaluate the "save it" step in the conservation process (*sensu* Janzen 1992). One approach focuses on intensive resource management in the existing reserve and supports restoration as a stewardship consideration (Medley 1994; see chapter 3). Forest restoration manipulates the landscape in order to increase forest area, establishes corridors between forest patches, and enriches food resources for primates. Research on the compositional characteristics of the forests and their potential for regeneration can be directly applied to the selection of sites and tree species (Medley 1994). Sites are ranked according to their ecological potential for regeneration (environmental conditions, existing evidence of tree saplings) and their probable contribution toward supporting primate populations. Trees are selected based on their success in preliminary nursery trials, tolerance of open (high-light) or forest (low-light) conditions, and value as a forest resource. A proposal for restoration was submitted as part of a Global Environmental Facility project in the reserve (Kiss 1993). The objectives are to rehabilitate and establish primate habitat such that a change in their ranging patterns toward the restored areas and/or population growth are indirect measures of project success (see chapter 2). Forest degradational losses in a small reserve are compensated by the cultivation of new forest sites.

A second approach broadens resource management to consider equally forests protected in the reserve and the whole forest landscape along the river corridor. Many forest patches downstream maintain populations of either or both primates. Butynski and Mwangi (1995) found in a 1994 survey that only 37% of the colobus groups and 56% of the mangabey groups reside in the reserve. They saw primates in 18 of the 26 forest patches on trust/government lands, and in 6 of the 14 forest patches managed by the Tana Delta Irrigation Project. Their survey of these forests reduces the decline in the colobus from 80% to $< 50\%$. Although it may be possible to translocate isolated and threatened populations to forests in the reserve, their findings emphasize that currently unprotected groups are important to long-term population viability in both species. If their suggestions are followed, then resource management would be expanded to include forests downstream and outside of the reserve.

Furthermore, research findings on plant distributions and forest ecosystems show that the reserve does not adequately capture the regional biogeographical patterns of plant diversity along the river corridor. One can project a probable loss in ecosystem diversity, decline in species richness, and greater species isolation with the loss and/or degradation of forests along the Tana River upstream and downstream from the reserve. A loss in total forest area and the isolation of forest patches in a protected area may be equally important factors jeopardizing the conservation of community diversity. In looking outside of the reserve, the primates are only one measure of resource value. The patterns of diversity in species, communities, and resources as represented by forests upstream and downstream from the reserve, contribute to heterogeneity in this biogeographical landscape and are also worthy of protection (Noss 1990).

Longitudinal linkages also occur along the river corridor. Conservation efforts in the Tana River Primate National Reserve justify greater consideration of environmental impacts by upstream operations. For example, the Three Forks Dam that will be constructed near the tributaries from Mount Kenya (map 1.1) is projected to drop groundwater levels, further decrease sediment loads, and reduce seasonal flooding (Butynski 1995). The impacts on riverine forest could be devastating unless mitigation measures, such as controlled water releases from the reservoir, are a part of the final plan (Butynski 1995).

Both "natural" and "human" disturbances directly influence the distribution and stature of forests along the Tana River. One might argue that governments originally established reserves for posterity and to protect them from *human* disturbances (McNeely 1989; Shafer 1990). Economic development in the region is redirected toward nonconsumptive uses with the hope that these activities will provide compensation for the loss of lands by the local people. The Tana River Primate National Reserve presents a different case scenario: resident communities and human activities continue inside the boundary and economic development remains difficult for the resident population and surrounding communities.

Most of the past research, as represented by my study, focused on quantifying human impacts, rather than explicitly investigating potential positive links between humans and the environment. The social and economic factors influencing resource conservation are less understood (Mangel et al. 1996). Agricultural demands for land resources in the reserve are probably not compatible with sustainable development of high-quality primate habitat. The ecological impacts of forest-product extraction that depend on the number and types of products require further investigation (Posey 1993). If one views pole-cutting as an unacceptable impact, can maintaining a 50 × 50 meter trail-grid system in four forest areas for the monitoring of primate populations be viewed as compatible? How does resource extraction differ in selection pressure and plant response from the disturbances imposed during earlier times by large mammal populations? Will access to knowledge about resource diversity be lost when extraction is strictly regulated?

Active participation by the Pokomo in research and management, irrespective of their resource needs, is necessary for conservation in the reserve and along the floodplain corridor. They are important "stakeholders" who have "property rights" and are partially accountable for future sustainability (Mangel et al. 1996: 349). My research project would have been impossible, and future research will be impossible, without their cooperation. They "know" the system and contribute greatly to the compilation of data on resource diversity, land-use history, and landscape dynamics. The presence of the reserve can potentially contribute to their subsistence or development. To date, tourism is not well-established, wage labor in association with the research camp provides little employment (usually fewer than 50 persons), shifta/bandit raids continue to threaten local communities, and agricultural yields are unpredictable as determined by annual precipitation and the flooding regime. A major shift in the river just south of the reserve in 1989 collapsed the irrigation scheme and repositioned the village of Mnazini to the less secure east bank (former locations shown on map 1.2). Just as it may be necessary to extend the conservation of biological resources beyond the boundaries of protected areas, it is equally important to better support the local community as part of the conservation process (Western 1989, 1994). Humans rely on, and are very much a part of, this biogeographical landscape (Bell 1987).

Conclusion: Insights on Sustainable Development

Developing countries that recognize the potential threats to resources in dynamic landscapes, the importance of resources both inside and outside reserve boundaries, and the need for community participation now address rural development as a major component of their management activities (Alpert 1996; MAB 1987; Western 1989, 1994). The boundaries between protected areas and surrounding landscapes are becoming more diffuse and the linkages among landscape components are viewed as a vital component of any conservation plan (Arnold 1995; Mangel et al. 1996; Western 1989). In this setting, a framework for long-term conservation is being developed by the Kenya Wildlife Service for the Tana River Primate National Reserve. An important challenge is to work toward a balance between the protection and restoration of forests, and the "natural" and "human" disturbances that promote forest fragmentation. Biodiversity protection requires the maintenance of regional-to-landscape disturbance processes (Arnold 1995), and is often a scientific, economic, and social problem (Mangel et al. 1996). Accordingly, the future holds tremendous potential for collaboration in the conservation of forest resources as the landscape continues to change along the Tana River, and throughout developing regions.

Acknowledgments

Wildlife Conservation International (now The Wildlife Conservation Society) sponsored the field research; additional funds and language training were provided by the Michigan State University African Studies Center through a Department of Education National Resource Fellowship. The study was affiliated with the National Museums of Kenya and the Institute of Primate Research, under permission granted by the Office of the President (Permit No. OP.13/001/17 C 24/9). It was completed as part of my doctoral program in the Department of Botany and Plant Pathology, Michigan State University. I thank the staff at the East African Herbarium, especially Steven Rucina and Ann Robertson, for assistance with my plant identifications, Bakari Mohammed Garise for his assistance during the field research, my academic advisor John Beaman, and Kamau Wakanene Mbuthia and Tom Butynski for their reviews of the manuscript and guidance on the status of the Global Environmental Facility project.

References

Allaway, J. D. 1979. *Elephants and Their Interactions with People in the Tana River Region of Kenya*. Ph. D. dissertation, Cornell University.

Alpert, P. 1996. Integrated Conservation and Development Projects: Examples from Africa. *BioScience* 46(11): 845–855.

Andrews, P., C. P. Groves, and J. F. M. Horne. 1975. Ecology of the Lower Tana River Floodplain (Kenya). *Journal of East African Natural History Society and National Museums* 151: 1–31.

Arnold, G. 1995. Incorporating Landscape Pattern into Conservation Programs. In *Mosaic Landscapes and Ecological Processes*, eds. L. Hansson, L. Fahrig, and G. Merriam, pp. 309–337. London: Chapman and Hall.

Beentje, H. J. 1994. *Kenya Trees, Shrubs and Lianas*. Nairobi: National Museums of Kenya.

Bell, R. 1987. Conservation with a Human Face: Conflict and Reconciliation in African Land Use Planning. In *Conservation in Africa: People, Policies, and Practice*, eds. D. Anderson and R. Grove, pp. 79–101. Cambridge: Cambridge University Press.

Botkin, D. B. 1990. *Discordant Harmonies: A New Ecology for the Twenty-first Century*. New York: Oxford University Press.

Butynski, T. M. 1995. Report Says Dam Could Threaten Kenya's Endangered Primates. *African Primates* (IUCN) 1(1): 14–17.

Butynski, T. M., and G. Mwangi. 1995. Census of Kenya's Endangered Red Colobus and Crested Mangabey. *African Primates* 1(1): 8–10.

Decker, B. S. 1989. *Effects of Habitat Disturbance on the Behavioral Ecology and Demographics of the Tana River Red Colobus* (Colobus badius rufomitratus). Ph. D. dissertation, Emory University.

Decker, B. S. 1994. Effects of Habitat Disturbance on the Behavioral Ecology and Demographics of the Tana River Red Colobus (*Colobus badius rufomitratus*). *International Journal of Primatology* 15: 703–737.

Decker, B. S., and M. F. Kinnaird. 1992. Tana River Primates: Results of Recent Censuses. *American Journal of Primatology* 26: 47–52.

Hamilton, A. 1974. The History of Vegetation. In *East African Vegetation*, eds. E. M. Lind and M. E. S. Morrison, pp. 188–209. London: Longman.

Holdridge, L. R. 1967. *Life Zone Ecology*. Revised edition. San Jose, Costa Rica: Tropical Science Center.

Homewood, K. 1976. *The Ecology and Behavior of the Tana Mangabey* (Cercocebus galeritus galeritus). Ph. D. dissertation, University of London.

Hughes, F. M. R. 1985. *The Tana River Floodplain Forest: Ecology and the Impact of Development.* Ph. D. dissertation, University of Cambridge.

Hughes, F. M. R. 1988. The Ecology of African Floodplain Forests in Semiarid and Arid Zones: A Review. *Journal of Biogeography* 15: 127–140.

Hughes, F. M. R. 1990. The Influence of Flooding Regimes on Forest Distribution and Composition in the Tana River Floodplain, Kenya. *Journal of Applied Ecology* 27: 475–491.

Hughes, F. M. R. 1994. Environmental Change, Disturbance, and Regeneration in Semiarid Floodplain Forests. In *Environmental Change in Drylands: Biogeographical and Geomorphological Perspectives,* eds. A. C. Millington and K. Pye, pp. 321–345. New York: John Wiley and Sons.

Janzen, D. 1992. A South-North Perspective on Science in the Management, Use, and Economic Development of Biodiversity. In *Conservation of Biodiversity for Sustainable Development,* eds. O. T. Sandlund, K. Hindar, and A. H. D. Brown, pp. 27–52. Oslo: University Scandinavian Press.

Kinnaird, M. F. 1990. *Behavioral and Demographic Responses to Habitat Change by the Tana River Crested Mangabey* (Cercocebus galeritus galeritus). Ph. D. dissertation, University of Florida.

Kiss, A. 1993. The Global Environmental Facility and the Project for Conservation of the Tana River Primate National Reserve. *East African Natural History Society Bulletin* 22(3): 34.

MAB. 1987. *A Practical Guide to the MAB.* Programme on Man and the Biosphere. Paris: UNESCO.

Mangel, M., L. M. Talbot, and G. K. Meffe et al. 1996. Principles for the Conservation of Wild Living Resources. *Ecological Applications* 6(2): 338–362.

Marsh, C. W. 1976. *A Management Plan for the Tana River Game Reserve.* Report to the Kenya Department of Wildlife Conservation and Management, Nairobi. New York: New York Zoological Society.

Marsh, C. W. 1978. *Ecology and Social Organization of the Tana River Red Colobus,* Colobus badius rufomitratus. Ph. D. dissertation, University of Bristol.

Marsh, C. W. 1986. A Resurvey of Tana River Primates and Their Forest Habitat. *Primate Conservation* 7: 72–81.

McNeely, J. A. 1989. Protected Areas and Human Ecology: How National Parks Can Contribute to Sustaining Societies. In *Conservation for the Twenty-First Century,* eds. D. Western and M. C. Pearl, pp. 150–157. New York: Oxford University Press.

Medley, K. E. 1990. *Forest Ecology and Conservation in the Tana River National Primate Reserve, Kenya.* Ph. D. dissertation, Michigan State University.

Medley, K. E. 1992. Patterns of Forest Diversity along the Tana River, Kenya. *Journal of Tropical Ecology* 8: 353–371.

Medley, K. E. 1993a. Extractive Forest Resources of the Tana River National Primate Reserve, Kenya. *Economic Botany* 42: 171–183.

Medley, K. E. 1993b. Primate Conservation Along the Tana River, Kenya: An Examination of the Forest Habitat. *Conservation Biology* 7: 109–121.

Medley, K. E. 1994. Identifying a Strategy for Forest Restoration in the Tana River National Primate Reserve, Kenya. In *Beyond Preservation: Restoring and*

Inventing Landscapes, eds. D. Baldwin, J. DeLuce, and C. Pletsch, pp. 154–167. Minneapolis: University of Minnesota Press.

Medley, K. E., and F. M. R. Hughes. 1996. Riverine Forests. In *Ecosystems and Their Conservation in East Africa,* eds. T. McClanahan and T. Young, pp. 361–383. New York: Oxford University Press.

Medley, K. E., M. F. Kinnaird, and B. S. Decker. 1989. A Survey of the Riverine Forests in the Wema/Hewani Vicinity with Reference to Development and the Preservation of Human Resources. *Utafiti* 2(1): 1–6.

Muchena, F. N. 1987. *Soils and Irrigation of Three Areas in the Lower Tana Region, Kenya.* University of Wageningen.

Njue, A. 1992. *The Tana River Floodplain Forest, Kenya: Hydrologic and Edaphic Factors as Determinants of Vegetation Structure and Function.* Ph. D. dissertation, University of California–Davis.

Noss, R. F. 1990. Indicators for Monitoring Biodiversity: A Hierarchical Approach. *Conservation Biology* 4(4): 355–364.

Odhiambo, O. W. 1990. The Tana River Red Colobus (*Colobus badius rufomitratus*). *Utafiti* 3(1): 1–5.

Pickett, S. T. A., and J. N. Thompson. 1978. Patch Dynamics and the Design of Nature Reserves. *Biological Conservation* 13: 27–37.

Posey, D. A. 1993. Indigenous Knowledge in the Conservation and Use of World Forests. In *World Forests for the Future: Their Use and Conservation,* eds. K. Ramakrishna and G. M. Woodwell, pp. 59–77. New Haven: Yale University Press.

Richards, P. W. 1973. Africa, the "Odd Man Out." In *Tropical Forest Ecosystems in Africa and South America: A Comparative Review,* eds. B. J. Meggers, E. Ayensu, and W. D. Duckworth, pp. 21–26. Washington, D. C.: Smithsonian Institution Press.

Seal, U. S., R. C. Lacy, K. Medley, R. Seal, and T. J. Foose, eds. 1991. *Tana River Primate Reserve Conservation Assessment Workshop Report.* Captive Breeding Specialist Group (CBSG/SSC/IUCN) in collaboration with the Kenya Wildlife Service. Apple Valley, Minnesota: Captive Breeding Specialist Group.

Shafer, C. L. 1990. *Nature Reserves: Island Theory and Conservation Practice.* Washington, D. C.: Smithsonian Institution Press.

Western, D. 1989. Conservation Without Parks: Wildlife in the Rural Landscape. In *Conservation for the Twenty-First Century,* eds. D. Western and M. C. Pearl, pp. 158–165. New York: Oxford University Press.

Western, D. 1994. Conservation and People. *Swara* 17(2): 35–36.

White, P. S., and S. P. Bratton. 1980. After Preservation: Philosophical and Practical Problems of Change. *Biological Conservation* 18: 242–255.

Zimmerer, K. S. 1994. Human Geography and the "New Ecology": The Prospect and Promise of Integration. *Annals of the Association of American Geographers* 84(1): 108–125.

2

Forest Landscape Dynamics and Panda Conservation in Southwestern China

Alan H. Taylor and Qin Zisheng

Identifying the role of disturbance in controlling the dynamics and diversity of forest communities at stand and landscape scales is a central focus of ecological and biogeographical research (Connell 1978; Pickett and White 1985; Veblen 1992; Watt 1947; White 1979). Many species regenerate only after certain types, sizes, or severities of disturbance (Foster 1988; Franklin and Hemstrom 1981; Marks 1974; Runkle 1982; Taylor 1990; Veblen 1992; Veblen et al. 1981; Whitmore 1989), and these compositionally and structurally variable patches are often used by distinct wildlife assemblages (Hunter 1990; Morrison et al. 1992). The disturbance-patch dynamics perspective has led to a new framework for ecologically sustainable forestry that is based on the premise that species evolved with, and are adapted to, a range of disturbance conditions (Agee and Johnston 1988; Swanson et al. 1994). The probability of a species' persistence, at landscape scales, is presumed to be high when disturbance processes are within their range of historic variability, and low when patch-generating disturbances fall outside their historic range.

Identification of ecologically sustainable forestry practices in many developing countries is fraught with uncertainty because the relationships between natural and human-induced disturbance processes and forest regeneration are poorly known. Moreover, disturbance processes also affect the dynamics of forest understory plants (Halpern and Spies 1995), and these are crucial forage for wildlife in forested landscapes (Hunter 1990; Morrison et al. 1992). Yet little is known about the ecology of understory species in most forest ecosystems (Halpern and Spies 1995). Learning more about the role of disturbance on the regeneration dynamics of trees and understory plants is therefore crucial for the development of successful forest and wildlife conservation efforts.

In this chapter, we synthesize our research over the last decade on the dynamics of montane conifer forests in southwestern China. The findings are relevant to the preservation of forest-dwelling giant pandas (*Ailuropoda melanoleuca*). Pandas forage exclusively on understory bamboos in these forests (Schaller et al. 1985). We place particular emphasis on identifying how tree and bamboo species respond to different types of disturbances and their severities, such as treefalls, selective logging, clear-cut logging, and bamboo flowering.

In order to set our study in a broader context, we first describe regional- and landscape-scale changes in bamboo forest cover that have reduced and fragmented panda habitat in southwestern China, causing the panda conservation problem. Secondly, we describe panda food habits that relate to bamboo forest structure and distribution. Thirdly, the relationships between disturbances, forest and bamboo regeneration, forest structure and composition, and bamboo life cycles are described. Finally, we discuss how these data can be used to implement forest practices that are consistent with the dual objectives of timber production and panda conservation. The field data was collected in the Wolong, Wang Lang, and Tangjiahe Natural Reserves in Sichuan Province, China. Annual precipitation in these areas is about 1000–1200 millimeters and occurs mostly during the March to November wet season. Summer temperatures reach 20°C in August, and winter temperatures reach −11°C in January.

Bamboo Forest Fragmentation and the Panda Conservation Problem

The giant panda is perhaps the world's most recognized endangered species. Fewer than 1500 survive in a region of temperate, montane, and sub-alpine forests in the provinces of Sichuan, Gansu, and Shaanxi in southwestern China (map 2.1) (Schaller et al. 1985). Pandas once ranged over much of eastern China but are now confined to six forested tracts that cover a total of about 10000 square kilometers (map 2.1) (MacKinnon et al. 1989; MacKinnon and De Wulf 1994). Although the contraction of the panda's range is partly the result of climatic changes during the Pleistocene epoch, it has mostly been caused by people. Forests that were once occupied by pandas have been cleared for agriculture to feed China's burgeoning human population over the last 2000 years. More recently, loggers

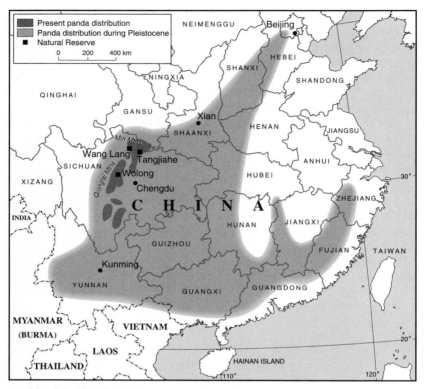

Map 2.1. Historic and current range of the giant panda and study area locations.

have moved into high, remote mountain areas where pandas still live to
cut forests to meet demand for saw timber (MacKinnon and De Wulf
1994; Richardson 1990; Schaller et al. 1985). The agricultural expansion
and logging have fragmented and isolated patches of forest landscape used
by pandas. Pandas now occur in 24 separate and isolated populations
within their range, and some populations have fewer than 20 individuals
(Schaller 1993). The isolation of small panda populations poses a long-
term threat to their survival because of inbreeding depression (O'Brien
and Knight 1987; Schaller et al. 1985). Inbreeding reduces reproduction,
fecundity, and survival of young. But poaching and habitat destruction
still remain the most acute, short-term threats to panda survival.

The plight facing the giant panda first drew widespread attention in the
mid-1970s. In 1975 and 1976, 138 pandas were found dead of starvation
in the Min Mountains (map 2.1). The panda starvation was caused by a
combination of factors: (1) habitat destruction; (2) its complete reliance

on bamboo for food; and (3) the unusual life cycle of bamboos in the panda range. At least three bamboo species flowered and died back synchronously in the Min Mountains, and pandas did not have alternative bamboos to use as forage at other landscape positions because of habitat destruction (Schaller et al. 1985).

About half the giant panda population lives within 13 nature reserves that were established between 1963 and 1979. There are plans to add 14 new reserves that will increase the size of the population in protected areas (MacKinnon et al. 1989; MacKinnon and De Wulf 1994; Pan et al. 1988; Schaller 1993; Schaller et al. 1985). Other pandas live on production forest land in the southwestern China. Panda habitat in many of the established and proposed natural reserves, and on forest production lands, has been altered by people through agriculture or timber-cutting. Moreover, there are large human communities within some of the reserves that continue to clear forests for fuelwood and building materials (De Wulf et al. 1988; Schaller et al. 1985). Forest managers who want to preserve panda habitat are rarely in the position of just maintaining pristine forest for pandas. Instead they are faced with the challenging tasks of restoring degraded panda habitat and developing forestry practices that can maintain forests for timber harvest and panda habitat.

Panda conservation measures that reduce the risk of extinction identify hazards to the populations and understand how pandas respond to these hazards (Botkin 1990; Holling 1978). Poaching has reduced panda populations by 50% in some areas and is an acute short-term threat (Schaller 1987). Logging and forest clearance have modified and reduced forest cover by 50% within the panda range over the last 15 years (MacKinnon and De Wulf 1994).

Panda Food Habits and Habitat Selection

Pandas are obligate bamboo feeders and evergreen bamboos comprise 99% of their diet (figure 2.1) (Schaller et al. 1985). A diet of plant stems and leaves is typical of herbivores—both the cell content and the cellulose and hemicellulose in cell walls are digested after microbial symbionts break these plant constituents down in their complex stomachs. Pandas, however, are carnivores with simple stomachs and can only digest the cell content and about 25% of the hemicellulose in plant cell walls (Schaller et al. 1985). Although the nutritive content of bamboo is low, its supply and level of nutrients are relatively constant year-round. Since the panda only digests up to 23% of the bamboo parts it eats, it spends over half of each day feeding to meet its nutritional needs (Schaller et al. 1985).

Pandas shift their feeding behavior and patterns of activity seasonally

Fig. 2.1. Mixed *Abies faxoniana–Betula utilis* forest with an understory of bamboo (*Fargesia denudata*) used by giant pandas (*Ailuropoda melanoleuca*) in the Wang Lang Natural Reserve, China (altitude 2940 meters).

to maximize the intake of digestible energy from different parts of the bamboo plant (Schaller et al. 1985). In the Qionglai Mountains in Wolong, the pandas' preferred food is the bamboo *Bashania fangiana*. It has a relatively continuous distribution beneath a mixed conifer canopy between the elevations of 2600 and 3200 meters. In winter, pandas select culms less than one year old, and in the spring they shift to older culms. Pandas eat mostly *B. fangiana* leaves during the summer months. They will also migrate downslope in spring to feed on the bamboo *Fargesia robusta*. It occurs below 2400 meters and produces shoots between April and June that are highly sought by pandas (Johnson et al. 1988; Reid and Hu 1991; Reid et al. 1989; Schaller et al. 1985; Taylor and Qin 1987). Pandas in the Min Mountains (Schaller et al. 1989) and Qin Ling Mountains (Pan Wenshi et al. 1988) also eat different bamboo species and parts in different seasons, but specific patterns vary by region.

The availability of panda forage can change suddenly (Johnson et al. 1988; Reid et al. 1989; Taylor and Qin 1987). The bamboos pandas eat are perennial monocarps: They grow vegetatively for set periods of several decades, flower, set seed, and die back simultaneously over wide areas (Taylor and Qin 1988a, 1993). Bamboo dieback reduces the standing crop and changes the spatial pattern of the panda's food source at landscape scales. In Wolong, 82% of the *B. fangiana* died back after it flowered in 1983 (Reid et al. 1989; Taylor et al. 1991a). Remaining live plants were distributed in a matrix of dead bamboo as live patches < 1.0 hectare in size (Reid et al. 1989; Taylor et al. 1991a).

Pandas responded conservatively to the bamboo dieback. They continued to eat *B. fangiana* parts in the same seasonal pattern they did before flowering, except they decreased selection of old shoots during the winter (preflowering diet of available stems 79%; postflowering diet 34%) and ate longer sections of each stem. They also increased use of bamboo stands above 3200 meters (Johnson et al. 1988; Reid et al. 1989). More important, however, pandas began to increase use of lower-altitude *F. robusta* by eating leaves during winter (Johnson et al. 1988; Reid et al. 1989). Clearly, preservation of two or more species of bamboo in areas with pandas is a top conservation priority. Daily activity patterns, home range sizes, and home range use by pandas did not change after the flowering (Johnson et al. 1988; Reid et al. 1989).

Pandas select habitats for the suitability of foraging, the palatability and nutritional value of the bamboo, and accessibility (Reid and Hu 1991; Schaller et al. 1985). During the winter in Wolong, pandas eat mostly *B. fangiana* stems from patches with tall thick culms. In summer, when they eat mostly leaves, pandas prefer dense bamboo stands which, after the 1983 bamboo dieback, are found mostly in clear-cuts less than 15 years old (Reid and Hu 1991). Live bamboo patches are more common in open clear-cut environments after flowering than beneath forest, perhaps because of stress that shifted carbohydrate allocation in bamboo plants to vegetative growth instead of flowering (Taylor et al. 1991a). Prior to the bamboo flowering, when bamboo was abundant, pandas rarely used *B. fangiana* in clear-cuts (Schaller et al. 1985). In all seasons, pandas prefer to forage on low-angled slopes (Reid and Hu 1991; Schaller et al. 1985). The amount of digestible nutrients in *B. fangiana* stands in Wolong are greatest under mixed conifer-broadleaved forests, followed by deciduous woods in forests below 3200 meters (Reid and Hu 1991).

Forest Overstory Influences on Bamboo Stand Structure

The composition and structure of subalpine forests (2700–3200 meters) in Wolong have a strong influence on understory bamboo (*B. fangiana*) stand structure and hence the spatial distribution of good foraging sites (Reid et al. 1991). Forests used by pandas in the subalpine zone are a mosaic of uncut, selectively cut, and clear-cut (old and new) stands. Forest composition varies with elevation (Taylor and Qin 1988b, 1989). The evergreen conifers *Tsuga dumosa* and *Abies faxoniana* are canopy dominants be-

Table 2.1. Average bamboo (*Bashania fangiana*) culm density (m^{-2}), average culm height (cm), average basal diameter (cm), and proportion of shoots by forest cover type; *n* is the number of stands sampled (adapted from Reid et al. 1991).

Cover type	*n*	Culm density (m^{-2})	Average height (cm)	Basal diameter (cm)	Percent shoots
Clear-cut	3	118.3	85.0	0.387	18.7
Deciduous	6	63.7	102.6	0.368	24.0
Mixed	9	64.4	110.1	0.411	21.7
Evergreen	4	45.5	94.2	0.395	15.5
High elevation (> 3200 m)	3	44.0	65.6	0.410	16.3
ANOVA *P*		0.001	0.002	0.083	0.028

tween 2700 and 2900 meters; above 2900 meters *A. faxoniana* dominates. Deciduous broadleaved *Betula albosinensis* and *B. utilis* are important subcanopy trees that vary widely in abundance. They regenerate in treefall gaps in uncut forests, and after selective and clear-cut logging (Taylor and Qin 1988b, 1988c, 1989; Taylor et al. 1991b).

Bamboo stand density, culm height, culm diameter, and the proportion of old shoots in a stand vary by forest cover type (table 2.1). Bamboo stands in young clear-cuts (< 15 years old) are the densest, and culms are short and thin. Bamboo stands that develop naturally or after selective cutting beneath mixed conifer-broadleaved forest (Taylor and Qin 1988b, 1989) have the tallest and thickest culms. Stands beneath broadleaved forest are similar in density but have shorter, thinner culms. Compared to mixed or broadleaved forests, bamboo stands beneath evergreen conifer canopies have lower densities and culms are shorter (Reid et al. 1991) (table 2.1).

Forest canopy effects on bamboo stand structure change when bamboos enter the die back-and-building phase of the bamboo growth cycle (mature/dieback/building). The germination and seedling establishment phase is the most vulnerable period in a plant's life cycle (Harper 1977). Seedling mortality is higher during these periods than when plants are mature because seedlings are unable to tolerate the same range of biotic and abiotic environmental fluctuations. Germination and establishment of *B. fangiana* over seven years in Wolong varied with forest cover (figure 2.2). Mean *B. fangiana* seedling density beneath uncut or older, selectively cut forest was higher than in clear-cuts. In fact, few bamboo seedlings remained in clear-cuts seven years after the flowering (figure 2.2) (Taylor and Qin 1988a, 1993). There are no similar data on bamboo seedling germination and establishment in other parts of the panda range. But one-time surveys of bamboo seedling density in clear-cuts and beneath forest showed lower

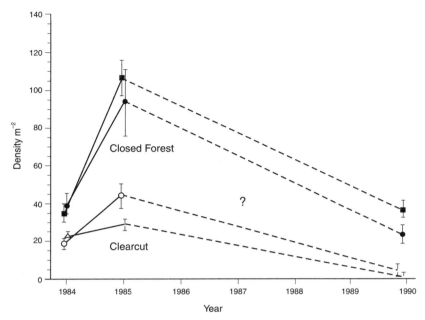

Fig. 2.2. Mean (±SE) density (m⁻²) of bamboo seedlings (*Bashania fangiana*) in 1984 ($n = 90$), 1985 ($n = 90$), and 1990 ($n = 64$), in permanent plots beneath forest and in clearcut in the Wolong Natural Reserve, China. Adapted from Taylor and Qin (1993).

seedling density in clear-cuts, suggesting that other bamboo species also regenerate poorly in clear-cuts (Taylor and Qin 1993).

The rate of bamboo stand recovery after flowering varies spatially within a forest stand and may be similar among bamboo species in the panda range (Taylor and Qin 1993; Taylor et al. 1995). Within forest, bamboo seedling cover and height is greater in treefall gaps than beneath adjacent closed forest in both Wolong and Wang Lang Natural Reserves (table 2.2). Estimates of the height recovery rate for three bamboo species in the Qionglai and Min Mountains are similar after 17–20 years. Pandas do forage on seedlings before they reach full height after about ten years, so recovering bamboo becomes increasingly available to pandas after that time.

Forest Regeneration Patterns

Uncut Forests

Forest canopy disturbances and bamboo abundance strongly influence tree regeneration patterns and forest structure in panda habitat. In uncut

Table 2.2. Average ($\pm$SE) bamboo seedling cover and height beneath 20 canopy gaps and adjacent closed forest.

	Wolong Natural Reserve	
Bashania fangiana	Canopy gap	Closed forest
Bamboo cover (%)	23.4 $\pm$ 3.3	6.9 $\pm$ 1.3
Maximum culm height (cm)	37.1 $\pm$ 2.6	21.1 $\pm$ 1.6
	Wang Lang National Reserve	
Fargesia denudata	Canopy gap	Closed forest
Bamboo cover (%)	34.4 $\pm$ 1.7	19.5 $\pm$ 1.3
Maximum culm height (cm)	150.0 $\pm$ 0.03	110.0 $\pm$ 0.03

Notes: Bamboo died back in Wolong in 1983 and culms were measured in 1990; dieback in Wang Lang occurred in 1976 and culms were measured in 1992 (adapted from Taylor and Qin 1992, 1995). Values in a row were significant ($P < 0.05$, t-test).

forests in Wolong and Wang Lang Natural Reserves, tree regeneration in subalpine conifer forests occurs in gaps 44 to 800 square meters in area in the forest canopy made by the death of one or more canopy trees (Taylor and Qin 1988b, 1988c, 1989, 1992; Taylor et al. 1995, 1996). The conifers *Abies faxoniana, Picea purpurea, Tsuga dumosa,* and hardwoods *Betula albosinensis* and *B. utilis,* can all persist in old-growth forests by regenerating in canopy gaps of this size, but gap regeneration patterns vary among species. *Betula* regenerates most frequently in larger multiple-tree gaps, while the conifers regenerate in gaps of all sizes. *Betula* is less shade tolerant, shorter-lived, produces more frequent seed crops, has smaller seeds that disperse further, and grows faster than the associated conifers. These life history traits promote the capture of larger gaps by *Betula* species. The coexistence of conifers and hardwoods in undisturbed subalpine forests is mediated by the tendency for conifers and hardwoods to regenerate in different size gaps. Overall, tree replacement patterns in mature and old-growth stands indicate that forests are compositionally stable (Taylor and Qin 1988b, 1988c, 1989, 1992; Taylor et al. 1995, 1996), but proportions of conifers and hardwoods are spatially variable because of disturbance history, chance, and the abundance of understory bamboos (table 2.3).

Selectively Cut Forests

Regeneration patterns of conifers and hardwoods in selectively cut forests in Wolong, where the basal area of *A. faxoniana* was reduced by about 50%, are broadly similar to those in uncut forests (Taylor and Qin 1988b, 1988c, 1989, 1992). *Betula* regenerates well in gaps created by loggers because it disperses quickly to gaps and grows faster than associated conifers

Table 2.3. Density of stems (ha⁻¹) in 43 forest plots (20 × 20 m) placed in clear-cut, selectively cut, and uncut forest with >40% bamboo cover in the Wolong Natural Reserve (summarized from Taylor and Qin 1988a, 1989; Reid et al. 1991).

Area type	Betula sp.	Abies faxoniana	Tsuga dumosa
Recent clear-cut			
stems <10 cm dbh	738	25	—
stems 10–40 cm dbh	—	—	—
stems >40 cm dbh	—	—	—
Old clear-cut			
stems <10 cm dbh	206	—	—
stems 10–40 cm dbh	219	—	—
stems >40 cm dbh	70	—	—
Selectively cut			
stems <10 cm dbh	100	56	32
stems 10–40 cm dbh	59	54	52
stems >40 cm dbh	3	49	42
Uncut			
stems <10 cm dbh	213	93	6
stems 10–40 cm dbh	98	76	6
stems >40 cm dbh	17	140	11

(Taylor and Qin 1988b, 1988c, 1989, 1992). The conifers *A. faxoniana* and *T. dumosa* also regenerate in logging-created gaps. New seedlings established in the gaps, and already established seedlings, saplings, and small trees, were released by selective logging. Tree replacement patterns indicate that conifers will be a dominant component of future stands as they develop (table 2.3).

Clear-Cut Forests

Tree regeneration patterns in clear-cuts are different than in uncut and selectively cut forest. In Wolong, stands of the conifers *A. faxoniana* and *T. dumosa* were clear-cut in the 1930s and are now dominated by hardwoods, particularly *Betula*. There are few conifer saplings in these stands or in the *Betula* thickets in recent clear-cuts (Schaller et al. 1985; Taylor et al. 1991b). Clear-cutting conifer forests, at least in Wolong, leads to development of persistent hardwood forests dominated by *Betula* species on many sites (table 2.3).

Bamboo Effects on Tree Regeneration

The lack of conifer regeneration in old and new clear-cuts is probably because of the bamboo understory. Clear-cut logging disturbs the bamboo understory, and tree seedlings establish and grow well in these disturbed areas. Since conifers are removed by loggers, conifer seed is scarce; thus most trees that establish right after clear-cutting are hardwoods, especially *Betula.* But recruitment of trees into the clear-cuts nearly stops after a few years because mature bamboos proliferate and impede further tree establishment. As the hardwood stands mature and thin, they become more open because few tree seedlings can establish in the bamboo (Reid et al. 1991; Schaller et al. 1985; Taylor et al. 1991b).

Bamboos also impede tree and shrub regeneration in uncut and selectively cut forests. In mixed conifer–hardwood forests in Wolong and Wang Lang, seedling and sapling density and woody plant diversity are negatively correlated with bamboo cover and culm density (Taylor and Qin 1988b, 1988c, 1989; Taylor et al. 1996). Gaps created by treefalls fill slowly because bamboo density and culm height increase in gaps, retarding seedling establishment (Taylor and Qin 1988b). A dense bamboo sward seems to favor establishment of *Betula* in gaps compared to the conifers, because there is no seedling-sapling pool in the forest understory than can release when a gap forms above it (Taylor and Qin 1988b, 1988c, 1992; Taylor et al. 1995). Instead, most gap-filling occurs by individuals that seed into gaps, and *Betula* disperses better and grows faster than associated conifers. Because conifers do establish in gaps and live longer than *Betula,* they remain codominant.

Conifer-hardwood dominance may drift because of the bamboo growth cycle (mature/dieback/building). In *A. faxoniana–Betula* forests in Wolong, tree seedling establishment increased in gaps from 1985 to 1990 after the bamboo *B. fangiana* died back in 1983 (Taylor and Qin 1992). The frequency of *A. faxoniana* regeneration in gaps increased during the dieback/building phases compared to *Betula* species (Taylor and Qin 1992). Consequently, *A. faxoniana* tends to increase relative to *Betula* during the dieback and building phases, and *Betula* increases during the mature phase of the approximately 50-year bamboo growth cycle (Taylor and Qin 1988b, 1992).

Bamboo dieback also synchronizes tree regeneration in conifer-hardwood forests. Treefall gaps fill slowly when bamboos are mature, but they are colonized rapidly by trees when the bamboos die. In Wolong and Wang Lang most *A. faxoniana, Acer caudatum,* and *Betula* seedlings in gaps established after the bamboos *B. fangiana* (Wolong) and *Fargesia denudata* (Wang Lang) died back (figure 2.3). In fact, in Wang Lang tree

Fig. 2.3. Age structure of tree seedlings in canopy gaps in the Wolong Natural Reserve where bamboo (*Bashania fangiana*) died back in 1983 (*bottom*), and the Wang Lang Natural Reserve where bamboo (*Fargesia denudata*) died back in 1975–1976 (*top*). Seedlings in 20 gaps were aged in Wolong ($n = 369$) in 1990 and in Wang Lang ($n = 460$) in 1992. Adapted from Taylor and Qin (1992); Taylor et al. (1995).

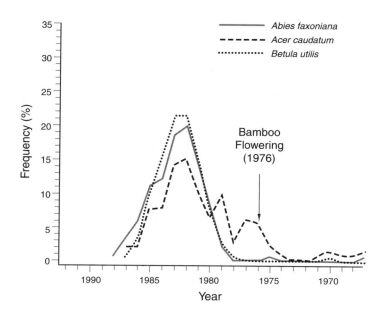

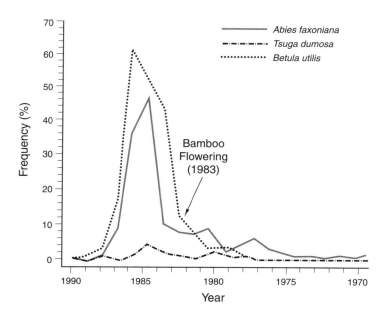

regeneration was delayed for several years. This delay could be a result of:
(1) dense ranks of dead culms that continued to impede tree regeneration
until they collapsed after three to four years; (2) low seed dispersal to gaps
because of poor seed crops; or (3) high seed predation by birds and ro-
dents. Seed predator populations frequently increase dramatically after a
massive bamboo seed crop is produced (Janzen 1976) and may have consumed dispersed seed.

The number of years bamboo stands can be invaded by tree seedlings
after they flower is uncertain because the rate of bamboo seedling development is influenced by forest canopy density and the initial density of bamboo seedling populations (Makita 1992; Makita et al. 1993; Nakashizuka
1988; Taylor and Qin 1993). The tree age structure data from Wang Lang
gaps where bamboos died back in 1975 and 1976 suggest that trees can
invade developing bamboo stands for at least eight to ten years (figure
2.3).

Discussion

The structure, composition, and dynamics of montane conifer forests in
panda habitat are strongly linked to different species responses to the
types and severities of disturbances that have occurred in these landscapes. In mature/old-growth forests, tree regeneration occurs in canopy
gaps created by small-scale, low-severity disturbances such as treefalls.
Species abundance patterns in gaps and stands indicate that both conifers
and hardwoods persist in these forests via gap-phase regeneration. However, hardwoods such as *Betula* regenerate best in large multiple-tree gaps.
Tree regeneration patterns after moderate-severity disturbances such as
selective logging are similar to those in uncut forests. Conifer and hardwoods both regenerate in gaps in selectively logged forests with hardwoods
responding more vigorously than conifers in large openings. In both forest
types, conifers will continue to be dominant because of their size and longevity. Patterns of tree regeneration after the large-scale severe disturbance
of clear-cutting are different than in uncut and selectively cut forests. Regeneration is virtually all by hardwoods and clear-cutting leads to persistent hardwood forests on many sites. Forests throughout panda habitat
are also influenced by understory bamboos that impede tree regeneration
and create more open forests than those on similar sites without bamboo
(Taylor and Qin 1988b, 1988c, 1989; Taylor et al. 1996). Moreover, tree
regeneration tends to be more synchronized in forests with bamboo because of the cyclic dieback/building/mature phases of the bamboo growth
cycle (Taylor and Qin 1992; Taylor et al. 1995).

Bamboo regeneration from seed also varies with severity of disturbance to the forest canopy. Bamboo stands replace themselves after flowering and dieback beneath uncut and selectively cut forest, but bamboo regeneration fails in recently clear-cut forests.

The different regeneration responses of trees and bamboos to the types and severities of disturbances can be used to guide forest management practices in panda habitat. Forest and wildlife managers should focus their activities to achieve three goals: (1) maintenance of high-quality panda habitat on forest lands still managed for timber production; (2) ensuring that clear-cuts and other severely disturbed areas inside protected areas provide future panda habitat; and (3) restoration of degraded habitat or areas devoid of forest or bamboo to create habitat corridors that will link habitat fragments (Schaller et al. 1985).

Our data on forest and bamboo regeneration in uncut, selectively cut, and clear-cut forest indicate that selective cutting, and not clear-cutting, should be used on forest production land in panda habitat. In Wolong and Wang Lang, mixed forests of hardwoods and conifers developed after selective cutting 30 to 60 years ago, and bamboo in these stands are favored by foraging pandas. In fact, selective logging seems to have improved the quality of panda habitat on some sites by locally increasing the deciduous components of stands that improves bamboo growth. Moreover, loggers left large diameter, hollow conifers in stands because these trees had no commercial value. These now serve as maternity dens (Schaller et al. 1985) and may be a critical structural attribute of forests needed by pandas. Bamboo seedlings also regenerate well beneath selectively cut forest canopies.

Clear-cuts, in contrast, are now poor habitat for pandas, and will be poor habitat in the future because bamboo seedlings regenerate poorly in them. In Wolong alone, bamboo failed to regenerate in about 20 square kilometers of recent clear-cut after it flowered in 1983. The area of bamboo forest lost by bamboo regeneration failure in Wolong would support eight to ten pandas (Taylor and Qin 1993a). Moreover, conifers regenerate poorly in dense mature bamboo stands that develop in clear-cuts; thus clear-cut lands will have little future commercial value as timberland.

Data from our studies on forest-bamboo interactions and bamboo ecology can also be used by managers to improve the success of panda habitat restoration efforts. Degraded panda habitat inside and outside reserves needs to be restored. Restoration activities should be concentrated in places where pandas prefer to forage, and completed at times when restoration plantings will succeed. Conifers need to be replanted in old clear-cuts now dominated by a sparse hardwood canopy, and in young ones with dense bamboo thickets to restore mixed forest. The best time to plant conifers is when bamboos die back (Taylor et al. 1991b). Light and other resources increase when bamboos die (Taylor and Qin 1988a), allowing newly planted tree seedlings to grow rapidly because there is little competition from slow-growing bamboo seedlings or other woody plants. Managers need to respond quickly to the restoration opportunity afforded by a bamboo flowering. It only takes bamboo seedlings 10 to 15 years to grow into stands that are dense enough to once again impede tree regeneration.

Trees and bamboo will have to be replanted in clear-cuts where bamboo failed to regenerate after flowering, and in proposed panda habitat corridors where bamboo forest has been cleared by agriculturalists. A widely-spaced conifer tree canopy should be established on restoration sites before bamboo clones are planted. Transplanted *B. fangiana* clones ($n = 20$) have low survival rates in open conditions (20%) compared to plantings beneath forest (100%), probably because of moisture stress. Moreover, bamboo clones should be planted in clumped patterns to reduce the risk of regeneration failure if they flower. Isolated flowering clones produce few seeds, perhaps because of poor pollination (Taylor and Qin 1988a). A clumped planting pattern increases the probability of fertilization by wind-borne pollen, and seed set. Conifers should not be planted as densely as they are on typical Chinese plantations (Richardson 1990) because dense conifer canopies retard bamboo growth.

Reintroduction of trees and bamboos should be focused on certain landscape positions. Pandas sit when they forage and prefer to feed in bamboo stands on low-angled slopes (Reid and Hu 1991; Schaller et al. 1985). Planners should focus their habitat restoration efforts on these sites first.

The disturbance patch dynamics perspective we used in our studies of forest and bamboo regeneration emphasized identifying the relationships between disturbance processes and the spatial and temporal patterns of tree and bamboo regeneration. Tree and bamboo regeneration patterns are broadly predictable across a disturbance gradient, and this knowledge can be used to guide forest management to achieve the conservation goal of preserving giant pandas. Understanding the relationship between disturbance and bamboo forest dynamics alone, however, will not assure panda survival. Poaching is the most acute short-term threat to pandas and enforcement of 1989 laws that ban the capture, killing, or trade of protected species is essential to stem the decline of the panda population (O'Brien et al. 1994; Schaller 1993). Panda numbers in Wolong between 1974 and 1986 declined by as much as 50% because of poaching (Schaller 1987). Rewards for poaching are great: A panda pelt can be sold for more than $10000 outside China (Schaller 1993). Pandas have a low reproductive rate, and poaching causes a rapid decline in a panda population (Schaller et al. 1985). Eradication of poaching, and changes in forest resource use by local people in panda habitat, are needed now to reduce the acute risks to panda survival. Only then will pandas benefit from improved habitats that result from advanced forestry practices and improved habitat restoration techniques.

Acknowledgments

The research reported in this chapter was funded by the World Wide Fund For Nature International (WWF), WWF-Canada, Wildlife Conservation International (WCI), the National Academy of Sciences Advanced Study in China Program, and the People's Republic of China. The work could not have been completed without the assistance of many individuals. We thank G. Schaller, D. Reid, K. Johnson, Hu Jinchu, Liu Jie, Wang Menghu, Hu Tieqin, Bi Fengzhou, Gon Tongyang, Fu Chengjun, Xiong Beirong, Lai Binghui, Li Chengyu, Shang Goujun, Zhang Kewen, Zhou Shoude, Yu Zhishun, Jiang Mingdao, Zhang Zhoumin, Gou Chencai, Mark Halle, Christopher Elliott, Andrew Laurie, and Pascale Moehrle for their assistance and support. K. Taylor, Cai Xucheng, Cheng Haizhou, He Xiangzhi, Huang Jinyan, Wang Sewei, Li Lianghua, Shen Heming, and Liu Mingchong assisted in the field.

References

Agee, J. K., and D. R. Johnson. 1988. *Ecosystem Management for Parks and Wilderness.* Seattle: University of Washington Press.

Botkin, D. B. 1990. *Discordant Harmonies: A New Ecology for the Twenty-first century.* New York: Oxford University Press.

Connell, J. S. 1978. Diversity of Tropical Forests and Coral Reefs. *Science* 199: 1302–1310.

De Wulf, R., R. Goosens, J. MacKinnon, and Wu Shencai. 1988. Remote Sensing for Wildlife Management: Giant Panda Habitat Mapping from LANDSAT MSS Images. *Geocarto International* 1: 41–49.

Foster, D. R. 1988. Disturbance History, Community Organization and Vegetation Dynamics of the Old-Growth Pisgah Forest, Southwestern New Hampshire, USA. *Journal of Ecology* 76: 105–134.

Franklin, J. F., and M. A. Hemstrom. 1981. Aspects of Succession in the Coniferous Forests of the Pacific Northwest. In *Forest Succession Concepts and Application,* eds. H. H. Shugart and D. B. Botkin, pp. 212–229. New York: Springer-Verlag.

Halpern, C. B., and T. A. Spies. 1995. Plant Species Diversity in Natural and Unmanaged Forests of the Pacific Northwest. *Ecological Applications* 5: 913–934.

Harper, J. L. 1977. *The Population Biology of Plants.* London: Academic Press.

Holling, C. S. 1978. *Adaptive Environmental Assessment and Management.* New York: Wiley and Sons.

Hunter, M. L. 1990. *Wildlife, Forests and Forestry.* Englewood, New Jersey: Prentice Hall.

Janzen, D. 1976. Why Bamboos Wait so Long to Flower. *Annual Review of Ecology and Systematics* 7: 347–391.

Johnson, K. G., G. B. Schaller, and Hu Jinchu. 1988. Response of Giant Pandas to a Bamboo Die-Off. *National Geographic Research* 4: 161–177.

MacKinnon, J., Bi Fengzhou, Qiu Mingjiang, Fan Chuandao, Wang Haiban, Yuan Shijun, Tian Anshun, and Li Jiangguo. 1989. *National Conservation Management Plan for the Giant Panda and Its Habitat.* Gland, Switzerland: World Wide Fund For Nature.

MacKinnon, J., and R. De Wulf. 1994. Designing Protected Areas for Giant Pandas in China. In *Mapping the Diversity of Nature,* ed. R. I. Miller, pp. 127–142. New York: Chapman Hall.

Makita, A. 1992. Survivorship of a Monocarpic Bamboo Grass, *Sasa kurilensis,* during the Regeneration Process After Mass-flowering. *Ecological Research* 7: 245–254.

Makita, A., Y. Komno, N. Fujita, K. Takada, and Hamabata, E. 1993. Recovery of a *Sasa tsuboiana* Population After Mass-Flowering and Death. *Ecological Research* 8: 215–224.

Marks, P. L. 1974. The Role of Pincherry (*Prunus pensylvanica*) in the Maintenance of Stability in Northern Hardwood Ecosystems. *Ecological Monographs* 44: 73–88.

Morrison, M. L., B. G. Marcot, and R. W. Mannon. 1992. *Wildlife-Habitat Relationships: Concepts and Applications.* Madison: University of Wisconsin Press.

Nakashizuka, T. 1988. Regeneration of Beech (*Fagus crenata*) After Simultaneous Death of Undergrowing Dwarf Bamboo *Sasa kurilensis. Ecological Research* 3: 21–35.

O'Brien, S. J., and J. A. Knight. 1987. The Future of the Giant Panda. *Nature* (London) 325: 758–759.

O'Brien, S. J., Pan Wenshi, and Lu Zhi. 1994. Pandas, People, and Policy. *Nature* (London) 369: 179–180.

Pan Wenshi, Gao Zhengsheng, Lu Zhi, Xia Zhengkai, Shang Miaodi, Meng Guangli, Shi Xiaoye, Liu Xushuo, Ma Laioling, and Chen Fengxiang. 1988. *The Giant Panda's Natural Refuge in the Qin Ling Mountains.* Beijing: Beijing University Press.

Pickett, S. T. A., and P. S. White. 1985. *The Ecology of Natural Disturbance and Patch Dynamics.* New York: Academic Press.

Reid, D. G., and Hu Jinchu. 1991. Giant Panda Selection between *Bashania fangiana* Bamboo Habitats in Wolong Reserve, China. *Journal of Applied Ecology* 28: 228–243.

Reid, D. G., Hu Jinchu, Dong Sai, Wang Wei, and Huang Yan. 1989. Giant Panda *Ailuropoda melanoleuca* Behaviour and Carrying Capacity Following a Bamboo Die-Off. *Biological Conservation* 49: 85–104.

Reid, D. G., A. H. Taylor, Hu Jinchu, and Qin Zisheng. 1991. Environmental Influences on *Bashania fangiana* Bamboo Growth and Implications for Giant Panda Conservation. *Journal of Applied Ecology* 28: 855–868.

Richardson, S. D. 1990. *Forests and Forestry in China.* Covelo: Island Press.

Runkle, J. R. 1982. Patterns of Disturbance in Some Old-Growth Mesic Forests of the Eastern United States. *Ecology* 63: 1533–1546.

Schaller, G. B. 1987. Bamboo Shortage Not Only Cause of Panda Decline. *Nature* 327: 562.

Schaller, G. B. 1993. *The Last Panda.* Chicago: University of Chicago Press.

Schaller, G. B., Hu Jinchu, and Zhu Jing. 1985. *The Giant Pandas of Wolong.* Chicago: University of Chicago Press.

Schaller, G. B., Teng Qitao, K. G. Johnson, Wang Xiaoming, Shen Heming, and Hu Jinchu. 1989. Feeding Ecology of Giant Panda and Asiatic Black Bear in Tangjiahe Reserve, China. In *Carnivore Behaviour, Ecology and Evolution,* ed. J. Gittleman, pp. 212–241. Ithaca: Cornell University Press.

Swanson, F. J., J. A. Jones, D. O. Wallin, and J. H. Cissel. 1994. Natural Variability and Implications for Ecosystem Management. In *Ecosystem Management Principles and Applications,* volume 2, eds. M. E. Jensen and P. S. Bourgeron, pp. 80–94. Washington, D. C.: USDA Forest Service.

Taylor, A. H. 1990. Disturbance and Persistence of Sitka Spruce (*Picea sitchensis* (Bong) Carr.) in Coastal Forests of the Pacific Northwest, North America. *Journal of Biogeography* 17: 47–58.

Taylor, A. H., and Qin Zisheng. 1987. Culm Dynamics and Dry-Matter Production of Bamboos in the Wolong and Tangjiahe Giant Panda Reserves, Sichuan, China. *Journal of Applied Ecology* 24: 419–433.

Taylor, A. H., and Qin Zisheng. 1988a. Regeneration from Seed of *Sinarundinaria fangiana,* a Bamboo, in the Wolong Giant Panda Reserve, Sichuan, China. *American Journal of Botany* 75: 1065–1073.

Taylor, A. H., and Qin Zisheng. 1988b. Regeneration Patterns in Old-Growth *Abies-Betula* Forests in the Wolong Natural Reserve, Sichuan, China. *Journal of Ecology* 76: 1204–1218.

Taylor, A. H., and Qin Zisheng. 1988c. Tree Replacement Patterns in Subalpine *Abies-Betula* Forests in Wolong Natural Reserve, China. *Vegetatio* 78: 141–149.

Taylor, A. H., and Qin Zisheng. 1989. Structure and Composition of Uncut and Selectively Cut *Abies-Tsuga* Forest in Wolong Natural Reserve and Implications for Panda Conservation in China. *Biological Conservation* 47: 83–108.

Taylor, A. H., and Qin Zisheng. 1992. Tree Regeneration After Bamboo Die-Off in *Abies-Betula* Forests, Wolong Natural Reserve, China. *Journal of Vegetation Science* 3: 253–260.

Taylor, A. H., and Qin Zisheng. 1993. Bamboo Regeneration After Flowering in the Wolong Giant Panda Reserve, China. *Biological Conservation* 63: 231–234.

Taylor, A. H., Qin Zisheng, and Liu Jie. 1995. Tree Regeneration in an *Abies faxoniana* (Rehder & Wilson) Forest After Bamboo Die-Off in the Wang Lang Natural Reserve, China. *Canadian Journal of Forest Research* 25: 2034–2039.

Taylor, A. H., Qin Zisheng, and Liu Jie. 1996. The Structure and Dynamics of Subalpine Forests in the Wang Lang Natural Reserve, China. *Vegetatio* 124: 25–38.

Taylor, A. H., D. G. Reid, Qin Zisheng, and Hu Jinchu. 1991a. Spatial Patterns and Environmental Associates of Live Bamboo (*Bashania fangiana*) After a

Mass Flowering in Southwestern China. *Bulletin of the Torrey Botanical Club* 118: 247–254.

Taylor, A. H., D. G. Reid, Qin Zisheng, and Hu Jinchu. 1991b. Bamboo Die-Off: An Opportunity to Restore Panda Habitat. *Environmental Conservation* 17: 166–168.

Veblen, T. T. 1992. Regeneration Dynamics. In *Plant Succession Theory and Prediction,* eds. D. C. Glenn Lewin, R. K. Peet, and T. T. Veblen, pp. 152–187. London: Chapman-Hall.

Veblen, T. T., C. Donoso, F. M. Schlegel, and B. Escobar. 1981. Forest Dynamics in South-Central Chile. *Journal of Biogeography* 8: 211–247.

Watt, A. S. 1947. Pattern and Process in the Plant Community. *Journal of Ecology* 35: 1–22.

White, P. S. 1979. Pattern, Process and Natural Disturbance in Vegetation. *Botanical Review* 45: 229–299.

Whitmore, T. C. 1989. Canopy Gaps and the Two Major Groups of Forest Trees. *Ecology* 70: 536–538.

3

Deforestation in Landscapes with Humid Forests in the Central Andes
Patterns and Processes

Kenneth R. Young

The distribution, extent, and composition of humid forests in landscapes located in the highland region of the central Andes of Ecuador, Peru and Bolivia are affected by both natural and anthropogenic processes. Understanding what these processes are, how long they have acted on the landscapes, and their relative importances in different humid forest types are key factors in interpreting the past, and predicting the future, of this important natural resource, just as they are in other contexts (Aiken and Leigh 1992; Balée 1992; Christensen 1989; Dodson and Gentry 1991; Groom and Schumaker 1993; Mather and Sdasyuk 1991; Rodgers 1993; Totman 1989; Williams 1989a; others in this volume). Using this kind of information, it is possible to propose regional policies and practices that could improve forest conservation.

Young and Valencia (1992a, 1992b) and Young and León (1995a, 1995b) reviewed what was known about these montane forests in highland Peru, and subdivided them in reference to the natural regions they occupy and their relative degree of connectivity. They concluded that two major classes of forest could be distinguished in terms of connectivity: naturally fragmented and naturally unfragmented forests. The unfragmented forest landscape is found in a continuous swath across eight degrees of latitude on the eastern slopes of the Peruvian Andes, overlooking the Amazon basin. Current deforestation along highways is resulting in anthropogenic fragmentation of the forest (Young 1992, 1994). The remainder of the highlands have montane forests that are naturally fragmented and isolated because of their distribution in complex mountainous terrain, with the presence of extensive tropical alpine ecosystems above timberline and numerous different arid and semiarid life zones located downslope (Sagástegui Alva 1988; Weberbauer 1945). In addition, these highland areas show evidence of millennia of human impact, as seen in the presence of archaeological sites and other cultural features, such as agricultural fields and terracing.

It is my goal in this essay to further explore the biogeographical implications of these landscape and regional patterns, extending my purview to include similar landscapes in Ecuador and Bolivia. After providing a brief description of the distribution and composition of these forests, I discuss them in reference to the type of human colonization that has or will engulf them, and outline the possible impacts on forests and forest biota. Finally, I speculate on the practical and theoretical implications of this new perspective.

Humid Forests of the Central Andean Region

According to the Holdridge life zone classification system (Holdridge 1947), as applied to the tropical Andean highlands (Cañadas Cruz 1983; Ewel et al. 1976; ONERN 1976; Tosi 1960), humid montane forests with closed canopies (those canopies formed of trees whose crowns touch or overlap) would generally be expected in areas with $\geq$ 700 millimeters of annual precipitation, as long as the dry season is not too extreme. These forests would fit into the moist (700–1100 mm/yr), wet (1100–2200 mm/yr), and rain (> 2200 mm/yr) divisions of the montane and lower montane life zones; I will refer to them as humid montane forests. Mean monthly temperatures would be on the order of 7 °C to 19 °C. The lower elevational limit of these forests is 1500–2000 meters, below which are found premontane and lowland vegetation types. The upper elevational limit of closed humid forest is variable, but often ends about 3500 meters, although forest patches with closed canopies can be found at 4200 meters.

The topography of the central Andes acts to naturally fragment this potential forest cover in some areas, juxtapositioning humid forests with drier vegetation types in rainshadow valleys and creating numerous different soil types. I have subdivided the potential humid montane forest zones for areas above 1500–2000 meters elevation in Ecuador, Peru, and tropical Bolivia in terms of this presumed degree of natural fragmentation (map 3.1). The naturally unfragmented montane forest would be found along the eastern slopes of the Andes of Ecuador, Peru, and Bolivia, in

Map 3.1. Naturally fragmented and unfragmented humid montane forest zones in Ecuador, Peru, and Bolivia.

addition to most of the interandean valleys of highland Ecuador and the western slope of the Ecuadorian Andes near the border with Colombia. Naturally fragmented montane forest zones can only be suggested at the scale of map 3.1, but would include the remainder of the highlands where humidity and elevation permit.

As a generalization, these forests have a canopy from five to as much as thirty meters in height. The taller the forest, the more complex the physiognomy, with some forests showing several strata and dense understories (Madsen and Øllgaard 1994; Young in press a). The forests with simpler structure tend to be dominated by several canopy-forming species from such genera as *Alnus, Buddleja, Escallonia, Gynoxys, Myrcianthes, Myrsine, Oreopanax, Polylepis, Vallea,* and *Weinmannia.* Most of the fragmented montane forests have this relatively simple composition and structure (for example, Yallico 1992).

More complex montane forests typical of naturally unfragmented forest zones can have canopy dominance shared by numerous species from many plant families, including Actinidiaceae (*Saurauia*), Aquifoliaceae (*Ilex*), Araliaceae (*Oreopanax*), Asteraceae (*Gynoxys, Vernonia*), Brunelliaceae (*Brunellia*), Chloranthaceae (*Hedyosmum*), Clethraceae (*Clethra*), Cunoniaceae (*Weinmannia*), Lauraceae (*Nectandra, Ocotea, Persea*), Melastomataceae (*Axinaea, Miconia*), Meliaceae (*Cedrela, Guarea, Ruagea*), Myrsinaceae (*Myrsine*), Myrtaceae (*Myrcianthes*), Podocarpaceae (*Podocarpus, Prumnopitys, Retrophyllum*), Rosaceae (*Prunus*), Rubiaceae (*Cinchona*), Solanaceae (*Saracha, Solanum*), Styracaceae (*Styrax*), Symplocaceae (*Symplocos*), and Theaceae (*Gordonia*).

The biota of Andean montane forests originated both from lowland tropical groups and from temperate latitude taxa that migrated along the cordilleras; at least this is the presumed origin of most tropical alpine plant groups in the high Andes (Ulloa Ulloa and Jørgensen 1995; van der Hammen and Cleef 1986). The biological diversity of the montane forests of the central Andes is notable, especially for vascular plants, birds, and amphibians. In the montane forests of the eastern Peruvian Andes, for example, Young and León (1997) reported the occurrence of 2400–2800 species of vascular plants, or about 14% of the national total. Dillon et al. (1995) found about 1100 plant species in five montane forest sites in northern Peru. Jørgensen et al. (1995) reported 200 botanical families, 1119 genera, and 4868 species of seed plants above 2400 meters in Ecuador; this would probably be equivalent to about 24% of the national total. Rodríguez et al. (1993) documented that most of Peru's frog species are found in eastern slope forests in Peru, and Leo (1995) showed high endemism there for mammals, birds, and amphibians.

Deforestation: Old and New Frontiers

As humans colonize a forested region, they typically convert it into a mixture of agricultural systems, grasslands, and scrub, leaving relictual forests along rivers or in inaccessible sites. However, there will never be a stable end point in terms of the mixture of vegetation types making up a particular landscape because land-use practices change for economic and social reasons. For example, forests of the eastern United States were cut as colonization proceeded in the eighteenth and nineteenth centuries; forest cover, however, has increased this century as the poorest lands were abandoned by agriculturists (Whitney 1994; Williams 1989b). The deforestation of Europe is generally accepted to have occurred over several millennia, and to have been quite variable temporally and spatially (Birks et al. 1988; Darby 1956). Currently much effort is going into planning for the technical and aesthetic challenges of restoring forest cover in Europe (Ferris-Kaan 1995; Peterken 1996).

The history of deforestation in the central Andes is a story yet to be told (Guillet 1985), but some general patterns can be delimited. The oldest archaeological sites come from high elevations and date back to the 2000-year period at the start of the Holocene (Bruhns 1994; Rick 1988). These sites are in areas that would have been only recently deglaciated and covered by tropical grasslands and other tropical alpine vegetation types (Colinvaux et al. 1988; Graf 1992; Hansen et al. 1984; Seltzer 1990). Thus, it is likely that human movements initially followed the open habitats of the high ridges of the north-south trending cordilleras.

Archaeological sites for the next seven thousand years (1000–8000 years B.P.) are found in two general regions: along Ecuador's and Peru's coast and in the highlands of Ecuador, Peru, and Bolivia. It is during this long period that evidence for the development of large settlements, social stratification, irrigation, and domestication of plants and animals is to be found (Bruhns 1994; Burger 1992; Lumbreras 1974; Zimmerer 1995). The interandean valleys and adjacent tropical alpine environments were home to most of the highland cultural change. Probably it was during this time that the original forest cover of the highlands was most modified (Smith 1980). The technology available included simple hand tools, such as stone axes and fire. Burning would have reduced the shrub component of tropical alpine vegetation types, expanded the area in grassland by lowering timberline, and converted some forest types into scrub. People might have burned in order to improve grazing conditions for domesticated camelids and possibly wild ungulates (deer and camelids), which would have made them more available for hunting.

Given the relatively high human population levels (Cook 1981; Denevan 1992), the continued development of more intensive agriculture, and the rise of the Huari and Inca empires (Bruhns 1994; Lumbreras 1974), it seems likely that by the time of contact with Europe there had already been massive changes in the forest cover of the central Andean highlands. Of course, following contact there were further dramatic changes in population densities, land-use practices, and the types of domesticated plant and animal species utilized (Gade 1992).

A major exception to this scenario is the case of the wettest montane forests, the true rain (cloud) forests found on the eastern slope of the Andes and on the western slope near what is now the border of Ecuador with Colombia, that were and are mostly uncolonized (although that is not the case for premontane and lowland forests; see Raymond [1988]). Known archaeological sites are much fewer and most are concentrated in northern Peru and associated with a poorly defined "Chachapoyan" cultural group. The earliest date is about 2400 years B.P. (Church 1991). It is likely that most deforestation occurred during the several centuries before European contact, and was limited to a few watersheds.

Ironically, the most visible deforestation frontier today is on the wet eastern slopes of the Andes (Myers 1993). Much of this colonization was initiated in the 1960s (Brown et al. 1994; Collins 1989; Rudel 1993; Young 1996). It largely bypassed the montane forest on its way down to the premontane and lowland ecological zones. However, the roads permit continued resource extraction from the montane forest and in some cases fire has been used to convert forest into scrub used as rangeland.

The landscape patterns that have resulted can be characterized in terms of the length of time that has passed since a deforestation and colonization frontier occupied the area (figure 3.1). The very humid montane forests are considered to be associated with "new frontiers" because their forest cover has only being markedly affected in the last forty years. As shown in figure 3.1C, the elevational limit to tree growth and the altitudinal gradient down to premontane and lowland forests are the two most important landscape-organizing features before deforestation. Following deforestation (figure 3.1D), the elevational limit of forest shifts downward (often about 500 meters), rangeland and agricultural fields are established in place of forest at middle and low elevations, and corridors representing the roads divide the remaining montane forest into linear-shaped blocks. Figure 3.2 shows a road cutting through an eastern-slope montane forest in southern Peru.

The landscapes long occupied by people are more complicated conceptually because they include naturally fragmented montane forests. As humans converted these landscapes (figures 3.1A and 3.1B), the original hu-

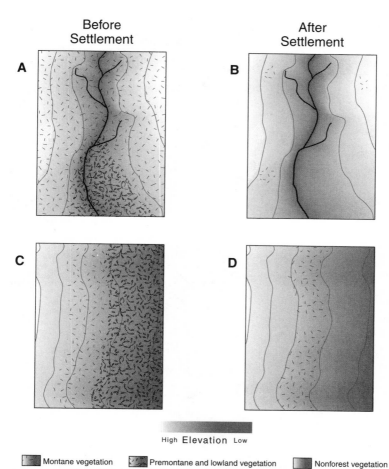

Fig. 3.1. Landscape forest patterns in the central Andes. Shown are a naturally fragmented montane forest landscape before (A), and after (B) "old frontiers" settlement and deforestation, and a naturally unfragmented montane forest landscape before (C), and after (D) "new frontiers" road construction and settlement.

mid montane forests were left behind as remnants. There might also have been conversion of this forest to scrub. The result is that landscapes today, including both the humid valleys of Ecuador with originally unfragmented forest and all the rest of the highlands with originally fragmented montane forests, have been converted into landscapes with little forest cover. Figure 3.3 shows an example of such a landscape in the central highlands of Peru.

Fig. 3.2. A road in tropical montane forest in southern Peru at 2600 meters. Photograph by the author.

Fig. 3.3. Landscape in central Peru, department of Ancash, at 2500 meters. Photograph by Blanca León.

Biogeographical Processes in Forested Landscapes

The landscape patterns discussed are the result of numerous processes acting over time. The processes of concern are those that affect the sizes, locations, and connectivity of these forests in relation to elevation and humidity. Also of concern is how the occurrence of plant and animal species, and their relative abundances, change along environmental gradients and in relation to forest size, isolation, and continuing human impacts. These are biogeographical processes that are ecological, evolutionary, and anthropogenic in nature (Young 1995). Evolutionary changes take place in the absence of human influence over many generations; anthropogenic changes directly or indirectly involve humans. Most important are the anthropogenic processes, although others that could be confused, confounded, or synergistic in effect cannot be ignored. In the central Andes, these processes involve the introduction of exotic species, the elimination or favoring of particular native species, and the modification of forest cover.

By identifying the processes involved, it becomes possible to evaluate their relative importances and to search for ways to test the ideas presented in this essay. Below I examine them in relation to their expected importance on old and new deforestation and colonization frontiers.

Deforestation Effects on Forest Distribution

The humid forests that remain on landscapes with ancient settlement and deforestation are isolated and small (usually < 10 hectares). They are often linear in shape because they tend to be found in ravines or near streams. These forest patches will have maximal amounts of edge habitat—forest bordering on more open habitat types. Trees growing under these conditions are short, often multiple-stemmed, and shrub-like in appearance (Crawford et al. 1970). Tree regeneration under these conditions can be asexual, as described in high-elevation *Polylepis* forests (Kessler and Driesch 1993). Tree populations may also require chronic disturbances to open up forest canopies; this is suggested by the rarity of seedlings and saplings of some species in closed forests (Valencia and Jørgensen 1992).

Because of human needs for firewood and polewood, there have been numerous private and state-sponsored reforestation projects in the central Andes. Most of these have used eucalypts such as *Eucalyptus globulus* (Dickinson 1969; Wiggins 1946), with pines (*Pinus* species) a distant second. These forests are monospecific, with little evidence of native plants colonizing the understory. The eucalypt is maintained by allowing it to coppice following cutting, or by planting seedlings grown in tree nurseries. In many areas the most visible trees on the landscape are from eucalypt plantings.

In contrast, the new frontiers and their accompanying roads divide what was once continuous montane forest into irregularly shaped blocks. The only natural edge habitats in these forests were along timberline, rivers, and landslides (Young 1993a). Road construction and subsequent habitat alteration dramatically increase the amount of edge at the expense of interior forest. Although the amount of surface area deforested is impressive given the few decades that these forests have been under direct human pressure, there are still large tracts of intact forest in watersheds not crossed by roads.

The small and isolated forest patches in many places in the central Andes that characterize old frontiers are under less direct human pressure than they were several decades ago. The reason is that the highlands are urbanizing as people move to larger towns and cities (Donoso de Baixeras 1992; Southgate and Whitaker 1994). In contrast, the rate of change in forest cover on the new frontiers has never been greater.

Impacts of Deforestation on Flora

Numerous plant species are intolerant of edge habitat because they require the shaded conditions of a tall closed forest. It seems likely that at least some of these species have gone locally extinct on the anciently deforested landscapes of the central Andes. Table 3.1 contrasts some commonly found tree and shrub genera on old and new deforestation frontiers. There are fewer taxa on the older landscapes: They tend to be species that resprout and develop multiple stems after being cut or burned (for example *Escallonia*), are shade intolerant (Hensen 1995), or have relatively small seeds that are dispersed by the wind (*Alnus, Gynoxys*) or small birds (*Berberis, Miconia, Myrica, Myrsine*). Exceptions such as *Schmardaea microphylla* (Meliaceae) can be highly endangered (Pennington et al. 1990; Styles and Bennett 1992). The forests in the new frontier areas have these same kinds of species, but also contain numerous additional taxa (table 3.1). These include interior forest genera with massive single trunks and large seeds (*Ocotea, Podocarpus, Prumnopitys, Symplocos*) that are dispersed by large-bodied frugivores such as cracids (*Crax, Ortalis, Penelope*), toucanets (*Aulacorhynchus*), and trogons (*Pharomarcus, Trogon*).

If extinction in the past seems plausible in the old frontiers, then wholesale extinction in the future seems likely for localized, endemic species on the new frontiers. Gentry (1986) reported that at least 38 (and possibly 90) endemic plant species occurred in one isolated premontane forest left on a ridgetop in western Ecuador; this forest was later cut down, so presumably these species are now extinct. The most susceptible species would be those with small ranges and a dependence on interior forest habitat.

Some species are subjected to extraction for human needs. People living nearby revisit trees and shrubs in old frontiers to harvest branches and trunks for firewood, charcoal production, or timber (Young 1993b). This chronic disturbance has, of course, a much longer history on the old frontiers than the new. However, the roads that create new deforestation frontiers also soon tie the forest resources to market demands, and thus may facilitate a rapid and more devastating extraction than in old frontiers where many people are only partially connected to market forces. Tree species valuable for timber can be threatened with local extinction when forests are accessible to markets.

Impacts of Deforestation on Fauna

The hunting of animals has a long and even ancient history in and around the montane forests of old landscapes in the central Andes. The principal game animal is the white-tailed deer (*Odocoileus virginianus*), although

Table 3.1. Presence of tree and shrub taxa common in humid montane forests on the old and new frontiers of the central Andes.

Botanical family and genus	Old frontiers	New frontiers
Aquifoliaceae		
Ilex		X
Araliaceae		
Oreopanax	X	X
Schefflera		X
Asteraceae		
Baccharis	X	X
Gynoxys	X	X
Pentacalia	X	X
Vernonia		X
Berberidaceae		
Berberis	X	X
Betulaceae		
Alnus	X	X
Brunelliaceae		
Brunellia		X
Buddlejaceae		
Buddleja	X	
Cecropiaceae		
Cecropia		X
Chloranthaceae		
Hedyosmum		X
Clethraceae		
Clethra	X	X
Cunoniaceae		
Weinmannia	X	X
Elaeocarpaceae		
Vallea	X	X
Ericaceae		
Cavendishia		X
Disterigma		X
Gaultheria	X	X
Vaccinium		X
Grossulariaceae		
Escallonia	X	X
Lamiaceae		
Satureja	X	
Lauraceae		
Nectandra		X
Ocotea		X
Persea		X
Loganiaceae		
Desfontainea		X
Loranthaceae		
Gaiadendron		X

86

Table 3.1. *Continued*

Botanical family and genus	Old frontiers	New frontiers
Melastomataceae		
Brachyotum	X	X
Miconia	X	X
Meliaceae		
Cedrela		X
Guarea		X
Ruagea		X
Moraceae		
Ficus		X
Morus		X
Monimiaceae		
Siparuna		X
Myrsinaceae		
Myrsine	X	X
Myrtaceae		
Myrcianthes	X	X
Podocarpaceae		
Podocarpus		X
Prumnopitys		X
Retrophyllum		X
Polygalaceae		
Monnina	X	X
Proteaceae		
Oreocallis	X	
Rosaceae		
Hesperomeles	X	X
Polylepis	X	X
Prunus		X
Rubiaceae		
Cinchona		X
Palicourea		X
Psychotria		X
Rudgea		X
Sabiaceae		
Meliosma		X
Sapindaceae		
Allophylus		X
Sapotaceae		
Pouteria		X
Solanaceae		
Cestrum		X
Saracha		X
Solanum	X	X
Styracaceae		
Styrax		X
Symplocaceae		
Symplocos		X

(continued)

Table 3.1. *Continued*

Botanical family and genus	Old frontiers	New frontiers
Theaceae		
Freziera		X
Gordonia		X
Ulmaceae		
Lozanella		X
Urticaceae		
Myriocarpa		X
Pilea		X

Notes: Compiled from Smith (1988), Valencia (1990), Young and León (1990), Young (1991, 1993b), Gentry (1992), Killeen et al. (1993), Ulloa Ulloa and Jørgensen (1995), Cano et al. (1995), and personal observation.

other species are taken opportunistically. Hunters also search for predatory species such as puma (*Puma concolor*) and Andean fox (*Pseudalopex culpaeus*) that threaten domesticated animals. The Andean bear (*Tremarctos ornatus*) can be a nuisance because it sometimes feeds on mature maize (Peyton 1980) and is potentially predatory on domesticated animals (Goldstein 1991); it is also eagerly hunted because the hide and fat, rendered into lard, can be sold.

These and many other large animal species are also found near or in the montane forests on new frontier landscapes. It is much more common there to find game offered in restaurants and consumed locally. Roads probably facilitate the rapid and often unnoticed overhunting of game species. Rare species threatened by the accessibility offered by road construction include the mountain tapir (*Tapirus pinchaque*) in the wet montane forests of Ecuador and northernmost Peru (Pulido 1991) and the yellow-tailed woolly monkey (*Lagothrix flavicauda*), endemic to montane forests in northern Peru (Leo 1984).

Monkeys, bear, tapir, and large, frugivorous birds are an important faunistic element in the intact montane forests of the central Andes. They soon disappear when hunters arrive and as interior forest is eliminated. Presumably the plant species whose seeds they disperse (Young 1990) are affected by their disappearance (Bond 1995). Various authors (Glanz 1991; Mittermeier 1987; Redford 1992; Terborgh et al. 1986) have documented the effect hunting has had on creating "empty" forests in some tropical lowlands—forests that have intact-appearing vegetation but are missing the large animal component. Something similar probably happens in the decades after a road creates a new deforestation frontier.

The fauna found in the isolated forest patches on old frontiers must be adapted (or preadapted) to those conditions by: (1) occupying edge habitat; (2) using the forests for only part of their food and shelter needs; or (3) being highly mobile and moving from patch to patch. Patterson et al. (1992) used DNA techniques to show that populations of fruit bats (*Artibeus*) were interbreeding because individuals had crossed mountain ranges and inhospitable habitat in northern Peru that had been presumed to be biogeographical barriers. These and other vagile species thus maintain population levels and genetic diversity in a patchwork landscape and across a heterogenous region through "metapopulation processes" (Gilpin and Hanski 1991).

On the other hand, Franke (1992) has shown a dramatic decline in species richness, and in trophic and behavioral diversity of birds, along a north-to-south gradient in the fragmented montane forests found on the western slopes of the Peruvian Andes from 7 to 12 degrees south latitude (Valencia 1990, 1992); further south the forests are smaller and more isolated from each other by large river valleys. Others have found bird diversity and composition to be affected by the size, location, and degree of isolation in patches of high-elevation Andean forests (Fjeldså 1992; Vuilleumier and Simberloff 1980). Arango-Vélez and Kattan (1997) showed that small Andean forest patches suffered higher rates of nest predation than did large patches.

The work of Silva (1992) should send a cautionary message concerning the more sedentary and localized species that constitute the montane forest fauna. She inventoried spiders in eight localities in Peruvian montane and premontane forests and found a total of approximately 750 species. More than 90% of these species were found in only one of the study sites and most of these were species new to science. Destruction of any of these forests will result in the extinction of unknown and unchampioned species.

Implications for Forest Conservation

The potential loss of endemic and interior-forest species to current deforestation is a powerful argument for the establishment, expansion, and maintenance of large nature reserves and national parks where unfragmented montane forests are found. Although Ecuador, Peru, and Bolivia have all made progress in these directions, much remains to be done (Fundación Natura 1992; Marconi 1992; Mares 1986; Mena Vásconez 1995; Pacheco et al. 1994). The exclusion of roads from protected watersheds is probably the most intractable issue politically, although good alternatives do not exist.

Management of forest resources on old frontiers is more complicated because it requires an unprecedented degree of collaboration between institutions involved with watershed protection and others concerned with nature reserves, plus local communities, nongovernmental organizations (NGOs), and regional governments (Morris 1985; Young in press b; Young and León 1995b). However, there are some technical issues that clearly could be addressed by biogeographers, ecologists, foresters, and other researchers. The increased use of native tree and shrub species in reforestation plantings is hindered by the lack of information on nursery care and silviculture (Brandbyge and Holm-Nielsen 1986; CESA 1991, 1992a), and the great effort needed to acquire and distribute the appropriate seeds. Although management of native forests in the central Andes is at a nascent stage (CESA 1992b, 1992c), this should not prove impossible, given that local Andean people have sophisticated and ancient land-use systems that include trees and shrubs (Ansión 1986; Johannessen and Hastorf 1990; Reynel 1988; Reynel and Felipe-Morales 1987; Torres et al. 1992). Because of the recognized need for reforestation projects, new restoration projects will be proposed that use the plantings of native species in ways that recreate what are presumed to be the original landscape patterns (Sarmiento 1995; Werner and Santisuk 1993).

Restoration would also increase forest connectivity and lessen the negative impacts of isolation on the plants and animals. The establishment of habitat corridors for the northern Andes in Colombia has been discussed by Kattan and Alvarez-López (1996), and by Yerena (1994) for Venezuela. In some cases this could be done within the respective national park systems; in others this would not be necessary.

Restoration of montane forests in the landscapes categorized here as "old deforestation frontiers" would require a series of experiments to determine suitable species and to identify and mitigate the factors that prevent or inhibit arboreal growth and forest formation (Smith 1977; Uhl 1988). This means that the very practical goals of restoration would also provide information on more theoretical concerns that involve attempts to understand the role that anthropogenic processes have played in creating current biotic landscapes. Ellenberg (1958, 1979) questioned the "naturalness" of the grassland and scrub vegetation that characterize so much of the tropical Andean highlands. Many others (Budowski 1968; Gade 1975; Jørgensen and Ulloa Ulloa 1994; Kessler 1995; Laegaard 1992; Seibert 1994; White 1985) have followed his lead, including myself. Although a refined paleorecord would help to distinguish anthropogenic impacts from those due to climate (Horn, chapter 5), even the best proxies from fossil material are still correlative. The most direct evidence will come from experiments that try to recreate the original fragmented and unfragmented forests.

Since restoration practices would also help protect rare and endangered species, and could dovetail with watershed programs and the rehabilitation of degraded landscapes, this approach may start a new era in the relations of humans with humid montane forest in the highland region of the central Andes. Implementation can begin with local projects and in a piecemeal fashion. The inclusion of restoration in conservation practices for the central Andes provides a more realistic and balanced selection of management tools for landscapes that include rural people. Uncertainties in this approach arise from the lack of knowledge of how biogeographical processes acted in the past (for example, what was the nature of prehistoric fire regimes?) and what changes are irreversible (for example, have species gone extinct?). However, the risks in experimenting are minimal as long as environmental and social concerns are addressed, and the changes that might result warrant the intent.

In a more general context, the landscape transformations discussed in this essay are to be found in other forested regions. With time total area of forest decreases, the area of nonarboreal vegetation types increases, the number of small, isolated forest patches increases, exotic species become more common, and some native species become rarer or go extinct. An additional complication in the Andes is the amount of preexisting environmental heterogeneity, although this is probably typical of most mountainous regions in the world. The role of environmental heterogeneity is often overlooked in discussions of deforestation, but is of critical importance.

Another feature of general importance is the need to clarify the time scales involved. Long-term impacts, accumulated over centuries, have transformed many Andean landscapes, and other long-settled regions, in ways that make them useful to subsistence agriculturalists. Very different are the new frontiers, where current economic processes and road construction are driving deforestation and other land-cover modifications over decadal time periods. Responses by conservation planners and development workers need to be calibrated accordingly. Prompt action is required to protect intact landscapes from deforestation and to reconnect them regionally. In contrast, deforestation is not the critical issue on old frontiers. Instead, the sustainable use of forest products and the implementation of restoration practices are more appropriate goals. The history and nature of human impact are among the most important variables affecting conservation policy decisions concerning forest distribution and associated biological diversity.

Acknowledgments

For financial support I am grateful to the John D. and Catherine T. MacArthur Foundation, Pew Charitable Trusts, the National Science Foundation (SES-8713237 and EAR-9422423), and the Danish Environmental Research Programme for the DIVA project (1994–1998). I thank Peter M. Jørgensen, Blanca León and Karl Zimmerer for comments on the manuscript.

References

Aiken, S. R., and C. H. Leigh. 1992. *Vanishing Rain Forests: The Ecological Transition in Malaysia.* Oxford: Clarendon Press.

Ansión, J. 1986. *El Arbol y el Bosque en la Sociedad Andina.* Lima: Proyecto Food and Agriculture Organization/Holanda/Instituto Nacional Forestal.

Arango-Vélez, N., and G. H. Kattan. 1997. Effects of Forest Fragmentation on Experimental Nest Predation in Andean Cloud Forest. *Biological Conservation* 81: 137–143.

Balée, W. 1992. Indigenous History and Amazonian Biodiversity. In *Changing Tropical Forests: Historical Perspectives on Today's Challenges in Central and South America,* eds. H. K. Steen and R. P. Tucker, pp. 185–197. Durham: Forest History Society.

Birks, H. H., H. J. B. Birks, P. E. Kaland, and D. Moe, eds. 1988. *The Cultural Landscape: Past, Present and Future.* Cambridge: Cambridge University Press.

Bond, W. J. 1995. Assessing the Risk of Plant Extinction due to Pollinator and Disperser Failure. In *Extinction Rates,* eds. J. H. Lawton and R. M. May, pp. 131–146. Oxford: Oxford University Press.

Brandbyge, J., and L. B. Holm-Nielsen. 1986. Reforestation of the High Andes with Local Species. *AAU Report 13.* Aarhus, Denmark: Botanical Institute, Aarhus University.

Brown, L. A., S. Digiacinto, R. Smith, and R. Sierra. 1994. Frameworks of Urban System Evolution in Frontier Settings and the Ecuador Amazon. *Geography Research Forum* 14: 72–96.

Bruhns, K. O. 1994. *Ancient South America.* Cambridge: Cambridge University Press.

Budowski, G. 1968. La Influencia Human en la Vegetación Natural de Montañas Tropicales Americanas. *Colloquium Geographicum* 9: 157–162.

Burger, R. L. 1992. *Chavin and the Origins of Andean Civilization.* New York: Thames and Hudson.

Cañadas Cruz, L. 1983. *El Mapa Bioclimático y Ecológico del Ecuador.* Quito: Banco Central del Ecuador.

Cano, A., K. R. Young, B. León, and R. B. Foster. 1995. Composition and Diversity of Flowering Plants in the Upper Montane Forest of Manu National Park, Southern Peru. In *Biodiversity and Conservation of Neotropical Montane Forests,* eds. S. P. Churchill, H. Balslev, E. Forero, and J. L. Luteyn, pp. 271–280. New York: New York Botanical Garden.

CESA. 1991. *Usos Tradicionales de las Especies Forestales Nativas en el Ecuador. Tomo 1: Informe de Investigación.* Quito: Central Ecuatoriana de Servicios Agrícolas.

CESA. 1992a. *Usos Tradicionales de las Especies Forestales Nativas en el Ecuador. Tomo 2: Catálogo de Especies.* Quito: Central Ecuatoriana de Servicios Agrícolas.

CESA. 1992b. *El Deterioro de los Bosques Naturales del Callejón Interandino del Ecuador.* Quito: Central Ecuatoriana de Servicios Agrícolas.

CESA. 1992c. *Explotación de Bosques Campesinos con Herramientas Manuales.* Quito: Central Ecuatoriana de Servicios Agrícolas.

Christensen, N. L. 1989. Landscape History and Ecological Change. *Journal of Forest History* 33: 116–125.

Church, W. B. 1991. La Occupación Temprana del Gran Pajatén. *Revista del Museo de Arqueología* (Universidad Nacional de Trujillo) 2: 7–38.

Colinvaux, P. A., K. N. Olson, and K.-b. Liu. 1988. Late-Glacial and Holocene Pollen Diagrams from Two Endorheic Lakes of the Inter-Andean Plateau of Ecuador. *Review of Palaeobotany and Palynology* 55: 83–99.

Collins, J. 1989. Small Farmer Responses to Environmental Change: Coffee Production in the Peruvian High Selva. In *The Human Ecology of Tropical Land Settlement in Latin America,* eds. D. A. Schumann and W. L. Patridge, pp. 238–263. Boulder: Westview Press.

Cook, N. D. 1981. *Demographic Collapse: Indian Peru, 1520–1620.* Cambridge: Cambridge University Press.

Crawford, R. M., D. Wishart, and R. M. Campbell. 1970. A Numerical Analysis of High Altitude Scrub Vegetation in Relation to Soil Erosion in the Eastern Cordillera of Peru. *Journal of Ecology* 58: 173–191.

Darby, H. C. 1956. The Clearing of the Woodland in Europe. In *Man's Role in Changing the Face of the Earth,* ed. W. L. Thomas, Jr., pp. 183–216. Chicago: University of Chicago Press.

Denevan, W. M. 1992. Native American Populations in 1492: Recent Research and a Revised Hemispheric Estimate. In *The Native Population of the Americas in 1492,* 2d edition, ed. W. M. Denevan, pp. xvii–xxix. Madison: University of Wisconsin Press.

Dickinson, J. C. 1969. The Eucalypt in the Sierra of Southern Peru. *Annals of the Association of American Geographers* 59: 294–307.

Dillon, M. O., A. Sagástegui Alva, I. Sánchez Vega, S. Llatas Quiro, and N. Hensold. 1995. Floristic Inventory and Biogeographic Analysis of Montane Forests in Northwestern Peru. In *Biodiversity and Conservation of Neotropical Montane Forests,* eds. S. P. Churchill, H. Balslev, E. Forero, and J. L. Luteyn, pp. 251–269. New York: New York Botanical Garden.

Dodson, C. H., and A. H. Gentry. 1991. Biological Extinction in Western Ecuador. *Annals of the Missouri Botanical Garden* 78: 273–295.

Donoso de Baixeras, S. 1992. La Población Rural en Bolivia. In *Conservación de la Diversidad Biológica en Bolivia,* ed. M. Marconi, pp. 181–196. La Paz: Centro de Datos para la Conservación and United States AID Mission to Bolivia.

Ellenberg, H. 1958. Wald oder Steppe? Die natürliche Pflanzendecke der Anden Perus. *Umschau* 1958: 645–648, 679–681.

Ellenberg, H. 1979. Man's Influence on Tropical Mountain Ecosystems in South America. *Journal of Ecology* 67: 401–416.

Ewel, J. J., A. Madriz, and J. Tosi. 1976. *Zonas de Vida de Venezuela (Memoria Explicativa sobre el Mapa Ecológico).* Segunda edición. Caracas: Ministerio de Agricultura y Cría.

Ferris-Kaan, R., ed. 1995. *The Ecology of Woodland Creation.* Chichester: John Wiley and Sons.

Fjeldså, J. 1992. Un Análisis Biogeográfico de la Avifauna de los Bosques de Queñoa (*Polylepis*) de los Andes y su Relevancia para Establecer Prioridades de Conservación. In *Biogeografía, Ecología y Conservación del Bosque Montano en el Perú,* eds. K. R. Young and N. Valencia, pp. 207–222. Memorias del Museo de Historia Natural UNMSM 21. Lima: Universidad Nacional Mayor de San Marcos.

Franke, I. 1992. Biogeografía y Ecología de las Aves de los Bosques Montanos del Perú Occidental. In *Biogeografía, Ecología y Conservación del Bosque Montano en el Perú,* eds. K. R. Young and N. Valencia, pp. 181–188. Lima: Universidad Nacional Mayor de San Marcos. Memorias del Museo de Historia Natural UNMSM 21.

Fundación Natura. 1992. *Acciones de Desarollo y Areas Naturales Protegidas en el Ecuador.* Quito: Fundación Natura.

Gade, D. W. 1975. *Plants, Man and the Land in the Vilcanota Valley of Peru.* The Hague: Dr W. Junk Publishers.

Gade, D. W. 1992. Landscape, System, and Identity in the Post-Conquest Andes. *Annals of the Association of American Geographers* 82: 460–477.

Gentry, A. H. 1986. Endemism in Tropical versus Temperate Plant Communities. In *Conservation Biology: The Science of Scarcity and Diversity,* ed. M. E. Soulé, pp. 1532–1581. Sunderland, Massachusetts: Sinauer Associates.

Gentry, A. H. 1992. Diversity and Floristic Composition of Andean Forests of Peru and Adjacent Countries: Implications for Their Conservation. In *Biogeografía, Ecología y Conservación del Bosque Montano en el Perú,* eds. K. R. Young and N. Valencia, pp. 11–29. Memorias del Museo de Historia Natural UNMSM 21. Lima: Universidad Nacional Mayor de San Marcos.

Gilpin, M., and I. Hanski, eds. 1991. *Metapopulation Dynamics: Empirical and Theoretical Investigations.* New York: Academic Press.

Glanz, W. E. 1991. Mammalian Densities at Protected versus Hunted Sites in Central Panama. In *Neotropical Wildlife Use and Conservation,* eds. J. G. Robinson and K. H. Redford, pp. 163–173. Chicago: University of Chicago Press.

Goldstein, I. 1991. Spectacled Bear Predation and Feeding Behavior on Livestock in Venezuela. *Studies on Neotropical Fauna and Environment* 26: 231–235.

Graf, K. 1992. Pollendiagramme aus den Anden: eine synthese zur Klimageschichte und Vegetationsentwicklung seit der letzten eiszeit. *Physische Geographie* (Universität Zürich) 34: 1–138.

Groom, M. J., and N. Schumaker. 1993. Evaluating Landscape Change: Patterns of Worldwide Deforestation and Local Fragmentation. In *Biotic Interactions and Global Change,* eds. P. M. Kareiva, J. G. Kingsolver, and R. B. Huey, pp. 24–44. Sunderland, Massachusetts: Sinauer Associates.

Guillet, D. 1985. Hacia una Historia Socio-Económica de los Bosques en los Andes Centrales del Perú. *Boletín de Lima* 38: 79–84.

Hansen, B. C. S., H. E. Wright, Jr., and J. P. Bradbury. 1984. Pollen Studies in the Junin Area, Central Peruvian Andes. *Geological Society of America Bulletin* 95: 1454–1465.

Hensen, I. 1995. Die Vegetation von Polylepis-waldern der Ostkordillere Boliviens. *Phytocoenologia* 25: 235–277.

Holdridge, L. R. 1947. Determination of World Plant Formations from Simple Climatic Data. *Science* 105: 367–368.

Johannessen, S., and C. A. Hastorf. 1990. A History of Fuel Management (A.D. 500 to the Present) in the Mantaro Valley, Peru. *Journal of Ethnobiology* 10: 61–90.

Jørgensen, P. M., and C. Ulloa Ulloa. 1994. Seed Plants of the High Andes: A Checklist. *AAU Report* 34. Aarhus, Denmark: Botanical Institute, Aarhus University.

Jørgensen, P. M., C. Ulloa Ulloa, J. E. Madsen, and R. Valencia. 1995. A Floristic Analysis of the High Andes of Ecuador. In *Biodiversity and Conservation of Neotropical Montane Forests,* eds. S. P. Churchill, H. Balslev, E. Forero, and J. L. Luteyn, pp. 221–237. New York: New York Botanical Garden.

Kattan, G. H., and H. Alvarez-López. 1996. Preservation and Management of Biodiversity in Fragmented Landscapes in the Colombian Andes. In *Forest Patches in Tropical Landscapes,* eds. J. Schelhas and R. Greenberg, pp. 3–18. Washington, D. C.: Island Press.

Kessler, M. 1995. Present and Potential Distribution of *Polylepis* (Rosaceae) Forests in Bolivia. In *Biodiversity and Conservation of Neotropical Montane Forests* eds. S. P. Churchill, H. Balslev, E. Forero, and J. L. Luteyn, pp. 281–294. New York: New York Botanical Garden.

Kessler, M., and P. Driesch. 1993. Causas y Historia de la Destrucción de Bosques Altoandinos en Bolivia. *Ecología en Bolivia* 21: 1–18.

Killeen, T. J., E. García, and S. G. Beck. 1993. *Guía de Arboles de Bolivia.* La Paz: Herbario Nacional de Bolivia and Missouri Botanical Garden.

Laegaard, S. 1992. Influence of Fire in the Grass Páramo Vegetation of Ecuador. In *Páramo: An Andean Ecosystem under Human Influence,* eds. H. Balslev and J. L. Luteyn, pp. 151–170. London: Academic Press.

Leo, M. 1984. *The Effect of Hunting, Selective Logging and Clear-Cutting on the Conservation of the Yellow-Tailed Woolly Monkey* (Lagothrix flavicauda). M. A. thesis, University of Florida–Gainesville.

Leo, M. 1995. The Importance of Tropical Montane Cloud Forest for Preserving Vertebrate Endemism in Peru: the Rio Abiseo National Park as a Case Study. In *Tropical Montane Cloud Forests,* eds. L. S. Hamilton, J. O. Juvik, and F. N. Scatena, pp. 198–211. New York: Springer-Verlag.

Lumbreras, L. G. 1974. *The Peoples and Cultures of Ancient Peru.* Washington, D. C.: Smithsonian Institution Press.

Madsen, J. E., and B. Øllgaard. 1994. Floristic Composition, Structure, and Dynamics of an Upper Montane Rain Forest in Southern Ecuador. *Nordic Journal of Botany* 14: 403–423.

Marconi, M. 1992. El Sistema Nacional de Areas Protegidas y las Areas Bajo Manejo Especial. In *Conservación de la Diversidad Biológica en Bolivia,* ed. M. Marconi, pp. 321–370. La Paz: Centro de Datos para la Conservación and United States AID Mission to Bolivia.

Mares, J. V. 1986. Conservation in South America: Problems, Consequences, and Solutions. *Science* 233: 734–739.

Mather, J. R., and G. V. Sdasyuk, eds. 1991. *Global Change: Geographical Approaches*. Tucson: University of Arizona Press.

Mena Vásconez, P. 1995. Las Areas Protegídas con Bosque Montano en el Ecuador. In *Biodiversity and Conservation of Neotropical Montane Forests*, eds. S. P. Churchill, H. Balslev, E. Forero, and J. L. Luteyn, pp. 627–635. New York: New York Botanical Garden.

Mittermeier, R. 1987. Effects of Hunting on Rain Forest Primates. In *Primate Conservation in the Tropical Rain Forest*, eds. C. Marsh and R. Mittermeier, pp. 109–148. New York: Alan R. Liss.

Morris, A. 1985. Forestry and Land-Use Conflicts in Cuenca, Ecuador. *Mountain Research and Development* 5: 183–196.

Myers, N. 1993. Tropical Forests: The Main Deforestation Fronts. *Environmental Conservation* 20: 9–16.

ONERN. 1976. *Mapa Ecológico del Perú y Guía Explicativa*. Lima: Oficina Nacional de Evaluación de Recursos Naturales.

Pacheco, L. F., J. A. Simonetti, and M. Moraes. 1994. Conservation of Bolivian Flora: Representation of Phytogeographic Zones in the National System of Protected Areas. *Biodiversity and Conservation* 3: 751–756.

Patterson, B. D., V. Pacheco, and M. V. Ashley. 1992. On the Origins of the Western Slope Region of Endemism: Systematics of Fig-Eating Bats, Genus *Artibeus*. In *Biogeografía, Ecología y Conservación del Bosque Montano en el Perú*, eds. K. R. Young and N. Valencia, pp. 189–205. Memorias del Museo de Historia Natural UNMSM 21. Lima: Universidad Nacional Mayor de San Marcos.

Pennington, T. D., S. C. Diaz, M. Timaná de la Flor, and C. R. Reynel. 1990. Un Raro "Cedro" Redescubierto en el Perú. *Boletín de Lima* 67: 41–46.

Peterken, G. F. 1996. *Natural Woodland: Ecology and Conservation in Northern Temperate Regions*. Cambridge: Cambridge University Press.

Peyton, B. 1980. Ecology, Distribution, and Food Habits of Spectacled Bears, *Tremarctos ornatus*, in Peru. *Journal of Mammalogy* 61: 639–652.

Pulido, V. 1991. *El Libro Rojo de la Fauna Silvestre del Perú*. Lima: Instituto Nacional de Investigación Agraria y Agroindustrial.

Raymond, J. S. 1988. A View from the Tropical Forest. In *Peruvian Prehistory*, ed. R. W. Keatinge, pp. 279–300. Cambridge: Cambridge University Press.

Redford, K. H. 1992. The Empty Forest. *BioScience* 42: 412–422.

Reynel, C. 1988. *Plantas para Leña en el Sur-occidente de Puno*. Puno, Peru: Proyecto Arbolandino.

Reynel, C., and C. Felipe-Morales. 1987. *Agroforestería Tradicional en los Andes del Perú*. Lima: Proyecto Food and Agriculture Organization/Holanda/Instituto Nacional Forestal.

Rick, J. W. 1988. The Character and Context of Highland Preceramic Society. In *Peruvian Prehistory*, ed. R. W. Keatinge, pp. 3–40. Cambridge: Cambridge University Press.

Rodgers, W. A. 1993. The Conservation of the Forest Resources of Eastern Africa: Past Influences, Present Practices and Future Needs. In *Biogeography and Ecol-*

ogy of the Rain Forests of Eastern Africa, eds. J. C. Lovett and S. K. Wasser, pp. 283–327. Cambridge: Cambridge University Press.

Rodríguez, L. O., J. H. Córdova, and J. Icochea. 1993. Lista Preliminar de los Anfibios del Perú. *Publicaciones del Museo de Historia Natural UNMSM Serie A* 45: 1–22.

Rudel, T. K. 1993. *Tropical Deforestation: Small Farmers and Land Clearing in the Ecuadorian Amazon.* New York: Columbia University Press.

Sagástegui Alva, A. 1988. *Vegetación y Flora de la Provincia de Contumazá.* Trujillo, Peru: Consejo Nacional de Ciencia y Technología.

Sarmiento, F. O. 1995. Restoration of Equatorial Andes: The Challenge for Conservation of Trop-Andean Landscapes in Ecuador. In *Biodiversity and Conservation of Neotropical Montane Forests,* eds. S. P. Churchill, H. Balslev, E. Forero, and J. L. Luteyn, pp. 637–651. New York: New York Botanical Garden.

Seibert, P. 1994. The Vegetation of the Settlement Area of the Callawaya People and the Ulla-Ulla Highlands in the Bolivian Andes. *Mountain Research and Development* 14: 189–211.

Seltzer, G. O. 1990. Recent Glacial History and Paleoclimate of the Peruvian-Bolivian Andes. *Quaternary Science Review* 9: 137–152.

Silva, D. 1992. Observations on the Diversity and Distribution of the Spiders of Peruvian Montane Forests. In *Biogeografía, Ecología y Conservación del Bosque Montano en el Perú,* eds. K. R. Young and N. Valencia, pp. 31–37. Memorias de Museo de Historia Natural UNMSM 21. Lima: Universidad Nacional Mayor de San Marcos.

Smith, A. P. 1977. Establishment of Seedlings of *Polylepis sericea* in the Páramo (Alpine) Zone of the Venezuelan Andes. *Bartonia* 45: 11–14.

Smith, C. E. 1980. Vegetation and Land Use near Guitarrero Cave. In *Guitarrero Cave: Early Man in the Andes,* ed. T. Lynch, pp. 65–83. New York: Academic Press.

Smith, D. N. 1988. *Flora and Vegetation of the Huascarán National Park, Ancash, Peru, with Preliminary Taxonomic Studies for a Manual of the Flora.* Ph. D. dissertation, University of Iowa–Ames.

Southgate, D., and M. Whitaker. 1994. *Economic Progress and the Environment: One Developing Country's Policy Crisis.* New York: Oxford University Press.

Styles, B. T., and S. T. Bennett. 1992. Notes on the Morphology, Chemistry, Ecology, Conservation Status and Cytology of *Schmardaea microphylla* (Meliaceae). *Botanical Journal of the Linnean Society* 108: 359–373.

Terborgh, J., L. H. Emmons, and C. Freese. 1986. La Fauna Silvestre de la Amazonía: el Despilfarro de un Recurso Renovable. *Boletín de Lima* 46: 77–85.

Torres, H., R. Borel, N. Bustamente, and M. I. Centeno. 1992. *Usos Tradicionales de Arbustos Nativos en el Sur de Puno.* Puno, Peru: Proyecto Arbolandino.

Tosi, J., Jr. 1960. *Zonas de Vida Natural en el Perú.* Lima: Organización de Estados Americanos.

Totman, C. 1989. *The Green Archipelago: Forestry in Preindustrial Japan.* Berkeley: University of California Press.

Uhl, C. 1988. Restoration of Degraded Lands in the Amazon Basin. In *Biodiversity,* eds. E. O. Wilson and F. M. Peter, pp. 326–332. Washington, D. C.: National Academy Press.

Ulloa Ulloa, C., and P. M. Jørgensen. 1995. *Arboles y Arbustos de los Andes del Ecuador.* 2nd edition. Quito, Ecuador: Ediciones Abya-Yala.

Valencia, N. 1990. *Ecology of the Forests on the Western Slopes of the Peruvian Andes.* Ph. D. thesis, University of Aberdeen.

Valencia, N. 1992. Los Bosques Nublados Secos de la Vertiente Occidental de los Andes del Perú. In *Biogeografía, Ecología y Conservación del Bosque Montano en el Perú,* eds. K. R. Young and N. Valencia, pp. 155–170. Memorias del Museo de Historia Natural UNMSM 21. Lima: Universidad Nacional Mayor de San Marcos.

Valencia, R., and P. M. Jørgensen. 1992. Composition and Structure of a Humid Montane Forest on the Pasochoa Volcano, Ecuador. *Nordic Journal of Botany* 12: 239–247.

van der Hammen, T., and A. M. Cleef. 1986. Development of the High Andean Páramo Flora and Vegetation. In *High Altitude Tropical Biogeography,* eds. F. Vuilleumier and M. Monasterio, pp. 153–201. New York: Oxford University Press.

Vuilleumier, F., and D. Simberloff. 1980. Ecology versus History as Determinants of Patchy and Insular Distributions in High Andean Birds. *Evolutionary Biology* 12: 235–379.

Weberbauer, A. 1945. *El Mundo Vegetal de los Andes Peruanos.* Estación Experimental de la Molina. Lima: Ministerio de Agricultura.

Werner, W. L., and T. Santisuk. 1993. Conservation and Restoration of Montane Forest Communities in Thailand. In *Restoration of Tropical Forest Ecosystems,* eds. H. Lieth and M. Lohmann, pp. 193–202. Dordrecht: Kluwer Academic.

White, S. 1985. Relations of Subsistence to the Vegetation Mosaic of Vilcabamba, Southern Peruvian Andes. *Conference of Latin Americanist Geographers Yearbook 11:* 3–10.

Whitney, G. G. 1994. *From Coastal Wilderness to Fruited Plain: A History of Environmental Change in Temperate North America, 1500 to the Present.* Cambridge: Cambridge University Press.

Wiggins, I. L. 1946. The Australian Blue Gum (*Eucalyptus globulus* Labill.) in Ecuador. *Lloydia* 9: 310–314.

Williams, M. 1989a. Deforestation: Past and Present. *Progress in Human Geography* 13: 176–208.

Williams, M. 1989b. *Americans and Their Forests: A Historical Geography.* Cambridge: Cambridge University Press.

Yallico, E. 1992. *Distribución de* Polylepis *en el Sur de Puno.* Pomata, Peru: Arbolandino.

Yerena, E. 1994. Corredores Ecológicos en los Andes de Venezuela. *Parques Nacionales y Conservación Ambiental* 4: 1–186.

Young, K. R. 1990. Dispersal of *Styrax ovatus* Seeds by the Spectacled Bear (*Tremarctos ornatus*). *Vida Silvestre Neotropical* 2: 68–69.

Young, K. R. 1991. Floristic Diversity on the Eastern Slopes of the Peruvian Andes. *Candollea* 46: 125–143.

Young, K. R. 1992. Biogeography of the Montane Forest Zone of the Eastern Slopes of Peru. In *Biogeografía, Ecología y Conservación del Bosque Montano en* el Perú, ed. K. R. Young and N. Valencia, pp. 119–140. Memorias del Museo de Historia Natural UNMSM 21. Lima: Universidad Nacional Mayor de San Marcos.

Young, K. R. 1993a. Woody and Scandent Plants on the Edges of an Andean Timberline. *Bulletin of the Torrey Botanical Club* 120: 1–18.

Young, K. R. 1993b. National Park Protection in Relation to the Ecological Zonation of a Neighboring Human Community: An Example from Northern Peru. *Mountain Research and Development* 13: 267–280.

Young, K. R. 1994. Roads and the Environmental Degradation of Tropical Montane Forests. *Conservation Biology* 8: 972–976.

Young, K. R. 1995. Biogeographical Paradigms Useful for the Study of Tropical Montane Forests and Their Biota. In *Biodiversity and Conservation of Neotropical Montane Forests,* eds. S. P. Churchill, H. Balslev, E. Forero, and J. L. Luteyn, pp. 79–87. New York: New York Botanical Garden.

Young, K. R. 1996. Threats to Biological Diversity Caused by Coca/Cocaine Deforestation in Peru. *Environmental Conservation* 23: 7–15.

Young, K. R. In press a. Composition and Structure of a Timberline Forest in North-Central Peru. In *Forest Biodiversity in North, Central and South America and the Caribbean: Research and Monitoring,* ed. F. Dallmeier. Pearl River, New York: Parthenon Press.

Young, K. R. In press b. Wildlife Conservation in the Cultural Landscapes of the Central Andes. *Landscape and Urban Planning* 38.

Young, K. R., and B. León. 1990. Catálogo de las Plantas de la Zona Alta del Parque Nacional Río Abiseo, Perú. *Publicaciones del Museo de Historia Natural UNMSM Serie B* 34: 1–37.

Young, K. R., and B. León. 1995a. Distribution and Conservation of Peru's Montane Forests: Interactions between the Biota and Human Society. In *Tropical Montane Cloud Forests,* eds. L. S. Hamilton, J. O. Juvik, and F. N. Scatena, pp. 363–376. New York: Springer-Verlag.

Young, K. R., and B. León. 1995b. Connectivity, Social Actors, and Conservation Policies in the Central Andes: The Case of Peru's Montane Forests. In *Biodiversity and Conservation of Neotropical Montane Forests,* eds. S. P. Churchill, H. Balslev, E. Forero, and J. L. Luteyn, pp. 653–661. New York: New York Botanical Garden.

Young, K. R., and B. León. 1997. Eastern Slopes of Peruvian Andes, Peru. In *Centres of Plant Diversity: A Guide and Strategy for their Conservation,* volume 3, eds. S. D. Davis, V. H. Heywood, O. Herrera MacBryde, and A. C. Hamilton, pp. 490–495. London: World Wildlife Fund and the World Conservation Union.

Young, K. R., and N. Valencia, eds. 1992a. *Biogeografía, Ecología y Conservación del Bosque Montano en el Perú.* Memorias del Museo de Historia Natural UNMSM 21. Lima: Universidad Nacional Mayor de San Marcos.

Young, K. R., and N. Valencia. 1992b. Introducción. In *Biogeografía, Ecología y Conservación del Bosque Montano en el Perú,* eds. K. R. Young and N. Valencia, pp. 5–9. Memorias del Museo de Historia Natural UNMSM 21. Lima: Universidad Nacional Mayor de San Marcos.

Zimmerer, K. S. 1995. The Origins of Andean Irrigation. *Nature* 378: 481–483.

4

Monitoring Forests in the Andes Using Remote Sensing
An Example from Southern Ecuador

Fernando R. Echavarria

Introduction

Remote sensing is a procedure for collecting data about features and targets on the earth's surface, often in the form of photographs and images. It is one of the most powerful tools available today for collecting and analyzing data on environmental change (Botkin et al. 1984; Greegor 1986; Myers 1988a, 1988b; Price 1986; Sader et al. 1990). Its technology allows for the monitoring, quantification, and mapping of both natural and human-induced changes on the environment—two key themes of this volume.

This study used remote sensing techniques to measure rates of montane forest clearing in a national park landscape: Ecuador's Podocarpus National Park. The analysis concentrates on the eastern side of the park and looks at two distinct parameters: (1) quantification of area loss, and (2) mapping the spatial patterns of forest fragmentation. Fragmentation of biogeographical landscapes, or the breaking up of large habitat or land areas into smaller parcels, is a significant worldwide issue for numerous disciplines (Forman 1995). These two parameters are critical in environmental assessment because they influence habitat loss for flora and fauna (Meffe and Carroll 1994; Schelhas and Greenberg 1996), and other land-

Table 4.1. FAO assessment of world deforestation rates for the tropics between 1981 and 1990 (FAO 1993: 9; Aldhous 1993: 3211).

Forest zone	Annual deforestation 1981–1990	
	Million hectares	% per annum
Rain forest	4.6	0.6
Moist deciduous	6.1	1.0
Dry deciduous	2.2	0.9
Hill and mountain	2.5	1.1
Total	15.4	0.9

scape processes including soils (Hamilton 1985; Harden 1991), hydrology (Hamilton and King 1983; Salati 1987) and climate (Salati and Nobre 1991).

This analysis focuses on the importance of the spatial constructs "landscape" and "region" for understanding land cover change and designing conservation strategies. Despite general attention given to tropical deforestation in the popular and scientific press (Dale et al. 1994; Simons 1989; Skole and Tucker 1993), few reports focus on the almost 5000 kilometers of montane forest along the Andean cordillera in Venezuela, Colombia, Ecuador, Peru, and Bolivia, or on specific sites designated for conservation (Hamilton et al. 1995).

Recently investigators have begun to document that regions containing tropical montane forests are being subjected to rapidly increasing rates of clearing (Aldous 1993; Churchill et al. 1995; Hamilton et al. 1995; Young 1992; Young and León 1995). The recent forest resource assessment by the Food and Agricultural Organization (FAO) gives world estimates for annual tropical deforestation rates from 1981 to 1990 (table 4.1). Figures indicate that montane forest zone depletion is almost twice as high as that for lowland forest (1.1% for montane versus 0.6% for lowland) and therefore warrants greater attention. The FAO study provides broad, general estimates of forest cover, but does not provide specific data on rates or patterns of change for specific regions or landscapes, including forest patch characteristics (FAO 1993). As a result, its usefulness for forest management, especially at a national or local level, is limited.

In fact, specific knowledge about the extent of forest clearing, analysis of forest patch structure and fragmentation, and information on the nature and degree of human change has been restricted by the absence of baseline data and information on rates of change (Echavarria 1991, 1993; Goodman 1993; Keating 1995). With remote sensing procedures, the potential to provide resource managers and policymakers specific information about landscape change at a scale that can foster both the development of appropriate conservation strategies and the ability to evaluate their effectiveness is greatly enhanced. It provides the best available spatial and temporal information about natural and anthropogenic change. This chapter illustrates through a case study how remote sensing techniques can provide baseline data on forest resources, quantify rates of change, and map spatial patterns of montane forest degradation for the forest landscape in eastern Podocarpus National Park. Some of the advantages and obstacles to using the technology in this type of landscape are discussed. The findings are then placed in the context of southern Ecuador and the Andean region.

Study Area

The landscape examined in this study is located in southern Ecuador, on the eastern slopes of the Andes. Over half of the site is located within the boundaries of Podocarpus National Park (map 4.1). Established southeast of the city of Loja in 1982, it is the southernmost national park in Ecuador. It falls within a region that is crossed from north to south by the main Andean cordillera, straddling the border between Ecuador's provinces of Loja and Zamora-Chinchipe. The mountain spurs crossing the region (which in places have a local relief of 3000 meters) are dissected by the headwaters of many rivers, all of them draining into the Amazon watershed, including the Zamora, Loja, Vilcabamba and Malacatos. The study area contains a portion of the eastern border of the park and the valleys of two small streams, the Bombuscara and Jambue. Both rivers flow into the Zamora River, which then joins the Santiago River before it meets the Marañon/Amazon mainstem at a point along the disputed Peru-Ecuador border (map 4.1). Hermessen (1917) gives a detailed description of the physical geography of this region.

Podocarpus National Park, named after the last remaining patches of the tree *Podocarpus oleifolius,* encompasses an area of 146280 hectares and includes some lowland forest, various types of montane forests between the altitudes of 1000 meters and 3500 meters, and some páramo vegetation (Apolo 1984). Lowland forest occurs below 1000 meters in elevation and can only be found in a few valley bottoms of the study area. Montane and premontane forests occur between 900 meters and 3200 meters above sea level and dominate the study area. Páramo, mostly grassland vegetation, occurs at elevations exceeding 3200 meters in the Andes (Acosta-Solís 1977; Troll 1968). Climate data for southern Ecuador is limited (Landívar 1989). Mean daily temperatures fluctuate between 23°C and 13°C at elevations from 1000 meters to 3000 meters above sea level. Precipitation is

Map 4.1. Podocarpus Nationak Park, Ecuador. Source: Fundacion Natura-DINAF A-12, 1989.

highest along the eastern slopes of the Andes with annual rates between 2000 millimeters and 6000 millimeters (Apolo 1989).

Despite the park's official protected status, forests continue to be cleared inside its boundaries (ARCOIRIS 1989; Espinosa et al. 1992). The embattled situation of park managers attempting to protect the park's natural resources with minimal enforcement or financial support from the government, is well recognized (Apolo 1984; Encalada 1982; Grenager 1991). The montane forests in the upper reaches of the Bombuscara and Nangaritza watersheds are being degraded by both small-scale gold mining (the mining cooperatives of "Cooperativa de Mineros de San Luís") and industrial-scale gold mining (the joint venture between multinational Rio Tinto Zinc, Inc. and the Ecuadorian company Cumbinamasa S.A.) (Fundación Natura 1993; Heylmun 1994; Sepulveda 1991). The lack of government action is because of a shortage of resources for conservation and the lack of baseline data on the extent and patterns of resource exploitation. No initial inventory of forests in the park exists and the amount of clearing is not being monitored.

Analysis

Remote sensing involves recording and analyzing electromagnetic energy (from the visible, infrared, or microwave portion of the electromagnetic spectrum) in the form of airphotos and satellite images (Greegor 1986: 429). From satellite data, it is possible to derive repeatable, verifiable information about montane forests over extensive areas. This information can then be used to calculate rates of forest conversion, map spatial patterns of fragmentation, and be incorporated into geographic information systems (GIS) for modeling purposes (Echavarria 1994; Scott et al. 1987; Scott et al. 1993). In this study, digital change detection techniques were used to examine changes in land cover between 1980 and 1992 in a small portion of the Bombuscara and Jambue watersheds. The information derived from the remote sensing data, in the form of maps and tables, was then verified with ground reconnaissance observations made during two field seasons in 1990 and 1991.

Sources of Data

Accuracy in land change detection calls for anniversary-date data that are collected on the same date but in different years (Jensen 1996). Unfortunately, in the tropics, cloud- or smoke-free imagery is rare. The cloud-free season over many parts of the Andes is the dry season, the time of year when farmers burn fields and forests. Because the skies are covered with smoke and dust, the spectral data is scattered, making the satellite images less useful. For this project, it was possible to obtain near-anniversary date satellite scenes that were generally, but not completely, free of both smoke and clouds. A Landsat Multispectral (MSS) image (path and row 10/63) with 30% cloud cover on December 19, 1980, was compared to a Landsat Thematic Mapper (TM) image (same path/row) for September 13, 1991. A 441 × 441 pixels subset of unrectified data, covering a 225-square-kilometer geographic area, was extracted from the 1980 MSS and 1991 TM images.

Processing of Remotely Sensed Data

Processing remotely sensed data requires the analyst to (1) extract the same subscene from the different satellite images, (2) correct the data for systematic and unsystematic geometric distortions, a process known as rectification, and (3) ensure that the images are registered to each other such that they represent the same geographic area. Several challenges were encountered in this study that represent typical problems when doing remote sensing in the tropics.

For example, although map sheets compiled by Ecuador's Instituto Geográfico Militar (IGM) in Quito were available for the study area, map coverage from which to derive coordinates for ground control points on the entire satellite image was incomplete. The 1980 imagery was rectified to 1:50000-scale maps using Universal Transverse Mercator (UTM) coordinates. At least half of the area covered by the image had never been mapped before. Thus, only portions of the image with available map coverage were used for the rectification. Where map coverage did exist, there was the additional problem that distinct features potentially useful as ground control points were not available, and, because of remoteness, the study area had few cultural features. This made it necessary to use node points in the stream network as ground control points.

Furthermore, cloud cover and topographic shadows increased the difficulty of selecting ground control points. The stream intersections were not always possible to locate at a subpixel level due to topographic shadows of the terrain. After rejecting 54 of the initial 123 extracted ground control points, it was possible to complete this image-to-map rectification (ERDAS 1990; Jensen 1996). Following rectification, perfect registration between the images had to be achieved for the change detection to be successful (Woodwell et al. 1986, 1986a, 1987).

Spectral normalization is also desirable in digital change detection of satellite data to ensure that observed changes in brightness value between the two images were because of changes in land cover and not due to

changes in atmospheric conditions (Eckhardt et al. 1990). The criteria suggested by Eckhardt et al. (1990) for normalization could not be met in this
case because of the high topography, smoke-filled skies, and the changes
in water levels in the streams of the study area.

Classification of 1980 MSS Data and 1991 TM Data

Classification of a satellite image requires the categorization of all pixels
in the image into land cover classes. Six land cover information classes
were utilized for this study: (1) forest in topographic slope, (2) intact forest
canopy, (3) degraded forest, (4) urban and roads, (5) hydrography, (6)
clouds and haze, and (7) clouds and topographic shadow.

The "forest in topographic slope" and "intact forest canopy" land cover
classes are basically the same, but have different illumination conditions.
Because of the high local topography, two land cover classes had to be
used to describe them: forests on the shadow slopes of mountains, and
forests facing the sun. Colby (1991) and Keating (1995) discuss the influence of topography on the classification of satellite data. These first two
land cover classes actually contain lowland forest, montane forest, and
páramo (Apolo 1989; Cifuentes et al. 1989; Holdridge 1974).

The third class, or "degraded forest," includes anthropogenic areas
where forests have been cleared for the planting of crops and African pasture grasses (Parsons 1972, 1975, 1976). This class contains the most important information for this study area since it represents patches that are
the end product of most deforestation. This class includes areas that had
not yet been cultivated but that clearly had been denuded of their original
forest cover. The land cover class "urban and roads" refers mainly to the
town of Zamora, located just upstream of the confluence of the Zamora
and Bombuscara Rivers, and its access roads.

The final two classes, "clouds" and "cloud shadows," unfortunately had
a higher presence in the images than desired; the two images used in this
investigation, however, had the least amount of cloud cover of any available satellite data over southern Ecuador.

1980 MSS Data
This survey used all four bands of Landsat MSS sensors (figure 4.1). The
procedure used to classify the 1980 MSS data is known as an unsupervised
maximum likelihood "cluster busting" classification. Echavarria (1993),
Jensen et al. (1993), and Jensen (1996) explain this procedure in detail.

Each pixel is assigned to a cluster based on its statistical characteristics
in n-dimensional feature space (where n is the number of spectral bands).
"Feature" or "spectral space" is defined as a two-dimensional scatter diagram where pixel reflectance from two different spectral bands is plotted.

STUDY AREA
BOMBUSCARA AND JAMBUE WATERSHEDS

RECTIFIED LANDSAT MSS DATA
19 December 1980
RGB = Bands 4,2,1

UTM Coordinates Zone 17 South

2000   0   2000   4000 m

Fig. 4.1. This 1980 Landsat MSS sub-scene illustrates the problems of clouds and topographic shadows in the remote sensing of forests in the tropical Andes.

Theoretically, pixels representing a given land cover type will cluster to-gether in a unique region of spectral space. The pixels assigned to the various land cover class clusters were then labeled according to the position of a cluster in spectral space and their relationship to the analyst's "real world" knowledge of the landscape (Jensen 1996; Jensen et al. 1987).

One problem encountered in the classification of the 1980 image was

Table 4.2. Inventory of land cover maps for eastern Podocarpus National Park, using 1980 (MSS) and 1991 (TM) Landsat data.

Class	1980 MSS data		1991 TM data	
	Area (ha)	%	Area (ha)	%
Forest in topographic shadow	5849.3	35.6	4958.8	30.2
Intact canopy forest	4475.2	27.2	9899.3	60.3
Degraded forest	474.6	2.9	907.4	5.5
Urban and roads	143.9	0.9	172.9	1.1
Hydrology	618.0	3.9	251.8	1.5
Clouds and haze	809.9	4.9	173.9	1.1
Cloud and topographic shadow	4049.1	24.6	55.9	0.3
Total	16420.0	100.0	16420.0	100.0

confusion between hydrology and topographic shadow features, because water in the forested upstream valleys was always under shadow. As a result, hydrologic features did not have a single spectral signature when plotted in feature space. A summary of the final classification for the 1980 MSS data is shown in table 4.2.

1991 TM Data

A slightly different classification approach was used for the 1991 TM data. Because of the TM sensor's higher spectral and spatial resolution, the final classification map had better separation between land cover classes on the basis of reflectance values and was therefore more accurate (Jensen 1996). For example, the final TM classification map (figure 4.2) had a more complete description of the area's hydrology than the MSS map (figure 4.1).

Because of all the forest clearing and burning, the skies over the study area were filled with smoke and dust. This caused severe atmospheric scattering in all the visible bands in the TM data set for 1991, making them unusable. The classification for the 1991 data set used different bands than those used in 1980. By using nonvisible bands 4, 5, and 7, it was possible to achieve excellent discrimination between forest and nonforest land cover types (figure 4.2). A summary of the final classification of the TM data is also included in table 4.2.

Accuracy Assessment of Classifications

Since the accuracy of postclassification change detection analysis depends on the accuracy of the initial classifications (Jensen 1996), it is necessary to apply the best possible accuracy assessment technique to the above two classification maps before change detection. Because no aerial photogra-

STUDY AREA
BOMBUSCARA AND JAMBUE WATERSHEDS

RECTIFIED LANDSAT TM DATA
13 September 1991
RGB = Bands 7,5,4

UTM Coordinates Zone 17 South

2000   0   2000   4000 m

Fig. 4.2. This 1991 Landsat TM sub-scene combining nonvisible bands provides good discrimination between forest and nonforest land cover types.

phy of the region exists for 1980 and 1991, it was impossible to verify the accuracy of the 1980 classification map. However, accuracy of the TM 1991 classification map could be determined using a digitized land cover map derived from the author's ground reconnaissance.

The 1991 classification error assessment was accomplished using field observations. Land cover types that were verified in the field were recorded over a topographic map sheet (1:50000 scale) identical to the one used to rectify the imagery. The land cover observations are limited to two watersheds of the Bombuscara and Jambue Rivers where ground reconnaissance could be completed. The sample points selected for the accuracy assessment were not random but biased to areas accessible from forest trails and dirt roads.

Classical remote sensing analysts use two different types of classification accuracies: "producer's accuracy" (errors of omission) and "user's accuracy" (errors of commission). The former measures the percentage of pixels that failed to be included in a certain land cover class; the latter measures the percentage of pixels that were incorrectly included in a particular land cover class (Congalton 1988, 1991). In this case, "producer's accuracy" and "user's accuracy" for the degraded forest class are 89.7% and 93.5%, respectively (Echavarria 1993).

Change Detection Results

Contemporary thinking on the management of natural resources encourages the recognition of uncertainty, randomness, and complexity in ecosystem processes (Botkin 1990: 155). Obviously, it is easier to think about absolute numbers rather than ranges or probability distributions. Absolute numbers, however, may present a false picture of the real world due to the inherent uncertainty of both natural processes themselves and the way we acquire, analyze, and interpret data. Without the use of varying assumptions to account for these uncertainties, it is unlikely that the analysis will help natural resource managers make effective decisions.

The results of this study are presented as a range of estimates of forest clearing that reflect some of the uncertainty inherent in modeling environmental problems. Many of these environmental problems have a characteristic indeterminacy, or an unknown probability, because of their complexity and randomness (Myers 1993). For example, there is considerable subjectivity in the labeling procedure of the spectral clusters generated by unsupervised classifications of satellite data. Because of this, different unsupervised classifications of the same satellite image are likely to produce slightly different classification maps.

In this study, several iterations of the unsupervised classification were made that are equally valid, given the land cover information available. In the computer algorithm, these iterations were accomplished by assigning small variations to the threshold settings for the unsupervised classification to yield slightly different classification maps. Multiple iterations based

Table 4.3. Maximum and minimum change from forest to nonforest from 1980 to 1991.

Land cover class	1980 MSS data		1991 TM data	
	High %	Low %	High %	Low %
Degraded forest	09.73	02.89	14.08	05.53
Urban	00.91	00.88	01.63	01.05
Total nonforest	10.64	03.77	15.71	06.58

on varying assumptions can provide a range of certainty that is expressed as an interval.

When these several equally valid classifications were compared for the two dates, slightly different overall rates of deforestation were obtained. A high and low estimate for each of the land cover classes in 1980 and 1991 was calculated. Since we are concerned with deforestation, it is only necessary to note the changes in two land cover classes: degraded forest and urban. Table 4.3 summarizes the high and low estimates for those two classes for the 1980 and 1991 classification maps. When the two maps with the high estimates for the two relevant land cover classes are compared, a maximum 5.1% and minimum 2.8% deforestation rate was calculated (table 4.3). The land patches associated with an increase in urban areas or degraded forest are visible on the 1991 classification map (figure 4.3).

Finally, a change detection matrix (table 4.4) was calculated that summarizes only the most conservative results of land cover change analysis for the two dates. The unsupervised classification of the 1980 subscene estimated 475 hectares of degraded forest (2.9% of the study area) and 144 hectares of urban/roads land cover (0.9%). The inventory for the 1991 classification estimated the degraded forest to have increased to 907 hectares, and the urban category to have increased to 173 hectares (5.5% and 1.1% of the study area, respectively). These results indicate a total degradation of 432 hectares during the eleven-year period, and an annual rate of forest loss of 39 hectares (0.2% of the study area). The total amount of urbanization for the 11-year period was 29 hectares.

Discussion

Implications for Podocarpus National Park

The results of this research document that the rate of forest degradation in the eastern Podocarpus National Park landscape differs markedly from other montane forest regions. Forest degradation for this landscape increased by a minimum of 2.8% and a maximum of 5.1% over an 11-year period (0.3%–0.5% per year), significantly lower than the FAO's Forest

BOMBUSCARA AND JAMBUE WATERSHEDS

CHANGE MAP

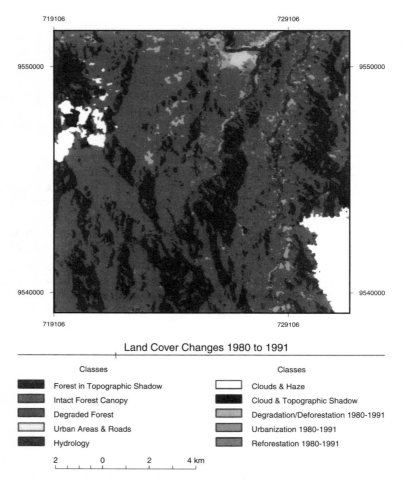

Fig. 4.3. Change detection map of Bombuscaro and Jambue watersheds for 1980–1991.

Resources Assessment figures for world annual deforestation during the same time period (table 4.1). According to the FAO report, the montane forests of the tropics suffered the highest deforestation rates during the 1980s (1.1% per annum). One explanation for the lower rate in the study landscape is that at least half of it falls inside Podocarpus National Park, which was established in 1981. Since then, agrocolonization in the landscape has been officially prohibited, even though poorly enforced.

Table 4.4. Change detection matrix for 1980 and 1991 satellite data (Matrix units = hectares).

1980	1991						
	Forest in topographic shadow	Intact forest canopy	Degraded forest	Urban and roads	Hydrology	Clouds and haze	Clouds and topographic shadow
Forest in topographic shadow	1914.7	3821.5	143.2	6.9	7.74	77.7	5.2
Intact forest canopy	194.9	4144.4	207.0	26.2	4.6	30.0	3.0
Degraded forest	13.3	247.8	128.8	40.8	7.5	0.0	0.0
Urban and roads	2.7	19.8	36.8	76.0	2.3	0.0	0.0
Hydrology	86.6	276.1	91.0	31.6	35.6	0.0	0.0
Clouds and haze	426.0	327.6	4.0	0.0	1.5	6.4	10.1
Clouds and topographic shadow	2399.9	1375.7	59.8	0.2	12.6	60.3	38.2
Total	5038.1	10212.9	670.6	181.7	71.8	174.4	56.5

Although one would expect no deforestation given the area's national park status, the satellite images reveal considerable forest clearing has occurred inside the park. This raises a number of questions: Has clearing occurred steadily over the study period or does the clearing rate observed only reflect a transition phase following the establishment of the park in 1981? Did deforestation in the immediate areas outside the park accelerate or halt with the establishment of the park? What impact has the rate of clearing had on the region's biological diversity? What are the edge effects due to forest clearing along the boundary of the park? How can the population living in the vicinity of the park be included in a conservation program that can ensure sustainable use of the area's resources, including its ecotourism potential? Is greater enforcement of the agrocolonization policy necessary? How are these policies to be defined and implemented? Finding the ultimate forces responsible for forest clearing will inevitably involve complex social, economic, and political issues that are far beyond the scope of this analysis (Brown and Pearce 1994; Rudel and Horowitz 1993; Sierra 1996).

The remote sensing study of the Podocarpus National Park landscape provides the baseline information for the formulation of these and other questions. Clearly, information derived from remote sensing data allows for more than mere quantification. The tables and maps for different points in time generated by this study provide information on, and raise new questions about, human impacts on the environment. Both help resource managers develop strategies that are effective and sustainable.

Implications for the Andean Region and Beyond

The application of remote sensing is relevant to organizations interested in forest conservation in the Andes and in other mountain regions of the world. The general FAO estimates on montane forest clearing do not provide adequate information to assist in the effort to manage montane forest resources, though they are important for distinguishing the severity of the problem compared to lowland forests.

Knowledge of the extent, spatial patterns, and rates of anthropogenic change and landscape modification in specific regions could lead to better-informed management decisions on the part of government and nongovernment institutions. This investigation provides specific information that could be used by Ecuador's Ministry of Agriculture Forestry Section to update monitoring programs for the country's forests. The methodology presented in this investigation could also be used to determine how fast forests are being cleared inside, or along the borders of, other protected areas in Ecuador and to what degree increasing fragmentation of forests threaten habitats. By replicating the analysis of Podocarpus National Park in other reserve areas, it may be possible to establish priorities of forest conservation. For instance, a list of Ecuador's most endangered protected areas could be generated. This in turn could be used to establish priorities for future government expenditures in forest conservation and habitat protection.

It might be expected that if forest clearing is occurring inside protected areas like Podocarpus National Park, much higher rates of clearing are taking place along the unprotected eastern slopes of the Andes. To prove this, it is necessary to apply the same methodology used in this study to much larger areas of Ecuador. The effort to obtain the information necessary for effective resource management, of course, will require an investment of resources to purchase satellite data and pay for its analysis and ground-truth verification.

Without access to this information, rational and effective natural resource policy by governments throughout the Andean region seems unlikely. Satellite-derived land-use and land cover maps will provide baseline data with which to pinpoint patches and areas that are most degraded and in greatest need of immediate protection. These could be used to draw buffers, or transition zones, with specific land use, around protected areas, similar to those in UNESCO-designated biosphere reserves, in order to improve conservation and encourage sustainable resource use (IUCN 1994).

Conclusion

This chapter has explained how remote sensing techniques can be used to quantify and analyze land cover change, specifically forest patches and forest fragmentation, in a montane forest landscape in eastern Podocarpus National Park. Two land cover maps were produced from the satellite images using an unsupervised classification. The two land cover maps for 1980 and 1991 were assessed for accuracy and then compared to quantify changes in forest cover (2.8% to 5.1% over an 11-year study period). The study revealed that the rates of deforestation for the landscape around Podocarpus National Park are lower than those reported by FAO (1993) for the region.

This case study, therefore, illustrates how landscape and region, as conceptual frameworks, help organize research questions at a level more useful than the FAO world estimates for those considering sustainable development strategies. Without the quantifiable information generated by the study, it could be assumed that no clearing was taking place in the park landscape because of its official protected status. It could also be assumed that the FAO world estimates for montane forest clearing apply to all of southern Ecuador. Policies to manage this natural resource, based on these incorrect assumptions or on a blanket policy covering all tropical forests, would most probably fail.

Remote sensing techniques contribute to our ability to monitor, quantify, and map both natural and human-induced changes on the environment. Both types of environmental change are the key themes of this volume. Due to the choice of landscape and absence of any major natural disturbance in the region, this chapter has focused on anthropogenic change. The ability to capture a synoptic snapshot of a landscape at different points in time reveals insights into the dynamics of the nonequilibrium conditions of these environments, that result from natural or human disturbances. In short, the techniques presented in this chapter can begin to uncover the essences of ecosystems—how they function on their own, and how they respond to human intervention.

Remote sensing techniques may not show the impact of forest patch disturbance on particular species or ecosystems, as evidenced by Podocarpus National Park, but they can actually map the size and frequency of landscape patches and monitor change.

Resource managers today face uncertainty about appropriate interventions because of limited knowledge about natural processes and the speed and impact of resource depletion. While analysis of remotely-sensed data begins to provide this missing information, the analysis should incorporate uncertainties inherent to natural processes and account for the ways we acquire, analyze, and interpret data. Although perhaps counterintuitive, an accurate analysis should therefore present a range, rather than an absolute number. Quantifiable estimates provide important information on rates of resource depletion with a level of certainty that will not mislead resource managers.

The findings alone from the research in this chapter will not solve the problem of montane forest degradation in the Andes of South America. In order to obtain a more accurate picture, the analysis needs to be repeated on a systematic and routine basis along the length of the montane forest belt that stretches for almost 5000 kilometers through Venezuela, Colombia, Ecuador, Peru, and Bolivia. These studies will lead to more effective natural resource policies that will allow countries in the tropics to pursue development and conservation strategies that are sustainable.

Acknowledgments

Professor John Winberry at the University of South Carolina kindly reviewed early versions of this chapter. Thanks are also due to Professor John R. Jensen in the USC Remote Sensing Laboratory where the satellite images were processed. The final version of this manuscript benefited from extensive review by Dr. Kenneth R. Young and Jane A. Loewenson.

References

Acosta-Solís, M. 1977. *División fitogeográfica y formaciones geobotánicas del Ecuador.* Quito: Casa de la Cultura Ecuatoriana.

Aldhous, P. 1993. Tropical Deforestation: Not Just a Problem in Amazonia. *Science* 259: 1390.

Apolo, W. B. 1984. *Plan de Manejo Parque Nacional Podocarpus.* Quito: Dirección Nacional Forestal, Ministerio de Agricultura y Ganadería.

Apolo, W. B. 1989. Influencias hidrológicas de los bosques. Ciencias Agricolas: Revista de Difusión Técnica y Científica de la Facultad de Ciencias Agrícolas, 16–17 (1–2): 10–19. Loja, Ecuador: Universidad Nacional de Loja.

ARCOIRIS. 1989. *Pérdida irreversible de áreas naturales claves en la región sur del país.* ARCOIRIS 1: 5.

Botkin, D. B. 1990. *Discordant Harmonies: A New Ecology for the Twenty-first Century.* Oxford: Oxford University Press.

Botkin, D. B., J. E. Estes, R. M. MacDonald, and M. W. Wilson. 1984. Studying the Earth's Vegetation from Space. *BioScience* 34(8): 508–514.

Brown, K., and D. W. Pearce, eds. 1994. *The Causes of Tropical Deforestation.* Vancouver: University of British Columbia Press.

Churchill, S. P., H. Balslev, E. Forero, and J. L. Luteyn, eds. 1995. *Biodiversity and Conservation of Neotropical Montane Forests.* Proceedings of the Neotropical Montane Forest Biodiversity and Conservation Symposium, The New York Botanical Garden, 21–26 June, 1993. New York: The New York Botanical Garden.

Cifuentes, M., A. Ponce, F. Alban, P. Mena, G. Mosquera, J. Rodrígues, D. Silva, L. Suáres, A. Tobar, and J. Torres. 1989. *Estrategia para el Sistema Nacional de Areas Protegidas del Ecuador, FASE II.* Quito: Dirección Nacional Forestal, Ministerio de Agricultura y Ganadería, y Fundación Natura.

Colby, J. D. 1991. Topographic Normalization in Rugged Terrain. *Photogrammetric Engineering and Remote Sensing* 57(5): 531–537.

Congalton, R. G. 1988. A Comparison of Sampling Schemes Used in Generating Error Matrices for Assessing the Accuracy of Maps Generated from Remotely Sensed Data. *Remote Sensing of Environment* 37: 45–46.

Congalton, R. G. 1991. A Review of Assessing the Accuracy of Classifications of Remotely Sensed Data. *Remote Sensing of Environment* 37: 35–46.

Dale, V. H., R. V. O'Neill, F. Southworth, and M. Pedlowski. 1994. Modeling Effects of Land Management in the Brazilian Amazonian Settlement of Rondônia. *Conservation Biology* 8(1): 196–206.

Echavarria, F. R. 1991. Cuantificación de la Deforestación en el Valle del Huallaga, Perú. *Revista Geográfica* (Instituto Panamericano de Historia y Geografía-IPGH) 114: 37–53.

Echavarria, F. R. 1993. *Remote Sensing of Montane Forest Degradation in Southern Ecuador.* Ph. D. dissertation, University of South Carolina.

Echavarria, F. R. 1994. GAP Analysis of the Montane Forests of Podocarpus National Park. In *Proceedings of the GIS/LIS 1994 Conference, October 23–28, Phoenix, Arizona,* pp. 235–244.

Eckhardt, D. W., J. P. Verdin, and G. R. Lyford. 1990. Automated Update of an Irrigated Lands GIS Using SPOT HRV Imagery. *Photogrammetric Engineering and Remote Sensing* 56(11): 1515–1522.

Encalada, M. 1982. *Evidencias del Deterioro Ambiental en el Ecuador.* Quito: Gangotera and Ruíz Editores S.A., and Fundación Natura.

ERDAS, 1990. *ERDAS Field Guide, Version 7.4,* p. 127. Atlanta: ERDAS, Inc.

Espinosa, D. A., E. Gallagher, P. A. Cueva, C. Gomez, and C. S. Bermeo. 1992. Diagnostico socio-económico del área de amortiguamiento occidental y sur oriental del Parque Nacional Podocarpus. *Boletín Informativo Sobre Biología, Conservación y Vida Silvestre* 3: 73–126.

Food and Agriculture Organization of the United Nations (FAO). 1993. *Summary of the Final Report of the Forest Resources Assessment 1990 for the Tropical World.* Eleventh Session of the Committee on Forestry, Rome, March 8–12. Rome: FAO.

Forman, R. T. T. 1995. *Land Mosaics. The Ecology of Landscapes and Regions.* Cambridge: Cambridge University Press.

Fundación Natura-DINAF. 1988. *Evaluación y revisión de la estratégia para el sistema de áreas protegidas del Ecuador. Mapas y Cartas Topográficas. II Fase.* Quito: Fundación Natura.

Fundación Natura. 1993. *Impacto ambiental de la minería en cuatro áreas protegidas del Ecuador. Estudios en Areas Protegidas,* Volumen 4. Quito: Fundación Natura.

Greegor, D. H. 1986. Ecology from Space. *BioScience* 36(7): 429–432.

Grenager, T. E. 1991. Estudio de los impactos de la extensión propuesta de la carretera Cajanuma en el Parque Nacional Podocarpus. *Boletín Informativo sobre Biología, Conservación y Vida Silvestre,* Facultad de Ciencias Veterinarias, Loja, Ecuador, Febrero, no. 2, pp. 99–125.

Goodman, B. 1993. Drugs and People Threaten Diversity in Andean Forests. *Science* 261: 293.

Hamilton, L. S. 1985. Overcoming Myths about Soil and Water Impacts of Tropical Forest Land Uses. In *Soil Erosion and Conservation,* eds. S. A. El-Swaify, W. C. Moldenhaur, and A. Lo, pp. 688–690. Ankeny: Soil Conservation Society of America.

Hamilton, L. S., J. O. Juvik, and F. N. Scatena, eds. 1995. *Tropical Montane Cloud Forests.* Ecological Studies 110. New York: Springer-Verlag.

Hamilton, L. S., and P. N. King. 1983. *Tropical Forested Watersheds: Hydrologic and Soil Response to Major Uses or Conversions.* Boulder: Westview Press.

Harden, C. 1991. Andean Soil Erosion. *National Geographic Research and Exploration* 7(2): 216–231.

Hermessen, J. L. 1917. A Journey on the Rio Zamora, Ecuador. *Geographical Review* 4(6): 434–449.

Heylmun, D. H. 1994. Gold in Ecuador. *International California Mining Journal* 63(8): 67–70.

Holdridge, L. R. 1974. Determination of World Plant Formations from Simple Climatic Data. *Science* 105: 367–368.

International Union for the Conservation of Nature (IUCN). 1994. *World Conservation Monitoring Center and the IUCN Commission on National Parks and Protected Areas. 1993 United Nations List of National Parks and Protected Areas.* New York: IUCN Publications.

Jensen, J. R. 1996. *Introductory Digital Image Processing. A Remote Sensing Perspective.* 2nd edition. Upper Saddle River, NJ: Prentice Hall, Inc.

Jensen, J. R., D. J. Cowen, J. D. Althausen, S. Narumalani, and O. Weatherbee. 1993. An Evaluation of the CoastWatch Change Detection Protocol in South Carolina. *Photogrammetric Engineering and Remote Sensing* 54(8): 1039–1046.

Jensen, J. R., E. W. Ramsey, H. E. Mackey, E. J. Christensen, and R. R. Sharitz. 1987. Inland Wetland Change Detection Using Aircraft MSS Data. *Photogrammetric Engineering and Remote Sensing* 53(5): 521–529.

Keating, P. L. 1995. *Disturbance Regimes and Regeneration Dynamics of Upper Montane Forests and Páramos in the Southern Ecuadorian Andes.* Ph. D. dissertation, University of Colorado—Boulder.

Landívar, C. B. 1989. *Análisis y Estudios Climatológicos en el Ecuador.* Quito: Instituto Panamericano de Geografía e Historia, Sección Nacional del Ecuador.

Meffe, G. K., and C. R. Carroll, eds. 1994. *Principles of Conservation Biology.* Sunderland, Massachusetts: Sinauer Associates, Inc.

Myers, N. 1988a. Threatened Biotas: "Hotspots" in Tropical Forests. *The Environmentalist* 8(30): 1–20.

Myers, N. 1988b. Tropical Deforestation and Remote Sensing. *Forest Ecology and Management* 23: 215–225.

Myers, N. 1993. Biodiversity and the Precautionary Principle. *Ambio* 22 (2/3): 74–79.

Parsons, J. 1972. Spread of African Pasture Grasses to the American Tropics. *Journal of Range Management* 25: 12–17.

Parsons, J. 1975. The Changing Nature of New World Tropical Forests since European Colonization. In *The Use of Ecological Guidelines for Development in the American Humid Tropics,* ed. Morges, pp. 28–38. New York: International Union for Conservation of Nature and Natural Resources Publications.

Parsons, J. 1976. Forest to Pasture: Development or Destruction? *Revista Biológica Tropical* 24(1): 121–138.

Price, M. 1986. The Analysis of Vegetation Change by Remote Sensing. *Progress in Physical Geography* 10(4): 473–491.

Rudel, T. K., and B. Horowitz. 1993. *Tropical Deforestation: Small Farmers and Forest Clearing in the Ecuadorian Amazon.* New York: Columbia University Press.

Sader, S. A., T. A. Stone, and A. T. Joyce. 1990. Remote Sensing of Tropical Forests: An Overview of Research and Application Using Non-photographic Sensors. *Photogrammetric Engineering and Remote Sensing* 56(10): 1343–1351.

Salati, E. 1987. The Forest and the Hydrological Cycle. In *Geophysiology of Amazonia: Vegetation and Climate Interaction,* ed. R. E. Dickinson, pp. 273–296. New York: Wiley Interscience.

Salati, E., and C. A. Nobre. 1991. Possible Climatic Impacts of Tropical Deforestation. *Climatic Change* 19 (1/2):177–196.

Schelhas, J., and R. Greenberg, eds. 1996. *Forest Patches in Tropical Landscapes.* Washington, D. C.: Island Press.

Scott, J. M., B. Csuti, J. J. Jacobi, and J. E. Estes. 1987. Species Richness: A Geographic Approach to Protecting Future Biological Diversity. *BioScience* 37: 782–788.

Scott, J. M., F. Davis, B. Csuti, R. Noss, B. Butterfield, C. Groves, H. Anderson, S. Caicco, F. D'erchia, T. C. Edwards, J. Ulliman, and R. G. Wright. 1993. Gap Analysis: A Geographic Approach to Protection of Biological Diversity. Wildlife Monograph No. 123. *The Journal of Wildlife Management* 57(1): 1–41.

Sepulveda, O. 1991. El frío color del oro. *Prisma Latinoamericano* 11: 41–143.

Sierra, R. 1996. *La deforestación en el noroccidente del Ecuador 1983–1993.* Quito: EcoCiencia-Fundación Ecuatoriana de Estudios Ecológicos.

Simons, M. 1989. In Amazon, Road is a Dream, but to Ecologists a Nightmare. *New York Times,* 19 February, p. 1.

Skole, D., and C. Tucker. 1993. Tropical Deforestation and Habitat Fragmentation in the Amazon: Satellite Data from 1978 to 1988. *Science* 260: 1905–1910.

Troll, C., 1968. The Cordilleras of the Tropical Americas. Aspects of Climatic, Phytogeographical and Agrarian Ecology. *Colloquium Geographicum* 9: 15–56.

Woodwell, G. M., J. E. Hobbie, R. A. Houghton, J. M. Melillo, B. J. Peterson, G. R. Shaver, and T. A. Stone. 1986a. *Deforestation Measured by LANDSAT: Steps Toward a Method* (DOE/EV/10468-1). Washington, D. C.: Office of Energy Research.

Woodwell, G. M., R. A. Houghton, and T. A. Stone. 1986b. Deforestation in the Brazilian Amazon Basin Measured by Satellite Imagery. In *Tropical Rain Forests and the World Atmosphere,* ed. G. T. Prance, pp. 23–32. Boulder: Westview Press.

Woodwell, G. M., R. A. Houghton, T. A. Stone, R. F. Nelson, and W. Kovalick. 1987. Deforestation in the Tropics: New Measurements in the Amazon Basin Using Landsat and NOAA Advance Very High Resolution Radiometer Imagery. *Journal of Geophysical Research* 92(D2): 2157–2163.

Woodwell, G. M., R. A. Houghton, T. A. Stone, and A. B. Park. 1986. Changes in the Area of Forests in Rondônia, Amazon Basin, Measured by Satellite Imagery. In *The Changing Carbon Cycle: A Global Analysis,* J. R. Trabalka and D. E. Reichle, pp. 45–59. New York: Springer-Verlag.

Young, K. R. 1992. Biogeography of the Montane Forest Zone of the Eastern Slopes of Peru. In *Biogeografía, Ecología y Conservación del Bosque Montano en el Perú,* ed. K. R. Young and N. Valencia, pp. 119–140. Memorias del Museo de Historia Natural UNMSM 21. Lima: Universidad Nacional Mayor de San Marcos.

Young, K. R., and B. Leon. 1995. Distribution and Conservation of Peru's Montane Forests: Interactions Between the Biota and Human Society. In *Tropical Montane Cloud Forests,* eds. L. S. Hamilton, J. O. Juvik, and F. N. Scatena, pp. 363–376. New York: Springer-Verlag.

PART 2

THE CONSERVATION CHALLENGES OF DISTURBANCES IN HIGH MOUNTAINS

Introduction

Mountains and hills cover perhaps a third of the earth's land mass. Highland regions are found at great heights in the world's largest mountain systems. Part 2 discusses sites in Nepal and Central and South America. Climate changes and natural disturbances have always caused shifts in the elevational limits of different ecological zones and have thus increased the heterogeneity of vegetation within each zone. These are also places that have long supported human populations, sometimes quite dense ones in the developing countries.

Sally Horn discusses these themes in the context of a páramo landscape in a national park in Costa Rica. Her analysis looks in detail at the history of fire in the park, both in the last several decades when fires have been discouraged, and also over many centuries as revealed in the charcoal particles preserved in lake sediment cores. She finds evidence of fires during the entire Holocene and suggests that park managers should not attempt to completely prevent fires if their goal is to maintain long-term conditions. Instead, reduced fire frequency and magnitude might be reasonable goals. An issue that Horn brings to the forefront is the possible conflict between expectations of park visitors, who often anticipate the experience of a sense of wilderness, and the visible manipulations that would be necessary to control fires. The documentation of disturbance regimes in this protected high Andean landscape allows policymakers to choose among possible trajectories of landscape change.

Francisco Pérez brings a long-term perspective to a protected páramo landscape in a drier portion of the Venezuelan Andes where fire is not an important disturbance type. Grazing by cattle has been a primary source of anthropogenic change over the last two centuries. Cattle increase soil erosion and selectively graze on plant species causing compositional and cover changes. A modern disturbance type that is particularly destructive to the cryptogamic crusts that dominate the landscape above 4600 meters is caused by the uncontrolled use of off-road vehicles. Pérez also shows that the utilization of native plants and the introduction of exotic plant species have had impacts in the lower elevations of the páramo landscape.

Barbara Brower and Ann Dennis conclude this part with their interpretation of livestock impacts in a narrow altitudinal belt of a high mountain region in Nepal. They evaluate vegetation patterns in relation to the amount of livestock grazing. The connections are not simple, and can best be understood as a function of the establishment requirements of fir trees and the kind and intensity of grazing. Contrary to many other interpretations, they find that forest cover is actually increasing in their research area. As in the other studies of this section, it is through an understanding of the details of complex biogeographical processes that the more difficult questions of what goals are relevant can be addressed. Their study is set in a large national park and the land use of the local people might well be in conflict with some park goals. However, they challenge underlying assumptions in a way that might permit the design of more flexible approaches that creatively link human needs with changing landscape patterns.

5

Fire Management and Natural Landscapes in the Chirripó Páramo, Chirripó National Park, Costa Rica

Sally P. Horn

Fires arise naturally from lightning and volcanism, and are set by people intentionally and by accident. The setting of fires is one of humankind's most potent tools for effecting landscape and environmental change. When weather and fuel conditions are appropriate, fires ignite readily, and without (or despite) further human effort can spread widely and quickly across the landscape. The degree of environmental modification wrought by fire depends on fire intensity, but even the lightest and spottiest of fires alter the appearance and functioning of ecosystems for intervals of time many orders of magnitude greater than the ignition time.

The potential of fire to so rapidly and greatly alter biotic landscapes makes fire management a critical issue in national parks and similar reserves throughout the world. This essay concerns fire management in the upper elevations of Chirripó National Park, a 50000-hectare protected area in southern Costa Rica. Chirripó National Park surrounds the highest peak in Costa Rica, Cerro Chirripó (3819 meters), and preserves the country's largest expanse of a treeless landscape of tropical alpine (páramo) vegetation, as well as a large area of relatively undisturbed montane forest. Fires have been a concern in this park since shortly after its establishment, and the management plan raises a number of issues related to fire. I examine these issues within the context of research I have conducted within and outside the park on páramo fire ecology and long-term fire and vegetation history. My objective is to highlight research findings, research and management uncertainties, and potential conflicts that might be usefully considered in designing fire management plans for the páramo landscape of Chirripó National Park. Although this essay focuses on one particular park, the questions and approaches considered are relevant to fire management efforts in other protected areas.

Characteristics of Chirripó National Park

The present boundaries of Chirripó National Park encompass about 44000 hectares of montane forest, and about 6000 hectares of treeless páramo vegetation. The focal point of the park is the páramo landscape surrounding Cerro Chirripó, the largest of a series of small islands of páramo vegetation that occupy the highest peaks and massifs of the Cordillera de Talamanca (map 5.1). The páramos and montane forests of Chirripó National Park, together with those of the adjacent La Amistad International Park, comprise the most biologically diverse area in Costa Rica (Boza 1987), and form the core of the larger region of the La Amistad Biosphere Reserve, declared in 1982 (Kappelle 1991).

Owing to extensive forest cover and heavy rainfall, Chirripó National Park serves an important hydrological function. Rainfall is highest within forests on the Atlantic side, where annual rainfall totals can reach 5000 millimeters or more (Coen 1983). Montane forests on both the Atlantic and Pacific slopes are dominated by oaks (*Quercus* species), in some areas with an understory of sprawling bamboos (*Chusquea* species) (Kappelle et al. 1995). The more compact dwarf bamboo *Chusquea subtessellata* dominates the cooler and somewhat drier páramos of the highest peaks, covering up to 60% of the surface (Kappelle et al. 1995) (figure 5.1). Shrub associates, none of which exceed 10% cover, include species in the Ericaceae, Hypericaceae, Compositae, and Rosaceae families. A diversity of herbs, ferns, club mosses, and bryophytes occur in the understory and in more open areas (Cleef and Chaverri 1992; Horn 1989a; Weber 1959). The páramo flora shows close floristic ties with that of the northern Andes (Cleef and Chaverri 1992; Weber 1959).

The Instituto Costarricense de Electricidad (ICE) installed a pluviometer in the Chirripó páramo in the early 1980s, but records are spotty and unreliable. Rainfall data from ICE's Repitidora Cerro de la Muerte station (3475 meters) in the Buenavista páramo (map 5.1) along the Inter-American Highway are probably broadly representative (ICE unpublished data). For the period 1971–1993, rainfall at this station averaged 2477 millimeters, of which less than 8% fell during the January–April dry season. Mean monthly temperatures (1971–1979) ranged from 6.8°C in January

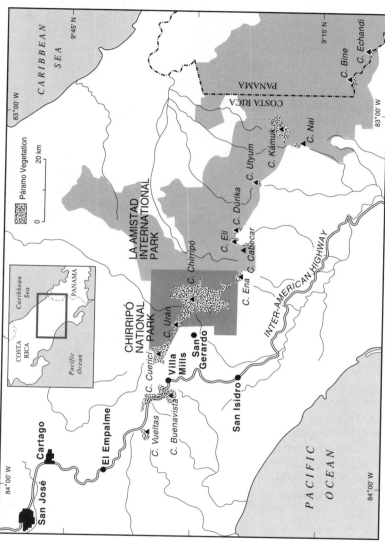

Map 5.1. Location of Chirripó National Park, La Amistad International Park, and peaks in the Cordillera de Talamanca of southern Costa Rica. Approximate extent of páramo vegetation is from 1:50000 topographic maps prepared by the Instituto Geográfico Nacional. National park boundaries within Costa Rica are from Magallón et al. (1987); park limits within western Panama are approximated from MINIREM and others (n.d.).

127

Fig. 5.1. The bamboo-dominated landscape of Chirripó National Park. The photograph shows the largest lake in the Valle de las Morrenas (3480 meters), source of the Morrenas 2 pollen and charcoal record.

to 8.2°C in March and April. Frosts are frequent in the Chirripó páramo but snowfall is unrecorded. However, cooler temperatures during intervals of the Pleistocene epoch led to the formation of small glaciers in the highest valleys. The last glaciers melted some 10000 years ago, leaving behind a picturesque, ice-carved landscape dotted by numerous glacial lakes (Horn 1990a; Horn and Haberyan 1993). Soils are generally thin, acidic, and stony.

Unlike the Buenavista páramo and some of the Andean páramos described by Pérez (chapter 6) and Keating (1995), the Chirripó páramo lacks both roads and nearby settlements, and is grazed only lightly (and within a discrete area) by horses used to transport supplies up the mountain. Several rough trails provide access to the páramo, the most travelled of which begins in San Gerardo (map 5.1), where the park maintains an administrative office.

Park Establishment and the 1976 Páramo Fire

Chirripó National Park was established by legislative decree on August 19, 1975. Cerro Chirripó had long been a favorite destination of hikers

(Kohkemper 1968), and was known for its spectacular glacial scenery and unique vegetation. The creation of the park apparently met with little controversy in the legislature (Wallace 1992). The objectives of park creation, as presented in the management plan published many years later (Bravo et al. 1991), were to conserve and protect biotic and abiotic resources, headwaters of important streams, endemic vegetation, representative glacial formations, and the glacial and postglacial geological history of Costa Rica (Bravo et al. 1991).

Less than a year after the park was created, a massive fire swept through the páramo landscape. Set by a young hiker during the driest March on record, the fire burned nearly the entire páramo area, along with some adjacent montane forest (Chaverri et al. 1976); its total extent was estimated at 5000 hectares. Coming as soon as it did after park establishment, the 1976 Chirripó fire generated front page headlines in Costa Rican newspapers. Journalists, and the officials and ecologists they interviewed, decried the fire as an unprecedented national disaster that had threatened a rare and fragile páramo ecosystem. The fear was expressed that the páramo landscape might never recover from the devastation (Horn 1990b).

Research on Fire Ecology and Fire History

The reports of the 1976 Chirripó fire caught my attention when I began in 1983 to investigate possible dissertation topics related to the biogeography of the Cordillera de Talamanca. I had read about and visited the more accessible Buenavista páramo along the Inter-American Highway, where fires had also occurred, and I was struck by the very different way fire was perceived in these two páramo landscapes. Along the highway, bamboo- and shrub-dominated vegetation were regarded as the *consequence* of fires and other human disturbances that had depressed the timberline (Janzen 1973a, 1983); in the more remote Chirripó highlands, similar vegetation was seen as natural (Hartshorn 1983), naturally fire-free, and in fact *menaced* by fires!

Alternative viewpoints were mentioned in some accounts, but none were backed by scientific research. Costa Rican forester Gerardo Budowski was quoted in a newspaper article as stating that the 1976 fire was only one of many fires on Cerro Chirripó, which burned every 25–30 years corresponding to exceptionally dry periods (Anon. 1976). In an article describing initial postfire reconnaissance in the 1976 burn site, Chaverri et al. (1976) stated that some scientists viewed páramo fires as necessary for the maintenance of treeless vegetation, and regarded plant species as adapted to fire. Boza and Bonilla (1978) indicated that "some researchers believe that the highland plains are adapted to fires, which can be set by lightning." In his book *Notes on Natural History in Costa Rica,* Valerio (1983) stated that the Chirripó páramo might owe its present composition to occasional fires over the past 1000 to 2000 years.

When I went to Costa Rica in 1984 to investigate páramo fire ecology and history, there was little in the literature aside from such second-hand speculation. Only one paper had been published on postfire regeneration of páramo plants, based on observations in the Buenavista páramo (Janzen 1973b). Chaverri et al. (1976) had established study plots within the Chirripó burn site following the 1976 fire, but had been unable to pursue the long-term monitoring they originally envisioned. The literature was similarly thin for the Andean páramos, and the sites that had been investigated were heavily grazed grass páramos that were not readily comparable to the shrubby páramos of Costa Rica, which are grazed lightly, if at all.

It was my good fortune that a second paper on postfire regeneration in the Buenavista páramo was in preparation (Williamson et al. 1986); the draft that the authors kindly shared provided a model for my research. Borrowing their methods, I spent the dry season of 1984–1985 locating and censusing old burns in the Chirripó and Buenavista páramos to determine rates of postfire regeneration and the impact of fires on species composition. I was helped in this work by the slow rate of wood decomposition in the páramo landscape, which made it possible to reconstruct prefire shrub composition from persistent dead stems.

To age the burn sites and reconstruct recent (post-1950) fire history over wider areas, I examined growth rings in stems of resprouting shrubs, interviewed park guards, and examined old photographs, newspapers, and reports that might show or mention fires. To determine longer-term fire history I recovered two sediment cores about one meter in length from Lago Chirripó, the largest of the 30-odd glacial lakes in the Chirripó páramo. In these cores—the first lake sediment cores recovered for paleoecological analysis in Costa Rica—I analyzed microscopic charcoal, focusing on the nearshore core, which turned out to span over 4000 years. Subsequent to this work the National Geographic Society generously funded an expedition to secure longer sediment cores from a second lake (Lago de las Morrenas); I analyzed pollen as well as charcoal at this site to try to resolve issues of long-term vegetation history. During the coring expedition I made additional observations on postfire regeneration in the Chirripó páramo, and since that time I have made repeated visits to the Buenavista páramo to monitor more recent fires and their impacts (Horn 1997). It is from this perspective that I examine issues related to fire and vegetation management in Chirripó National Park.

Table 5.1. Management objectives for Chirripó National Park.

1. Maintain this wildland in the most pristine state possible, and ensure the conservation of its biotic and abiotic characteristics.
2. Especially protect the páramo zone, which constitutes the greatest expanse of páramo in the country.
3. Protect the hydrographic basins to maintain water production and prevent erosion.
4. Protect the geological and geomorphological features, especially those formed by Pleistocene glaciations.
5. Permit and facilitate the biological recuperation of areas damaged by agricultural activities, fire, and tree-cutting.
6. Offer opportunities for scientific investigations, and recommend the study of certain aspects about which better knowledge is needed for optimal management of the park.
7. Offer an agreeable environment, pleasant and respectful of nature, for the visitors to the area.
8. Foster awareness among visitors to the park, and residents of the area, of the importance of conserving the park.

Source: Bravo et al. 1991.

Fire-related Issues in the Park Management Plan

The potential for fire, and issues of vegetation and biodiversity conservation related to fire, are mentioned frequently in the 1991 Chirripó management plan (Bravo et al. 1991). The possibility of fire is recognized as a key biophysical constraint for park management:

The possibility of forest fires within and outside of the park, ignorance about adequate control techniques, and the lack of equipment, make it imperative to prepare for this problem. This implies the need for financial and human resources for the prevention, fighting, control, and extinguishing of fires (Bravo et al. 1991: 17).

The potential for fire also figures prominently in the eight management objectives for the park (table 5.1). Fire is mentioned directly in objective 5, which concerns the need to facilitate the recovery of areas damaged by fire. This management objective is also reflected in the proposed park zoning, which includes areas kept off-limits to tourists and scientists to foster the regeneration of páramo vegetation burned in previous fires.

While not specifically mentioning fire, other management objectives are clearly related: for example, the first objective of maintaining the area in the most pristine state possible, and the second and third objectives of protecting the páramo and park watersheds. In this context I should note that the Spanish word that I have translated as pristine (pristino) appears to carry the same two meanings as the English word pristine does: the first connotes purity (untouched, unspoiled), and the second means original in the sense of being characteristic of the earliest, or an earlier, condition (McKechnie 1983; Ramondino 1969). General park management objectives regarding research facilitation, visitor experiences, and education also relate to fire management, as do specific objectives and responsibilities outlined for different subprograms. For example, objectives of the protection and vigilance subprogram are to "maintain the environment in the most natural state possible" and "plan for controlling fires in the park, especially in the páramo and in deforested areas near the south and west boundaries of the park."

The presentation of fire-related issues in the 1991 Chirripó management plan suggests that the authors of the plan share conceptions of fire similar to those highlighted in the media following the 1976 fire in the park. The emphasis on preventing fires, and on protecting vulnerable plants and animals from fire, suggests that fires are seen as incompatible with the management goals of keeping the area in (or returning it to) pristine and natural conditions. The listing of fire as a biophysical constraint for park management implies recognition that the seasonal páramo environment is conducive to burning. But only twentieth-century fires are mentioned, suggesting that the authors of the management plan discounted the possible importance of human- or lightning-set fires during earlier times.

Research Results and Management Implications

Results of research on páramo fire history and ecology suggest that fire may have a long-term role in the Chirripó páramo, as it does in other seasonally dry shrublands. Here I explore research findings and management implications with respect to five broad and overlapping themes relevant to fire management in Chirripó National Park and other protected areas. My objective is not to criticize the Chirripó management plan, which was completed before any of the research I describe was published. Rather, in keeping with the spirit of management objective 6 (table 5.1), I offer the following interpretations and ideas in the hope that they will provide some of the knowledge needed for optimal management of Chirripó National Park.

The Antiquity and Frequency of Fire in the Chirripó Páramo

The management plan mentions fires in the Chirripó highland in 1933, 1953, 1961, 1976, and 1985. Field and written evidence documents additional fires in the Chirripó páramo (some after the management report was completed), as well as fires in the Buenavista páramo and in the smaller páramos surrounding Cerro Cuericí (added to Chirripó National Park in 1982) and more remote peaks to the east in La Amistad National Park (Weston 1981a, 1981b; Horn 1989a, 1989b, 1990b, 1997; Sacchetti

1992). All twentieth-century páramo fires are known or suspected to have been set by people. While lightning-set fires have not been documented, lightning strikes have been observed within the Chirripó páramo and in forests at lower elevations in the Cordillera de Talamanca and elsewhere in Costa Rica (Horn 1989c; B. Middleton, personal communication).

Charcoal fragments in sediment cores from Lago Chirripó and Lago de las Morrenas extend the record of páramo fires into prehistoric times. Were fire an introduction of modern human society, I would expect charcoal to be rare in all but the most recent, uppermost sediments in the lake basins. I have found instead that macroscopic and microscopic fragments of charcoal occur throughout the length of the cores, attesting to periodic burning since deglaciation some 10000 years ago (figure 5.2) (Horn 1989c, 1993). People have been present in Central America since the end of the Pleistocene, and Piperno et al. (1990) have interpreted charcoal in 10000-year-old sediments in Panama to reflect human-set fires. Early fires recorded in the Morrenas sediments could also have been set by people, accidentally, or to improve hunting within the páramo or surrounding montane forests. Lightning also may have started fires; ignition sources cannot be determined from the charcoal. Regardless of origins, it is clear that fire has long been a component of the Chirripó landscape. The pristine, fire-free páramo that management seeks to recreate in Chirripó National Park may be without precedent in the postglacial period.

The abundance of charcoal in the Chirripó lake sediments suggests the need to reevaluate management goals. Rather than trying to eliminate all páramo fires, managers might want to direct efforts towards maintaining some target fire frequency. But just how often has the Chirripó páramo burned? This is a question that cannot be adequately answered from existing charcoal records, because I counted charcoal at intervals throughout the cores and likely missed evidence of other past fires. However, as part of a National Science Foundation Research Experience for Undergraduates project, Brandon League is now examining macroscopic charcoal at contiguous one-centimeter intervals in the Morrenas sediment record (League and Horn, in preparation). From this high-resolution charcoal record we hope to infer fire frequencies within different time periods and in this way place the recent fire history of the Chirripó páramo within its long-term context.

In the meantime, it is possible to make some inferences about *relative* fire frequencies as reflected by downcore changes in charcoal influx. Although relationships between fire history and sedimentary charcoal are complex (Clark 1983; Patterson et al. 1987), I assume that the influx of charcoal to the core sites is driven at least in part by fire frequency (it may also reflect fire intensity, or lake level). If the fire frequency signal is strong,

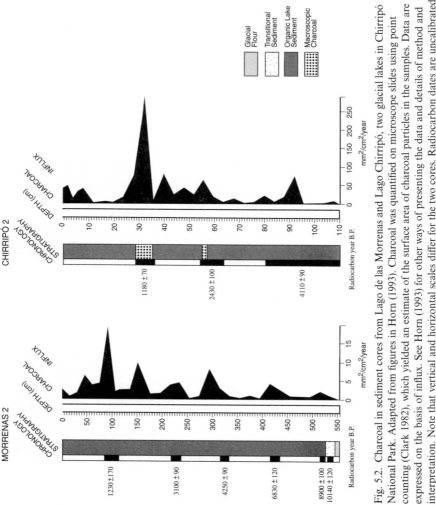

Fig. 5.2. Charcoal in sediment cores from Lago de las Morrenas and Lago Chirripó, two glacial lakes in Chirripó National Park. Adapted from figures in Horn (1993). Charcoal was quantified on microscope slides using point counting (Clark 1982), which yielded an estimate of the surface area of charcoal particles in the samples. Data are expressed on the basis of influx. See Horn (1993) for other ways of presenting the data and details of method and interpretation. Note that vertical and horizontal scales differ for the two cores. Radiocarbon dates are uncalibrated.

then the higher charcoal influx values in the upper part of the core suggest that fire frequencies were generally higher in the late Holocene (last 5000 years) than they were in the early Holocene, and that considerable variability existed in both time periods. This interpretation complicates management decisions, because it means that there is not one single pre-park or pre-European fire frequency that could be targeted by management, but many! That charcoal influx values are no higher for the tops of the sediment cores than they are in some deeper sections suggests that the post-1950 frequency of one fire every 8–15 years in the lake basins may not be unprecedented. A similar frequency of burning may have characterized some earlier time periods. Owing to slow rates of fuel accumulation, it is doubtful that fires ever occurred at frequencies much greater than those of recent decades. Under current conditions, a minimum of six years of postfire vegetation recovery seems to be required to generate enough fuel to carry a second fire. This interpretation is complicated, however, by the possibility that some core sections that are characterized by higher charcoal influx than at present may have been deposited during periods of lowered lake level, when core sites were closer to shore and more likely to receive charcoal directly, and through the erosion and redeposition of exposed charcoal-rich littoral silts (Horn and Sanford 1992). That some lower core sections seem to have received more charcoal per year than modern sediments may also reflect time lags in the delivery of charcoal from the most recent (1976) fire in the watershed. Some of this uncertainty in the interpretation of the charcoal records may be resolved by the high-resolution analysis of macroscopic charcoal now underway.

Natural Landscapes of the Chirripó Highlands

Pollen analysis of the sediment core from Lago de las Morrenas has given us a view of past vegetation in the Chirripó highlands (Horn 1993). Curves for individual pollen taxa (Horn 1993) show peaks and wiggles, but no dramatic shifts are apparent. Though fires have occurred since deglaciation some 10000 years ago, they have not carved páramo from forest. Pollen percentages of forest taxa increase upcore at the expense of grasses (including bamboo) and other páramo taxa (figure 5.3), indicating a general *upslope,* rather than downslope, movement of timberline and a shrinking rather than expanding páramo landscape. Further evidence of a glacial depression (and postglacial rise) of timberline in the Cordillera de Talamanca is provided by pollen records from bogs near El Empalme (map 5.1), which are presently located within the montane forest zone but were surrounded by páramo during the last glacial maximum (Hooghiemstra et al. 1992; Martin 1964).

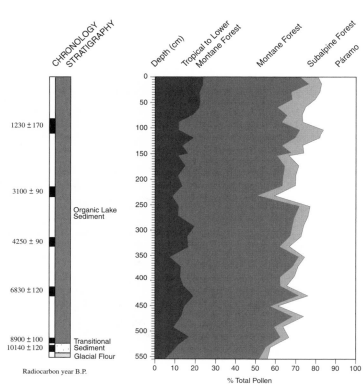

Fig. 5.3. Composite pollen diagram for Morrenas 2 sediment core. Pollen taxa have been grouped into categories that correspond to broad vegetation types of the Cordillera de Talamanca, as follows. Tropical to lower montane forest: *Cecropia,* other Urticales, *Acalypha, Alchornea, Piper,* Myrtaceae, *Ulmus, Alfaroa, Bocconia.* Montane forest: *Quercus, Alnus, Podocarpus, Myrica, Hedyosmum, Drimys, Ilex, Clethra, Cornus.* Subalpine forest: *Miconia*-type, *Weinmannia, Myrsine, Garrya,* Ericaceae, Compositae. Páramo: Gramineae, Cyperaceae, *Valeriana,* Umbelliferae, *Geranium, Acaena,* Caryophyllaceae. Oak *(Quercus)* and alder *(Alnus)* pollen dominate the montane forest category, and Gramineae pollen dominate the páramo category. Percentages are expressed on the basis of a pollen sum that consists of all pollen types included in the four categories. See Horn (1993) for curves for individual pollen and spore taxa and further discussion. Groupings of pollen types are based on Burger (1977), Hartshorn (1983), Kappelle (1991), and Kappelle et al. (1991).

That pollen assemblages in modern sediments do not differ greatly from those in sediments deposited thousands of years ago suggests that today's páramo communities are fairly similar to those of the past. The pollen data are thus consistent with the view that páramo is the "natural" vegetation of the high peaks of Cerro Chirripó (Hartshorn 1983), and not an artifact of human activity. In this sense, the pollen evidence validates management efforts to preserve the páramo: What is being preserved, although not untouched by humans, has a long history that is probably at least to some extent independent of human history. Further evidence of the antiquity of páramo vegetation in Chirripó National Park and elsewhere in the Talamancan region is provided by the modern flora, which includes two endemic páramo genera (Cleef and Chaverri 1992).

While the pollen record suggests that vegetation similar to that of the present has occupied the Chirripó highlands since deglaciation, there are possible indications of minor shifts in species composition within the páramo (Horn 1993). For example, the club moss *Lycopodium saururus* may have been more common during the past 5000 years that it was 5000–10000 years ago, judging from the distribution of its distinctive spores in the Morrenas sediments. The páramo herb *Valeriana,* on the other hand, may have been less common in recent millennia than in past. The changing importance of these páramo taxa suggests (as do the charcoal data) that in the Chirripó páramo there may not be one single "pristine" or original condition toward which management efforts might be directed, but many. Should we manage for the páramo of the late Holocene (more *L. saururus,* less *Valeriana,* intervals of more frequent burning) or for the páramo of the early Holocene (little or no *L. saururus,* more *Valeriana,* and possibly fewer fires)? Botkin (1990) asked similar questions for the Boundary Waters Canoe Area in northern Minnesota and Ontario, Canada. Paleoecological data from that protected area indicate a highly dynamic postglacial history that defies identification of a single original or natural state that could be the goal of managers.

The Ecological Impacts of Fire

Contrary to predictions made at the time, the 1976 Chirripó fire did not destroy the páramo. The botanist and intrepid páramo hiker and plant collector Arthur Weston reported that as of 1981, only one plant species known to have occurred in the páramo prior to the 1976 fire had not been found subsequently, and he believed that this *Xyris* species would ultimately be discovered if searched for during the wet season (Weston 1981b). However, while the known flora was essentially unchanged by the 1976 fire, vegetation patterns did change because of the varying ability of plant species to resprout or reseed following burning. I surveyed a site in the park in 1985, and found that 99% of the clumps of the dominant bamboo, *Chusquea subtessellata,* that had burned in 1976 had subsequently resprouted. The ericaceous shrub *Vaccinium consanguineum* showed a 90% rate of basal resprouting. In contrast, the shrub *Hypericum irazuense,* which prior to the fire was the second most common woody species, suffered very high mortality—only 6% of the shrubs resprouted. Unlike most other woody taxa in the Costa Rican páramos, which rely almost entirely on sprouting to recolonize after fires, *Hypericum irazuense* recolonizes burned areas via seeds. However, the large size and fairly complete coverage of the 1976 fire eliminated seed sources over wide areas, slowing recovery of this species. In the 1200-square-meter area that I surveyed, I found 80 shrubs of *Hypericum irazuense* that had been killed by the fire (and 5 that had resprouted), but only 1 seedling that had colonized since the fire. I saw more seedlings in 1989 than I did in 1985, but populations (and plants) were still much smaller than they were immediately prior to the 1976 fire. Postfire changes in herbaceous cover also occurred (Weston 1981b), but have not been systematically documented. (I collected data on postfire herb communities in 1985, but have no prefire data with which to compare them.)

The vigorous resprouting by the páramo bamboo suggests that the plant in part owes its present dominance to the recurrent fires of the latter twentieth century, which have favored the bamboo at the expense of more fire-sensitive woody associates. An interesting possibility is that the slow post-fire spread of bamboo, and the potentially more rapid spread of herbaceous grasses, may have increased the fuel load of the páramo, and hence influenced not just vegetation composition but also the likelihood of fire. This deserves study, both for its management implications and as a possible illustration of the complex ways in which human activity can alter disturbance regimes (Savage 1993).

The ability of *Chusquea subtessellata* to survive and thrive under frequent burning leads one to question the extent to which the plant might *depend* on fire. On the basis of observations made within the Buenavista páramo along the Inter-American Highway, Janzen (1983) suggested that populations of *C. subtessellata* may require burning at 15–100 year intervals to survive. Janzen projected that if the bamboo were not burned or otherwise cleared at long intervals, it would be shaded out by slower-growing dicotyledonous shrubs. If this holds true for the Chirripó páramo, it suggests that management efforts to prevent fires, if completely successful, might endanger the very plant that most typifies the páramo environment! Whether this would be desirable or acceptable depends on the nature of the páramo environment that managers seek to create or maintain.

Fire Climatology and the Inevitability of Fire

Although recent fires in the Chirripó páramo have been set by people, their occurrence appears to have been strongly affected by natural climate variability, which makes fires more likely in some years than in others (Horn, 1990a and unpublished). Between 1960 and 1993 there were three years (1961, 1976, 1985) in which rainfall records from in or near the Buenavista páramo showed less than 0.5 millimeters rainfall during the driest months of February or March (ICE unpublished data). In each of those years a large (> 100 hectares) fire occurred in the Chirripó páramo. A fourth major fire in the Chirripó páramo in 1992, and a 1964 large fire in the Cuericí páramo, occurred in years that were slightly wetter but still among the driest years of the past three decades. Smaller páramo fires, or no fires, characterized years of higher dry season precipitation. If rainfall trends for 1952–1993 are indicative of long-term trends, the data suggest that extremely dry years conducive to widespread burning may be expected to occur about once a decade in the Chirripó highlands, regardless of the fire management strategy adopted.

The available rainfall and fire data clearly support the recognition by authors of the management plan that the possibility of fire is a "biophysical constraint" for park management. With hundreds of visitors in the park each year, and agriculturalists burning forests and pastures along park boundaries, ignition sources are presumably not limiting. This, coupled with the periodic occurrence of extremely dry weather, make fires not just possible, but inevitable. Because the páramo largely lacks firebreaks—natural or human-made—we can expect páramo fires to be large and uncontrollable, given available resources.

The fire-free, pristine páramo may thus be unachievable, as well as unprecedented, in the Chirripó highlands. If this interpretation is correct, it presents managers with a difficult decision. Should they do nothing, and risk large fires? Or should they continue to try to prevent all fires, and risk failure? A more reasonable approach may be to change the focus of fire management efforts from the prevention of all fires to the reduction of fire frequency. While the charcoal record indicates that fire has a long history in the area, recent fire recurrence intervals may be shorter than they have been for some or perhaps most of the last 10000 years. While the data are not yet available to make precise comparisons between fire frequencies of the latter twentieth century and earlier periods, it is unlikely that fires were ever much *more* frequent than at present, owing to slow rates of fuel accumulation and to cycles of drought severity. Given existing uncertainties, it would seem prudent to err on the side of longer fire intervals, and to work towards reducing fire frequencies.

How do we lengthen fire recurrence intervals in a fire-prone ecosystem? The answer for the Chirripó páramo may lie in switching the emphasis from controlling ignitions to controlling fire size. This could be accomplished by clearing bamboo and shrubs to create a series of intersecting firebreaks or lanes. Firebreaks have been cleared *during* fires to control fire spread, but rarely in advance of fires, as a preventative measure. A series of interconnected fire lanes would need to be cleared to effectively limit fire size. These lanes would require periodic maintenance, but perhaps not more than every five years or so, owing to slow rates of shrub recovery and herb colonization in the páramos (Horn 1989a). Near the guard station and visitor cabins, grazing by horses used to transport gear to the páramo could assist in keeping fire lanes clear. Prescribed burning might also be used to create firebreaks that could limit the size of future fires.

That firebreaks will stop the spread of páramo fires is illustrated by the fire history of the Buenavista páramo, where intersecting roads, jeep trails, and electrical transmission line corridors contained recent fires. During the past fifty years, the number of fires that have occurred in the Buenavista páramo is the same or greater than the number of fires ignited in the Chirripó páramo, but the total area burned has been less, even though management efforts are minimal in the Buenavista páramo and fires usually are not fought. Observations of fire behavior in the Buenavista páramo could provide guidance on how wide firebreaks need to be to stop fires.

The existence of a network of firebreaks in the Chirripó páramo would make it possible to contain fires and would probably reduce fire frequencies within particular sites. Fires would still ignite, perhaps as often as they have in recent years, but they would not spread as far. The páramo as a whole might see fires as often, but particular sites within the páramo would tend to burn less frequently. The smaller fires that did occur within the páramo would themselves serve as firebreaks for several years, and thus reinforce a tendency towards smaller fires.

Keeping fires small ($\leq$ 25–50 hectares) would contribute to a number of other management goals, even if it didn't lengthen fire recurrence intervals. Small fires would be less of a threat to animal species (which could escape to unburned habitat), and to watersheds (since only small portions would burn). Keeping fires small would foster faster regeneration because of the proximity of seed sources. Managing for small fires would also tend to increase habitat diversity, because at any given time the páramo would consist of a patchwork of stands of different ages. The increased habitat diversity might help maintain species diversity.

However, active management of fire size through the establishment of firebreaks might be seen as conflicting with park objectives to maintain a natural environment, because under natural conditions the páramo would have lacked firebreaks. (Old burns would have limited the spread of subsequent fires, but such burns would not have been arranged in a pattern of intersecting corridors.) Conspicuous firebreaks might negatively impact visitor experience.

The Naturalness of the Landscape and Visitor Expectations

Just as with other management decisions, decisions about how to manage fire must take into consideration the characteristics and expectations of visitors. Who visits Chirripó National Park, and why? According to the management plan, foreign tourism in the park has increased, but Costa Rican citizens still comprise over three-quarters of all visitors. Most visitors come in groups during the dry season. Outdoor recreation is the principal goal, with the main objective being to climb Cerro Chirripó (accessible by trail). My own impressions, based on four 5–10 day visits to the park, are consistent with the observations in the management plan. Visitors seem to be attracted to the park primarily by the chance to experience a high montane environment that differs greatly from that of most of the rest of Costa Rica. Visitors come to stand atop Costa Rica's highest peak, and from there and other peaks enjoy panoramic vistas; to clamber over rocks shattered by frost or smoothed by glacial ice; to see and touch (and occasionally bathe in) the cold and clear glacial lakes; and to physically challenge themselves with difficult trails, the cold, and the altitude. Vegetation provides the backdrop for the geomorphic features but is otherwise probably of secondary interest to most visitors; animals such as tapir occur in the park but are too rarely seen to be a draw.

Visitors who are primarily interested in strenuous mountain hiking and cold weather may not be bothered by the existence of fire lanes and of recently burned areas. However, visitors who come to Chirripó with different or additional goals could find such obvious signs of human manipulation detract from their experiences. In considering what visitors to the Boundary Waters Canoe Area wanted from their experience, Botkin (1990) speculated that what many of them sought was a *sense of wilderness*—a feeling that they were in a place untouched by humans, where conditions were like those that the first European explorers faced. Visitors to Chirripó searching for such a sense of wilderness would likely react negatively to the sight of fire lanes and other obvious indications of vegetation management.

To my knowledge, visitor expectations have not been studied in Chirripó National Park; doing so may be critical to the success of management plans. Observations by Hofstede (1995) on outdoor recreation in the Andean páramos indicate that recreation in this area combines both Botkin's sense of wilderness and the physical exertion/rough terrain aspects that seem important to many visitors to Cerro Chirripó. Hofstede found that what people most enjoy in the Andean páramos is adventurous tracking in rough landscapes, or what he calls the wilderness experience. They are disappointed to encounter unnatural elements such as cattle and fences. Neither national nor international tourists care much about high species richness or other ecological characteristics of the landscape. What they want to see is a landscape that is as original as possible.

In considering management options for the higher elevations of Yosemite National Park, California, Vale (1987) stressed the need to keep in mind the recreation, as well as preservation, goals of the park. The interests and expectations of recreational users similarly deserve attention in the design of fire management strategies for the Chirripó páramo. If the wilderness aspect of the Chirripó páramo is an important part of its recreational appeal, options for managing fire size through the imposition of firebreaks may be limited. However, in thinking about what management tools visitors might find acceptable, we should compare visions of a páramo crisscrossed by fire lanes not just to that of an unbroken, unburned páramo, but also to that of an extensively burned páramo. A fire several thousand hectares in extent may negatively impact visitor experience more than a fire 50 hectares in extent. Actively managing for the latter outcome undeniably involves considerable uncertainty. But the uncertainty of managing for small fires may be preferable to the certainty of otherwise witnessing large ones.

Conclusion: Challenges, Uncertainties, and the Role of Science

The results of scientific research on fire history and fire ecology in páramo landscapes of Chirripó National Park and the surrounding region present managers with a number of difficult challenges. Some of these arise from uncertainties in the research, and others stem from findings that suggest a need to reevaluate the conception of fire that underlies the management plan. The following quote from Foresta (1991), made in connection with scientific research in the Amazon, seems equally relevant to the situation in Chirripó National Park:

[S]cience is seldom useful to policymakers in its pure form: it is seldom as conclusive as policy demands, and it frequently points to politically improvident actions.

Recognizing that fire may have a place in the Chirripó landscape, and sanctioning its role by working to control fire size rather than prevent all fires, represents a radical change in management approach. For political and philosophical reasons it may never take place. It is an approach that deserves consideration, but it is not without pitfalls. The creation of a landscape crisscrossed by fire lanes and containing a patchwork of younger and older burns may enhance habitat, species diversity, and opportunities for scientific research. Such a landscape appears more in line with fire history and climatology, and is likely a more easily obtainable outcome than the alternative goal of a fire-free páramo. But managing the páramo to control fire size via firebreaks conflicts with the goal of maintaining it in a pristine and natural condition, and may detract from human experiences in the park by reducing the feeling of wilderness. Whether the benefits and practicality of the former outweigh the importance of the latter is a question that scientific research alone can't answer.

This exploration of fire management in Chirripó National Park highlights management issues and uncertainties that confront protected areas worldwide. In many parks and reserves we know very little about the longer-term history and ecological interactions of fire and vegetation. Basic research on fire ecology, and fire and vegetation history, can provide useful perspectives for management. However, as this chapter demonstrates, translating research results into policy may prove difficult, owing to uncertainties in research results and implications, and to the possibility that findings or their management implications will contradict what managers and visitors expect and want from nature. As Zimmerer and Young point out in this volume, research on modern and past environments typically reveals a dynamism at odds with the concepts of long-term stability that underlie traditional approaches to the management of protected areas. Managing for change poses many dilemmas, not the least of which is a requisite sea change in the way scientists, conservationists, managers, park visitors, and the general public think about the ecosystems we hope to conserve in parks and other protected areas. Scientific research and environmental education on biophysical interactions can help broaden views of natural processes, and in this way facilitate the adoption of new management approaches that recognize the dynamic nature of ecosystems.

Acknowledgments

I am grateful to Marten Kappelle, Adelaida Chaverri, Bruce Williamson, Daniel Janzen, and Arthur Weston for sharing information and ideas about the Costa Rican páramos, and to Roger Horn and several present and former students for assisting with field work. I thank the National Park Service for permission to conduct my research, Carolyn Hall and Gilbert Vargas of the Department of Geography at the University of Costa Rica for logistical support, and Kenneth Young and Karl Zimmerer for improvements to the manuscript. The field and laboratory research that I draw upon here has been funded by the National Geographic Society, the National Science Foundation, the Association of American Geographers, the University of Tennessee, the University of California at Berkeley, and the U.S.

Information Agency (Fulbright program). I dedicate this essay to resource managers in Costa Rica who work under difficult political and economic conditions to protect the integrity of their country's natural landscapes.

References

Anonymous. 1976. Chirripó Se Quema Cada 25 Años. *Excelsior* (San José, Costa Rica), 12 June 1976, pp. 1, 3.

Botkin, D. B. 1990. *Discordant Harmonies: A New Ecology for the Twenty-first Century.* New York: Oxford University Press.

Boza, M. A. 1987. Los Parques Nacionales de Costa Rica 1987. Text on back of 1:500000 scale map *Mapa de los Parques Nacionales de Costa Rica.* San José, Costa Rica: Editorial Heliconia.

Boza, M. A., and A. Bonilla. 1978. *Los Parques Nacionales de Costa Rica.* Madrid: INCAFO.

Bravo, J., A. Chaverri, and G. Solano. 1991. *Plan de Manejo Parque Nacional Chirripó.* San José, Costa Rica: Instituto Geográfico Nacional, Universidad Nacional, and Servicio de Parques Nacionales.

Burger, W., ed. 1977. Flora Costaricensis. *Fieldiana: Botany,* new series, 40: 1–291.

Chaverri, A., C. Vaughan, and L. J. Poveda. 1976. Informe de la Gira Efectuada al Macizo de Chirripó al Raíz del Fuego Ocurrido en Marzo de 1976. *Revista de Costa Rica* 11: 243–279.

Clark, R. L. 1982. Point Count Estimation of Charcoal in Pollen Preparations and Thin Sections of Sediments. *Pollen et Spores* 24: 523–535.

Clark, R. L. 1983. Fire History from Fossil Charcoal in Lake and Swamp Sediments. Ph. D. dissertation, Australian National University.

Cleef, A. M., and A. Chaverri P. 1992. Phytogeography of the Páramo Flora of Cordillera de Talamanca, Costa Rica. In *Páramo: An Andean Ecosystem under Human Influence,* eds. H. Baslev and J. Luteyn, pp. 45–60. London and San Diego: Academic Press.

Coen, E. 1983. Climate. In *Costa Rican Natural History,* ed. D. H. Janzen, pp. 35–46. Chicago: University of Chicago Press.

Foresta, R. 1991. *Amazon Conservation in the Age of Development: The Limits of Providence.* Gainesville: University of Florida Press.

Hartshorn, G. S. 1983. Plants: Introduction. In *Costa Rican Natural History,* ed. D. H. Janzen, pp. 118–157. Chicago: University of Chicago Press.

Hofstede, R. G. M. 1995. Effects of Livestock Farming and Recommendations for Management and Conservation of Páramo Grasslands (Colombia). *Land Degradation and Rehabilitation* 6(3): 133–147.

Hooghiemstra, H., A. M. Cleef, G. W. Noldus, and M. Kappelle. 1992. Upper Quaternary Vegetation Dynamics and Paleoclimatology of the La Chonta Bog Area (Cordillera de Talamanca, Costa Rica). *Journal of Quaternary Science* 7(3): 205–225.

Horn, S. P. 1989a. Postfire Vegetation Dynamics in the Costa Rican Páramos. *Madroño* 36: 93–114.

Horn, S. P. 1989b. The Inter-American Highway and Human Disturbance of Páramo Vegetation in Costa Rica. *Yearbook of the Conference of Latin Americanist Geographers* 15: 13–22.

Horn, S. P. 1989c. Prehistoric Fires in the Chirripó Highlands of Costa Rica: Sedimentary Charcoal Evidence. *Revista de Biología Tropical* 37: 139–148.

Horn, S. P. 1990a. Vegetation Recovery After the 1976 Páramo Fire in Chirripó National Park, Costa Rica. *Revista de Biología Tropical* 38: 267–275.

Horn, S. P. 1990b. Timing of Deglaciation in the Cordillera de Talamanca, Costa Rica. *Climate Research* 1: 81–83.

Horn, S. P. 1993. Postglacial Vegetation and Fire History in the Chirripó Páramo of Costa Rica. *Quaternary Research* 40: 107–116.

Horn, S. P. 1997. Postfire Resprouting of *Hypericum irazuense* in the Costa Rican Páramos: Cerro Asunción Revisited. *Biotropica* 29(4): 529–531.

Horn, S. P., and K. A. Haberyan. 1993. Physical and Chemical Properties of Costa Rican Lakes. *National Geographic Research and Exploration* 9(1): 86–103.

Horn, S. P., and R. L. Sanford, Jr. 1992. Holocene Fires in Costa Rica. *Biotropica* 24: 354–361.

Janzen, D. H. 1973a. Sweep Studies of Tropical Foliage Insects: Descriptions of Study Sites, with Data on Species Abundances and Size Distributions. *Ecology* 54: 659–678.

Janzen, D. H. 1973b. Rate of Regeneration After a Tropical High Elevation Fire. *Biotropica* 5: 117–122.

Janzen, D. H. 1983. *Swallenochloa subtessellata* (chusquea, batamba, matamba). In *Costa Rican Natural History,* ed. D. H. Janzen, pp. 330–331. Chicago: University of Chicago Press.

Kappelle, M. 1991. Distribución Altitudinal de la Vegetación del Parque Nacional Chirripó, Costa Rica. *Brenesia* 36: 1–14.

Kappelle, M., J. G. Van Uffelen, and A. M. Cleef. 1995. Altitudinal Zonation of Montane *Quercus* Forests Along Two Transects in Chirripó National Park, Costa Rica. *Vegetatio* 119(2): 119–153.

Kappelle, M., N. Zamora, and T. Flores. 1991. Flora Leñosa de la Zona Alta (2000–3819 m) de la Cordillera de Talamanca, Costa Rica. *Brenesia* 34: 121–144.

Keating, P. L. 1995. *Disturbance Regimes and Regeneration Dynamics of Upper Montane Forests and Páramos in the Southeastern Ecuadorian Andes.* Ph. D. dissertation, University of Colorado—Boulder.

Kohkemper, M. 1968. *Historia de las Ascensiones al Macizo de Chirripó.* San José, Costa Rica: Instituto Geográfico Nacional.

Magallón, F., C. Segura, and J. C. Parreaguirre. 1987. *Mapa de los Parques Nacionales de Costa Rica* (1:500000 scale map). San José, Costa Rica: Editorial Heliconia.

Martin, P. S. 1964. Paleoclimatology and a Tropical Pollen Profile. In *Report on the VI International Congress on the Quaternary, Warsaw, 1961,* volume 2, pp. 319–323.

McKechnie, J. L., ed. 1983. *Webster's New Twentieth Century Dictionary.* 2nd edition. New York: Simon and Schuster.

MINIREM (Ministerio de Recursos Naturales, Energía, y Minas de Costa Rica), MIDEPLAN (Ministerio de Planificación Nacional y Política Económica), OEA (Organización de los Estados Americanos), and CI (Conservación Internacional). N.d. *Strategy for the Institutional Development of the La Amistad Biosphere Reserve: A Summary.* Costa Rica: MINIREM.

Patterson III, W. A., K. J. Edwards, and D. J. Maguire. 1987. Microscopic Charcoal As a Fossil Indicator of Fire. *Quaternary Science Reviews* 6: 3–23.

Pérez, F. L. 1998. Ecological Problems in a High Páramo of the Venezuelan Andes. In *Nature's Geography: New Lessons for Conservation in Developing Countries,* eds. K. S. Zimmerer and K. R. Young, pp. 147–183. Madison: University of Wisconsin Press.

Piperno, D. R., M. B. Bush, and P. A. Colinvaux. 1990. Paleoenvironments and Human Occupation in Late-Glacial Panama. *Quaternary Research* 33(1): 108–116.

Ramondino, S., ed. 1969. *The New-World Spanish-English and English-Spanish Dictionary.* New York: New American Library.

Sacchetti, M. 1992. After Fire, Rangers Try to Rebuild Chirripó. *The Tico Times* (San José, Costa Rica), 15 May 1992, p. 11.

Savage, M. 1993. Ecological Disturbance and Nature Tourism. *Geographical Review* 83(3): 290–300.

Vale, T. R. 1987. Vegetation Change and Park Purposes in the High Elevations of Yosemite National Park, California. *Annals of the Association of American Geographers* 77(1): 1–18.

Valerio, C. E. 1983. *Anotaciones sobre Historia Natural de Costa Rica.* San José, Costa Rica: Editorial Estatal a Distancia.

Wallace, D. R. 1992. *The Quetzal and the Macaw: The Story of Costa Rica's National Parks.* San Francisco: Sierra Club Books.

Weber, H. 1959. *Los Páramos de Costa Rica y su Concatenación Fitogeográfico con los Andes Suramericanos.* San José, Costa Rica: Instituto Geográfico Nacional.

Weston, A. 1981a. *Páramos, Cienagas, and Subpáramo Forest in the Eastern Part of the Cordillera de Talamanca* (unpublished). San José, Costa Rica: Tropical Science Center.

Weston, A. 1981b. *The Vegetation and Flora of the Chirripó Páramo* (unpublished). San José, Costa Rica: Tropical Science Center.

Williamson, G. B., G. E. Schatz, A. Alvarado, C. S. Redhead, A. C. Stam, and R. W. Sterner. 1986. Effects of Repeated Fire on Tropical Páramo Vegetation. *Tropical Ecology* 27: 62–69.

6

Human Impact on the High Páramo Landscape of the Venezuelan Andes

Francisco L. Pérez

Fast runs the wind; fast runs the water;
  fast runs the stone that falls from the mountain.
Run, warriors, fly against the enemy;
  run fast
  like the wind,
  like the water,
  like the stone that falls from the mountain.
Strong is the tree that resists the wind;
  strong is the rock that resists the river;
  strong is the snow of our páramos, that resists the sun.
Fight, warriors; fight, braves; be strong
  like the trees,
  like the rocks,
  like the snows of the mountains.
— War song of the Timotes Indians; translated by Tulio Febres
    Cordero (Cardozo 1965). English translation by the author.

The *páramos* are the alpine-equivalent regions of the tropical Andes. As in many other high-altitude areas, the landscape of the South American páramos has been significantly affected by different types of human activity. Several recent reports have focused on anthropogenic impact on páramos (Balslev and Luteyn 1992; Grubb 1970; Hofstede 1995; Pérez 1992a; Vargas and Rivera 1990). In contrast to the punas of the southern Andes, where human use can be traced back several millennia (Winterhalder and Thomas 1978; Ellenberg 1979), most north-Andean páramos were not heavily impacted by humans until recently, and vast páramo tracts still remain relatively untouched (Parsons 1982). However, recent anthropogenic disturbance is rapidly affecting many of the unusual vegetation types of the high páramo, which in dry Andean valleys is dominated by diverse giant caulescent rosette-plants (Espeletiinae) and by tiny cryptogamic crusts that cover the soil surface. This chapter examines the changing nature of landscape disturbance due to different past and present human-induced modifications of the high páramos in the Venezuelan Andes. In

147

this paper the "high páramo" corresponds with the "superpáramo" (Cuatrecasas 1968; Luteyn 1992); this last term seems to be restricted to Colombia. I will first discuss the historical antecedents of the human use of the high Venezuelan páramos, and then focus on some current aspects of environmental conservation in the largest high-páramo region, the Páramo de Piedras Blancas. Finally, some management and conservation recommendations are offered in light of the data and processes discussed.

Human Advance into the Páramo Region in Pre-Columbian and Colonial Times

As have other biogeographical landscapes discussed in this book, the Páramo de Piedras Blancas has gone through different phases of human influence and shifting disturbance regimes (Botkin 1990). The environmental history of humans in the Andes of present-day Venezuela during the past millennium has been one of settlement and land use in landscapes at progressively higher elevations. This vertical migration of humans has also been accompanied by intensification of their effects, as different cultures and technologies have impacted ever higher Andean belts.

The Andean foothills and valleys were settled by 650 A.D. (Wagner 1967). At this time, humans inhabited mostly the *tierra templada* (temperate land) below 2000 meters, where they practiced intensive maize and manioc agriculture (Wagner 1979). Human occupation of higher mountain belts did not occur until protohistoric times, when an extension of the central Andean cultures reached Venezuela through Colombia. Settlement of the *tierra fría* (cold land)—2000 to 3000 meters—by this new group began at about 1000 A.D. (Wagner 1973). Highland settlement during this Mucuchíes phase was made possible by the switch from maize—which grows below 2400 meters (Wagner 1978)—and manioc to a system centered on the production of tuber staples: potatoes (*Solanum* species), ulluco (*Ullucus tuberosus*), and cuiba (*Oxalis tuberosa*) (Wagner 1967, 1979). However, although agriculture was present as high as 3000 meters, it seems that the bulk of the human population was found below 2000 meters, primarily in dry intermontane valleys (Monasterio 1980a; Salgado-Labouriau 1975, 1979). At any rate, there were no permanent settlements in the Venezuelan páramos, which were visited only for brief periods for hunting, gathering wild plants, or practicing religious ceremonies. Páramos were also crossed during war expeditions or bartering and exploration trips (Wagner 1978).

The Spaniards reached the Venezuelan Andes in 1548–1549, when Diego Ruiz de Vallejo penetrated the Humocaro Mountains. Juan Rodríguez Suárez entered the Chama Valley (map 6.1) in 1558 and founded the city of Mérida on one prominent glaciofluvial terrace. The páramo proper was entered by the Spaniards next year, when an expedition led by Fernando Cerrada reached the site of Mucuchíes (2983 meters) (Wagner 1973). The native population was relatively dense in this region at the time of conquest by Europeans. Both the lands and the Indians inhabiting them were partitioned among the conquistadores in 1564, when Andrés Díaz Venero de Leiva granted the *encomiendas* covering the Chama Valley (Morón 1961). A census at the time gave a native population of 30000, but since many Indians had fled the conquerors and taken refuge in the páramos, a better total estimate is about 50000 (Brito 1962; Salas 1956, 1971; Wagner 1979).

After 1564, the páramos became a sanctuary for rebel Indians, who successfully resisted the Spaniards until at least 1575 (Cardozo 1965). Fighting with páramo Indians continued sporadically until the end of the sixteenth century, when the remaining rebels were reduced to the town of Mucuchíes and converted (Wagner 1979). In the early seventeenth century, Indians would occasionally still escape to the mountains. Several towns were founded in the high Andes in the early 1600s with the main goal of reducing and indoctrinating the native population (Burguera 1982). Colonization along the lower páramo ecotone started at this time (Pérez 1992a; Vila 1978), but the páramo itself appears to have remained mostly unsettled.

The changes in land use instituted by Europeans began to have an impact on páramo ecology. Wheat and cattle were introduced in high areas, and both gradually encroached upon the Andean highlands. Salgado-Labouriau and Schubert (1977) examined pollen from Laguna Victoria cores (3250 meters) and noted that the upper levels contained large amounts of *Rumex acetosella* (sheep sorrel) pollen, introduced with wheat. The upper core section also showed a drastic drop in tree pollen, suggesting that massive deforestation of the upper Andean forest took place at about the same time. Deforestation along the Andean forest ecotone did, in fact, help to extend the original boundaries of the páramos, which advanced downslope into cleared plots (Budowski 1968; Luteyn 1992). In the Santo Domingo area (map 6.1), rosettes of *Espeletia schultzii* can be seen today growing within cultivated fields at 2700 meters, while the surviving patches of high Andean forest in the same valley are found up to 3060 meters. Wheat spread rapidly along the Chama Valley during the mid-1500s (Brito 1972). Extensive wheat cultivation on even the steepest slopes produced widespread soil erosion, and deep gullies still scar the flanks of the lower Chama Valley (Crist 1943; Monasterio 1980a). Although agriculture became important after European settlement, it never

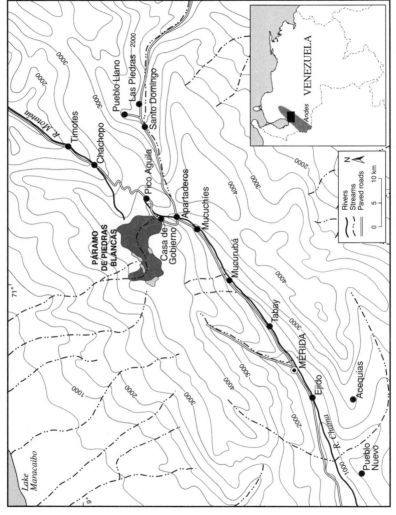

Map 6.1. Generalized map of the Mérida Andes and location of the Páramo de Piedras Blancas (*shaded area*). Contour interval is 500 meters.

150

Fig. 6.1. South view of the upper Chama Valley, near Casa de Gobierno, at the lower boundary of Piedras Blancas (about 3650 meters elevation). The peaks of the Sierra de Santo Domingo appear in the background. December 1977.

directly affected the high páramos. Today, wheat is only a marginal crop in the Andes, found up to about 3600 meters (figure 6.1).

In contrast to wheat, cattle were slow in occupying the mountains, but eventually reached the highest páramos. Domestication of native animals was not important in north-Andean cultures, and páramos lacked natural or domesticated grazers. Cattle were introduced in the upper Chama Valley during the late 1600s (Vila 1978), but were not significant until the 1760s, when several Andean towns below 2600 meters were reported to have "*mucho ganado*" (Vila 1978, 1980). The *Descripción de la Ciudad de Mérida* of 1783 also indicates how cattle rapidly multiplied throughout the region and were then present up to 3000 meters (Vila 1980). There are several possible reasons for the slow growth of cattle populations in the Andes. A period of gradual buildup of animals was probably needed, and cattle must have spread slowly at first during the colonial period. In addition, animals were not permitted in the Indian towns, and encomenderos were prohibited from keeping any cattle (*ganado mayor*) ". . . within less than a league [about 6.4 kilometers] of the indigenous towns and *repartimientos*" (Vila 1980: 107). Thus, growth of cattle had to await both the slow increase in European settlers and the abolition of the encomienda system in 1687 (Cardozo 1965). A significant problem was also posed by the fact that Mediterranean cattle needed a long period of adaptation to the highlands; Rouse (1977: 57) estimated that the acclimatization phase lasted about 50 years. For all these reasons, it appears that cattle were primarily found through most of the colonial period in landscapes well below 4000 meters (Pérez 1992a; Salgado-Labouriau 1975).

The colonial system collapsed in 1810, when Venezuela gained independence from Spain. The war resulted in a steep reduction of Andean cattle herds, and by 1832 few animals survived (Vila n.d.). Cattle regained pre-independence densities around 1847 (Vila 1962), but then several bloody revolutions took place, and by 1863 the livestock numbers had sharply dropped again (Brito 1972). National political instability lasted until 1870, when Guzmán Blanco's rise to power allowed for rapid cattle multiplication during the 1870s. Construction of the Panamerican Highway in 1925 accelerated human settlement along the lower páramo fringe (Horn 1989a) and increased cattle influence on the high páramo. Thus, grazing animals probably reached the highest areas of the Andes during the last three decades of the nineteenth century, or even as recently as the early 1900s (Pérez 1992a).

General Impact of Human Activities on the High-Páramo Flora

In addition to the more obvious impacts caused by these recent, intensive land uses, several activities have affected high Andean vegetation since human settlement in the region. These include native plant gathering, plant introduction (mainly accidental), disturbance of habitat, and alteration of the range of some native plants.

Early Venezuelan botanists and ethnographers took notice of the wide variety of uses that highland plants provided—first for the aborigines and later for settlers. A compilation from several sources indicates that more than 60 plant taxa were gathered from the páramo and its ecotone with the upper Andean forest for various purposes, but mainly as medicinal sources, firewood, lumber, and dietary supplements (table 6.1). Many páramo plants are still used for medicine or as stimulants; about half of the plants cited have some purported medicinal use for humans or for domestic animals, although, as Pittier (1939: 26) notes, ". . . for the majority of our medicinal plants, the action that is attributed to them remains to be demonstrated."

The occasional use of plants as food occupies a second place (some 23% of the taxa cited in table 6.1); the fruits of several shrubs and trees are consumed raw, and some plants are used to produce spices, butter and cheese, or preserves. Gade (1975) studied the use of wild plants in the Vilcanota Valley of Peru and found that only about 4% of the total food intake came from uncultivated plants, mostly during extraordinary occasions such as crop failures or famines. It seems probable that páramo food

Table 6.1. List of plant species and their local uses in the northern Andes.

Plant species (Family)	Life-form	Altitudinal range (m)	Plant use(s)
Acaena cylindrostachya R. et P. (Rosaceae)	H	2900–4200	Medicine: infusion taken as sedative.
Adipera jahnii Britt. et Rose (Caesalpiniaceae)	T	1500–3350	Medicine: today cultivated in the lower paramos as laxative.
Alnus acuminata Regel (Betulaceae)	T	1700–3200	Lumber, and possibly firewood; (= *A. jorullensis* H. B. K.)[1]
Alnus mirbellii Spach. (Betulaceae)	T	1950–3220	Its soft wood is used for firewood and lumber.[1] Vareschi (1970: 148) lists both species of *Alnus* as present in the Andes.
Arracacia spp. (Umbelliferae)	Su	2000–3450[2]	Spice: roots are also medicinal (laxative and emollient).
Azorella spp. (Umbelliferae)	H	3500–4600	Possibly used as fuel when plant is dried.[3]
Baccharis floribunda H. B. K. (Compositae)	T-S	2100–3300	Medicine: depurative and vulnerary (useful in healing wounds).
Baccharis tricuneata Pers. (Compositae)	S	2200–3800	Medicine: vulnerary. Other *Baccharis* species are used as well.
Bocconia frutescens L. (Papaveraceae)	T-S	1500–3000	Medicine: root is used against fungal infections. Also lumber (?).[1]
Calamagrostis effusa Steud. (Gramineae)	G	2700–4000	House construction (roofing) and basket weaving.
Calycopus moritzianus Burr. (Myrtaceae)	T	1200–3600	Food: aromatic, sweet fruits are consumed.
Cavendishia bracteata (Ruiz & Pavon ex J. St.-Hil.) Hoerold (Ericaceae)	S	2100–3500	Food: fleshy fruits are eaten: Recently used as ornamental (see [10]).
Chaetolepis alpestris Trian. (Melastomataceae)	T-S	3000–3900	Firewood: produces sparks when burned, thus name ("sparkler").[1]
Chaetolepis lindeniana Trian. (Melastomataceae)	S	3000–3900	Also used as firewood; wood produces sparks when burned.[1]
Cheilanthes myriophylla Desv. (Pteridaceae)	F	1400–3200	Medicine: an infusion used against common cold and chest pains. Also used as diuretic and sudorific.
Coriaria thymifolia H. et B. (Coriariaceae)	Su	1800–3000	Aborigines used it for dyeing. A source of ink in colonial times.

(continued)

153

Table 6.1. *Continued*

Plant species (Family)	Life-form	Altitudinal range (m)	Plant use(s)
Draba spp. (Cruciferae)	H	3500–4600	Food and spice: mixed with hot ají peppers. Also as medicine.[4]
Drymis winteri Forst. (Magnoliaceae)	T	2000–3000	Stimulant: bark is given to cattle mixed with salt.
Dyalianthera otoba (Humb. & Bompl.) Warb. (Myristicaceae)	T	Upper Andean forest	Medicine: its fruit is used against skin afflictions, ticks, and fleas; also against stomach pains, and as laxative.[5]
Echeverria venezuelensis Rose (Crassulaceae)	Su	2800–4000	Ground leaves are given to cattle as laxative.
Equisetum bogotense H. B. K. (Equisetaceae)	F	1500–3400	Medicine: antidiabetic.
Escallonia tortuosa H. B. K. (Saxifragaceae)	T	2750–3700	This, and other *Escallonia* species of the upper Andean forest (e.g. *E. floribunda* H. B. K.), are used as firewood, and for house construction (mainly as beams).[1]
Espeletia humbertii Cuatr. (Compositae)	T	~3500	Stems and resin are used as aromatic incense in local churches.
Espeletia neriifolia Sch. Bip. (Compositae)	T	2000–3600	Stems and resin are used as aromatic incense.
Espeletia schultzii Wedd. (Compositae)	R	2900–4300	Food (preserves): roots are used in butter production; leaves are also used to wrap cheese and butter.
Espeletia weddellii Sch. Bip. (Compositae)	R	3300–4250	Medicine: used to treat respiratory ailments.
Espeletia spp. (Espeletiinae, Compositae)	R	2000–4600[6]	Numerous uses: see text for details.
Eugenia triquetra Berg. (Myrtaceae)	S	2000–4000	Its wood may be used for firewood.[1]
Eupatorium neriifolium (?) Robinson (Compositae)[7]	S	1800–3200	Medicine: depurative and vulnerary.
Galium canescens H. B. K. (Rubiaceae)	H	500–3000	Roots are used in Mérida for dyeing.
Gnaphalium spp. (Compositae)	H	1500–4200[8]	Medicine: cure for pulmonary afflictions (?).
Grammadenia alpina Mez. (Myrsinaceae)	S	2500–3800	Medicine: its pungent bark is used to alleviate toothaches.
Hedyosmum glabratum H. B. K. (Chloranthaceae)	T	1250–3200	Medicine: its aromatic bark is reputed as tonic and invigorating.
Hesperomeles glabrata H. B. K. (Rosaceae)	T	2300–3600	Wood may be used for lumber (?), but tree is now rare.[1]

154

Species		Elevation	Uses/Notes
Hinterhubera columbica Sch. Bip. (Compositae)	S	3100–4000	Medicine (?).
Hypericum brathys Sm. (Guttiferae)	S	2600–4100	This plant has nearly disappeared from the páramos near Bogotá (Colombia), where its branches were used to produce brooms until the 1940s (Cuatrecasas and Torres 1988). No similar use has been ascertained for Venezuela, however.
Hypochoeris sessiliflora (?) H. B. K. (Compositae)[9]	H	2700–4200	Medicine: refreshing drink with sudorific properties.
Jamesonia canescens Kze. (Pteridaceae)	F	2900–4300	Medicine.
Macleania rupestris (Kunth) A. C. Smith (Ericaceae)[10]	S	2200–3900	Food: sweet, fleshy fruits are consumed.
Miconia theaezans Cogn. (Melastomataceae)	S	300–3200	Wood of this plant and other *Miconia* species (e.g., *M. jahnii* Pitt.) may be used as firewood.[1]
Oritrophium peruvianum Cuatr. (Compositae)	H	2700–4150	Medicine: used against diarrhea.
Oxalis medicaginea H. B. K. (Oxalidaceae)	H	~3200	Medicine (?).
Peperomia spp. (Piperaceae)	H	2000–3850[11]	Medicine: the boiled leaves are used as an anti-asthmatic.
Plantago major H. B. K. (Plantaginaceae)[12]	H	1300–3000	Medicine: all plant is astringent; seed decoction is diuretic, and tea of leaves is anti-hemorrhagic.
Podocarpus oleifolius var. *macrostachys* Buchh. et Gray (Podocarpaceae)	T	upper Andean forest	Food: the pericarp of its nuts is highly esteemed.
Polylepis sericea Wedd. (Rosaceae)	T	2400–4200	This tree has been heavily decimated, apparently for its excellent red wood.[1] A closely related species (*P. quadrijuga*) has become nearly extinct in Colombia, where it is exploited for fence posts.
Polymnia pyramidalis (?) Triana (Compositae)[13]	T	2000–3000	Medicine: vulnerary. Also used to prepare glue (?).
Polypodium spp. (Polypodiaceae)[14]	F	700–4200	Many species are used for different medicinal purposes.
Pteridium aquilinum Kuhn. (Pteridaceae)	F	1700–3000	Medicine: hernia remedy.
Puya aristeguietae L. B. Sm. (Bromeliaceae)	R	~3200	Rare and endangered plant: plants are burned and buds are eaten.
Puya venezuelana L. B. Sm. (Bromeliaceae)	R	~3200	Same use and impact as for *P. aristeguietae*.
Relbunium hypocarpium Hemsl. (Rubiaceae)	H	2000–4000	Roots are used to obtain a reddish dye (called "raicita": little root).
Rubus coriaceus (?) Poir. (Rosaceae)[15]	S	3100–3500	Food: its muriform fruits are edible.

(continued)

Table 6.1. *Continued*

Plant species (Family)	Life-form	Altitudinal range (m)	Plant use(s)
Salix humboldtiana Willd. (Salicaceae)	T	70–3000	Wood is a good but scarce source of paper.[1]
Salvia rubescens H. B. K. (Labiatae)	H	2500–3500	Used in popular medicine in Mérida, but purpose not reported.
Savastana mexicana Rupr. (Gramineae)	G	(High Sierra Nevada)	Medicine: root infusion is used against stomach cramps.
Senecio funckii (?) Sch. Bip. (Compositae)[16]	H	3500–4300	Medicine: leaves' infusion used to disinfect wounds and abscesses.
Senecio sclerosus (?) Cuatr. (Compositae)[17]	S	3900–4400	Leaves are used as food; chewing leaves is said to be refreshing.
Solanum columbianum Dun. (Solanaceae)	H	~3000	Tubers possibly consumed by humans (Bitter potato: "papa amarga") or fed to animals (pig potato: "papa de puerco").
Stipa ichu Kunth. (Gramineae)	G	2800–4000	House roofing and basket weaving.
Thamnolia vermicularis Schaer. (Lichenes Imperfecti)	EL	~2800–4600	Medicine: depurative.[18]
Vaccinium floribundum H. B. K. (Ericaceae)	S	3000–3900	Food: its globose fruits (blue berries) are eaten.

Notes:

1. Pittier (1939:29) identifies this species as "maderable" (lumber and/or firewood-producing).

2. Two species in the Venezuelan páramos. One (*Arracacia vaginata* Cout.) found in many páramos up to 3450 meters (Vareschi 1970:273).

3. Two species in the Venezuelan páramos: *A. crenata* Pers. and *A. julianii* Math. See Hodge (1960) and Ralph (1978) for plant uses.

4. Plant consumed is locally called "michiruy." Plant used as spice might be *D. bellardii* Blake, and perhaps also *D. cheiranthoides* HK. f., which is called "little michiruy." Vareschi (1970:225) implies that several *Draba* species are medicinal.

156

5. Pittier (1926:316) cites Father Gumilla (1741), a Spanish friar travelling in Venezuela in the 1730s, who mentioned how the Indians in the Andes " . . . go higher (towards the páramo) to find the trees that produce the Otova . . . which is not a resin or rubber, [but] like a white hazelnut" which they used in large amounts for their beneficial effects as medicine.

6. About 60 species of the Espeletiinae subtribe are found just in the Venezuelan Andes (Cuatrecasas 1979b).

7. Pittier (1926:307) lists this species used as *Eupatorium leuconyelum* (spp. ined) but this plant does not appear in later floras. Plant used appears to be *E. neriifolium* (Schnee 1973:498).

8. Eight or nine *Gnaphalium* species are present in the Venezuelan páramos (Aristeguieta 1964:351).

9. Actual species consumed uncertain. Pittier (1926:208) lists it as *H. acaulis*, but this is not present in Venezuela. Species used seems to be *H. sessiliflora* (Aristeguieta 1964:925).

10. The fruits of *Macleania rupestris* are consumed in Colombia (Cuatrecasas and Torres 1988). The vernacular name for *M. rupestris* in Venezuela is "cacagüito," which refers to its globose fruits (Schnee 1973). *Cavendishia bracteata* is also called "cacagüito."

11. About eight species of *Peperomia* present in the (low) Venezuelan páramos (Vareschi 1970:215).

12. This plant has been introduced in the páramo from Europe (see table 6.2).

13. Actual species used uncertain. Pittier (1926:113) indicates *P. eurylepis*, but the species used is probably *P. pyramidalis* (Schnee 1973:39).

14. At least seven species of *Polypodium* are found in the páramos; all grow at or above 3500 m elevation (Vareschi 1970:112).

15. Species used uncertain. Many *Rubus* species have globose fruits, and are called "mora" or "zarzamora" in Venezuela (Schnee 1973:778).

16. Pittier (1926) gives this plant as *Senecio crepidifolius*, but that species is from Peru, and is not present in Venezuela. Species utilized is, most likely, *S. funckii* (Aristeguieta 1964:780).

17. Pittier (1939:106) did not identify the species, but the vernacular name he cites ("salviecita de páramo") is that of *S. sclerosus* (Vareschi 1970:319).

18. This is a common "erratic" lichen in the páramo; it is normally found completely unattached to the substrate (Pérez 1994).

The list includes only plants present in the páramos or along their lower ecotone with the upper Andean forest. Compiled from Acosta (1954), Bauman and Young (1986), Cuatrecasas and Torres (1988), Guhl (1968), Jahn (1927), Lares (1907), Levi-Strauss (1950), Metraux and Kirchhoff (1948), Pittier (1926, 1939), Ramón y Rivera and Aretz (1963), Schnee (1973), Smith (1974, 1981), Vareschi (1970), Vila (1976), and Wagner (1967, 1979). Nomenclature and altitudinal ranges follow mainly Vareschi (1970), but see also Aristeguieta (1964) for taxonomic equivalents. Consult table notes for further details. Key to life-forms: EL: erratic lichen; F: fern; G: graminoid; H: herb; R: rosette plant; S: shrub; Su: suffruticose plant; T: tree.

sources were used in a similarly sparing manner, possibly for the most part during trips through the mountains.

In third place (with 18% of taxa in table 6.1) comes the utilization of woody plants as sources of firewood and, occasionally, lumber. Most woody sources are, of course, found below the páramo belt, so this use applies primarily to the lower páramo and its ecotone with the Andean forest. Finally, there is a diversity of other uses for páramo plants that include basket-weaving, house construction and insulation, cloth dyeing and ink production, and, recently, planting the more attractive species as landscape ornamentals.

Several rosette species from the Espeletiinae subtribe (Heliantheae Compositae), the most characteristic plant of the Venezuelan páramos, are used for numerous purposes. Rosette leaves have been utilized for house roofing, for filling wall interstices, and for insulation; they have also been used for clothing manufacture, to fill mattresses, and as overnight bedding while sleeping out. In the past, some species provided a source of printing ink, and an attempt to produce paper pulp from *Espeletia* leaves failed in Venezuela in the early twentieth century. The stems of some arborescent species are cut and burned as incense in the local churches because of their aromatic resins. The dry stems of the tall caulescent species may occasionally serve as fuel. The fleshy stem pith of some species is eaten raw, fried, or prepared as preserves. The roots of *E. schultzii* are utilized to produce butter, and its leaves then used to wrap butter and cheese. *E. weddellii* is used to cure some respiratory diseases. In Colombia, and possibly Venezuela, *Espeletia* plants are burned and their ash added to the soil as antiacid and fertilizer for potato cultivation in the lower páramos (Guhl 1968).

The most significant trends apparent from table 6.1 are that the majority of the plants gathered by humans have come from the lower páramos, and that only a few species from the higher elevations have been affected by human collection.

Páramo floras have also been altered by accidental introduction of several weed species which came adventitiously with crops imported by Spanish settlers (Sauer 1988). Table 6.2 indicates that at least 23 plant taxa have become naturalized in the páramo. Most are herbaceous plants or annuals, which are generally lacking in the native páramo flora (Vareschi 1970). There are also several introduced grasses. The main source of nonnative plants is, expectedly, Europe (about 75% of the taxa cited) with some species from Mexico, Africa, Asia, and the Middle East. Adventitious plants seem not to have altered natural páramo vegetation in a substantial manner—about half of them are restricted to cultivated fields and pastures, roads, and trails.

Table 6.2. List of adventitious plant species found in the Venezuelan páramos.

Plant species (Family)	Life-form	Altitude range (m)	Origin: main habitat in the Andes
Alchemilla vulgaris L. (Rosaceae)	H	~3300	Europe: in the páramos of Mérida state.
Anthoxanthum odoratum L. (Gramineae)	G	1000–3000	Europe: a forage plant, found in the páramos of Mérida and Táchira states.
Arabis turrita L. (Cruciferae)	H	~3200	Europe: found in the páramos near Pico Bolívar (Mérida).
Capsella bursa-pastoris Medc. (Cruciferae)	Th	1800–3500	Europe: crop weed, and along roads in páramos. In the area between Timotes and San Rafael de Mucuchíes (Mérida).
Conyza filoginoides Hier. (Compositae)	H	2000–3000	Mexico: in the Páramo Piñango (Mérida).
Erodium cicutarium L'Her. (Geraniaceae)	Th	2300–4080	Europe (Mediterranean): in the páramos of Mérida.
Erodium moschatum L'Her. (Geraniaceae)	Th	2300–4800	Europe (Mediterranean): in the páramos of Mérida.
Hierochloe mexicana Benth. (Gramineae)	G	3000–4000	Mexico (?): rare, in the páramos of Mérida.
Lappula echinata Gilib. (Borraginaceae)	Su	1700–3600	Europe: in the páramos of Mérida, Táchira, and Trujillo. Crop weed, often grown as ornamental. See note 1 in table 6.3.
Medicago denticulata Willd. (Papilionaceae)	Th	600–3500	Southern Europe: in páramos of Mérida, Táchira, and Trujillo.
Melilotus officinalis Willd. (Papilionaceae)	Th	1800–3000	Europe, Asia: in the páramos of Mérida.
Melinis minutiflora Beauv. (Gramineae)	G	800–3000	Africa (intr. from Brasil): forms masses in secondary vegetation throughout the Venezuelan Andes. Also found in pastures.
Papaver glaucum Boiss. et Hauss. (Papaveraceae)	Th	2600–3300	Middle East: along the edge of upper Andean forest, and in many Venezuelan páramos. Common in wheat fields.
Phytolacca octandra L. (Phytolacaceae)	Su	~3500	Asia (Japan?): along the edge of upper Andean forest and in the páramos of Mérida. Not very common.
Plantago major L. (Plantaginaceae)	H	1300–3000	Europe: common in the páramos of Mérida and Trujillo; also in crop fields and along roads (see table 6.1 also).

(continued)

Table 6.2. *Continued*

Plant species (Family)	Life-form	Altitude range (m)	Origin: main habitat in the Andes
Poa annua L. (Gramineae)	G	1600–3920	Europe: widely distributed in páramos of Mérida, Táchira, and Trujillo, from the upper edge of Andean forest up to the nival belt. In moist grasslands, and as crop weed.
Rhynchelitrum roseum Stapf et Hubb. (Gramineae)	G	up to 4200	South Africa: in secondary vegetation of páramos in Mérida, Táchira, and Trujillo.
Rumex acetosella L. (Polygonaceae)	H	1600–4000	Europe: páramos of Mérida, Táchira, and Trujillo. Common in *Espeletia* páramos, as well as in wheat and potato fields.
Rumex crispus L. (Polygonaceae)	Ge	1600–3500	Europe: throughout the Venezuelan Andes: most common in cultivated fields.
Silene inflata Sm. (Caryophyllaceae)	H	~3100	Europe: along the edge of upper Andean forest and in páramos.
Taraxacum officinale Web. (Compositae)	H	1500–4040	Europe: widely distributed in páramos of Mérida and Trujillo.
Trifolium repens L. (Papilionaceae)	H	1500–3800	Asia and Europe: in the páramos of Mérida and Táchira. Along roads and in cultivated fields.
Viola tricolor L. (Violaceae)	Th	2600–3200	Europe: in páramos all throughout the Andes. Common weed in wheat and potato fields.

Notes:

Compiled from Pittier (1926, 1939), Schnee (1973), and Vareschi (1970). Taxonomic nomenclature follows Vareschi (1970). Key to plant life-forms: G: graminoid; Ge: geophyte; H: herb; Su: suffruticose plant; Th: therophyte (annual).

Closely related is a third category of plants native to the páramo but whose distribution has been altered by human activities (table 6.3). These ruderal species come primarily from the montane forest and the low páramo, and are able to ascend to higher areas along roads and trails (Sauer 1988: 95). They also become weeds in croplands and pastures, where cattle manure and fertilizer help the proliferation of these nitrophilous plants (Lozano and Schnetter 1976; Vareschi 1970). I plotted all these taxa on a diagram of plant presence by 100-meter altitudinal belts (figure 6.2). It shows strikingly similar patterns for the three categories considered: the numbers of plant taxa affected by humans decrease steadily with increasing elevation, and fewer than 10% to 20% of them are found above 4250 meters. This trend fits appropriately with the historical narrative, which indicated that humans made only light use of the high páramo—at least until recently.

The Landscape of the Páramo de Piedras Blancas

The Páramo de Piedras Blancas is in the Sierra de La Culata, above 3700 m, at 8° 51′ N and 70° 53′ W (map 6.1). This arid páramo receives < 800 millimeters/year of precipitation: the dry period (December–March) averages < 20 millimeters/month, and the rainy season (May–August) averages > 100 millimeters/month. Due to the equatorial location and high altitude, temperatures show broad diurnal fluctuations. In the dry season, air temperatures reach highs of 15°C to 23°C and lows of −5°C to −12°C. The daily amplitude is reduced during the rainy season by persistent cloud cover, but some 350 freeze/thaw cycles occur annually at the highest elevations. Freezing is linked to cyclic, nightly formation of needle ice, a type of ground frost (Pérez 1991a).

Soils of Piedras Blancas are Inceptisols, Entisols, and Histosols (Soil Survey Staff 1994). They are mostly shallow, have low organic content, and have a coarse texture with a high percentage of gravel and sand. The fine fraction (≤ 0.05 mm) is mainly silt with a little clay (Pérez 1992b); this texture is highly conducive to needle ice growth. Recurrent freeze/thaw action creates a nearly continuous surface layer of small soil lumps ("nubbins" or "buds"; Pérez 1987a; Washburn 1980).

The high-páramo vegetation contains several species of caulescent rosettes mixed with many herbs and shrubs (Vareschi 1970). The most common rosette plant is *Coespeletia timotensis* Cuatr., found up to 4600 meters (figure 6.3A). The landscape above the upper limit of rosette growth is occupied by periglacial desert (Monasterio 1980b). This apparently barren belt contains only a few short vascular plants. Careful examination reveals, however, that soils are often covered by a thin crust of cryptogams

Table 6.3. List of plant species native to the Venezuelan páramos and the upper Andean forest, but which have been affected in their distribution by human activities.

Plant species (Family)	Life-form	Altitude range (m)	Main habitat in the páramos
Acaena cylindrostachya R. et P. (Rosaceae)	H	2900–4200	Common in overgrazed sites and in fertilized fields (nitrophylous plant). In páramos of Mérida, Táchira, and Trujillo.[1]
Acaena elongata L. (Rosaceae)	Su	3500–4000	Páramos of Mérida, Táchira, and Trujillo. Grows normally in shrub communities, but also invades same habitats as *A. cylindrostachya*.[1]
Argemone mexicana L. (Papaveraceae)	H	up to 3000	Normally found in the upper Andean forest, but it often ascends to the páramo along roads.
Bromus catharticus Vahl. (Gramineae)	G	1900–3300	Lower páramos of Mérida, Táchira, and Trujillo. Very common as crop weed and in grazing areas.
Calandrinia ciliata D. C. (Portulacaceae)	Su	2800–4200	Found along trails in the páramos of Mérida and Trujillo.
Chusquea spencei Ernst. (Gramineae)	G(B)	2000–3450	A plant of the upper Andean forest and moist areas near creeks, but also invades sites disturbed by humans.
Heliotropum spp. (Borraginaceae)	H	<3200	Páramos of Mérida. Plants of lower elevations, ascend to the páramos mainly as agricultural weeds.
Lachemilla fulvescens Rothm. (Rosaceae)	H	2500–2800	Páramos of Mérida. Nitrophylous, found in sites rich in nitrogen.
Lachemilla polylepis Wedd. (Rosaceae)	S	3600–4200	Páramos of Mérida. Extremely common in overfertilized pastures.
Lachemilla ramossisima Rothm. (Rosaceae)	H	3000–4160	Páramos of Mérida. On sites similar to those of *L. verticillata*.
Lachemilla verticillata Rothm. (Rosaceae)	H	3200–4300	Páramos of Mérida. Not very common, but prefers sites rich in nitrogen within the communities of *Espeletia* rosettes.
Lithospermum mediale Johnston (Borraginaceae)	Su	<3200	Páramos of Mérida. Plant of lower elevations, found in the páramos only sporadically or as a weed in crops.

Species (family)	Life-form	Elevation	Notes
Malvastrum acaule Gray (Malvaceae)	Ge	3200–4500	Páramos of Mérida. Prefers moist pasture sites with high nitrogen.
Moritzia lindenii (D. C.) Benth.	Su?	3000–3600	Páramos of Mérida and Táchira. Plant of lower elevations, ascends to the páramos sporadically or as an agricultural weed.
Oenothera cuprea Schl. (Onagraceae)	H	2000–4000	Páramos of Mérida and Táchira. Found along trails and roads.
Oenothera mollissima L. (Onagraceae)	H	1500–3300	Páramos of Mérida and Trujillo. Weed in cultivated fields and on the walls around these fields.
Ottoa oenanthoides Kunth. (Umbelliferae)	H	up to ~4350	In all Mérida páramos. Common in pastures with high nitrogen.
Oxalis spiralis R. et P. (Oxalidaceae)	H	2500–4200	Páramos of Mérida and Táchira. In communities of *Espeletia* rosettes, but also common as weed in cultivated fields.
Salix humboldtiana Willd. (Salicaceae)	T	70–3000	In the highest cultivated páramos, near inhabited areas, along roads.
Soliva anthemidifolia (Juss.) R. Brown (Compositae)	H	3000–3400	Páramos of Mérida (and Táchira ?). Common in cultivated fields.
Tournefortia scabrida (?) H. B. K. (Borraginaceae)	S	<3200	Páramos of Mérida, especially around Tabay. Plant of lower elevations, only found in páramos sporadically or as a weed in crops.
Veronica tourneforti Gmel. Th. (Scrophulariaceae)	H	3000–3300	In the páramos of Mérida (especially along Chama Valley and around the town of Timotes), associated with crops and roads.

Note:
1. These plants are able to spread widely because of the glochidia (tiny barbed hairs or bristles) that cover their seeds. These adhere easily to animals (mainly cattle) and people's clothes. In Venezuela, a common name for glochidiate plants is "cadillo" (Pittier 1926).

Compiled from Pittier (1926, 1939), Schnee (1973), and Vareschi (1970). Taxonomic nomenclature follows Vareschi (1970). Key to plant life-forms: B: bamboo; G: graminoid; Ge: geophyte; H: herb; S: shrub; Su: suffruticose plant; T: tree; Th: therophyte (annual).

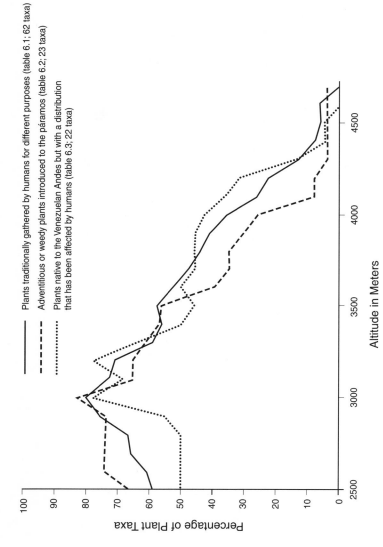

Fig. 6.2. Altitudinal distribution of plants associated with human activities in the Venezuelan páramos. Data for plant taxa compiled by 100-meter altitudinal belts.

including mosses, lichens, and hepatics (Pérez 1994, 1996). In contrast to the well-known vascular páramo flora, the microbiotic crust has been largely ignored by páramo researchers (Smith and Young 1987). These tiny plants add substantial amounts of organic material to otherwise poor soils, protect soil from aeolian erosion, rain, and runoff, and help capture fine particles that are gradually added to the substrate. Some cryptogams also produce unattached forms that can migrate when carried by wind, water, and gravity (Pérez 1991a, 1991b). This erratic habit may facilitate plant colonization of bare sites.

Environmental Human Impact on the Páramo de Piedras Blancas

Impact on Vascular Vegetation

Cattle are by far the most significant disturbance agent of vascular vegetation in Piedras Blancas (Pérez 1987b, 1992a, 1992b, 1993). This region has been held as *ejido* (communal property) by Andean peasants since the sixteenth century, when King Philip II granted the use of public lands to Andean towns (Troconis 1962). Although Piedras Blancas was included in 1989 as part of Sierra La Culata National Park, this páramo is still used for extensive grazing by cattle herds and horses.

Cows concentrate their feeding on low-lying, moist areas, where they normally eat short ground cover, particularly the soft, fleshy species of cushion plants. Cattle often kick the sward to detach palatable plants or parts (Grubb 1970). This may actually limit the establishment of caulescent rosettes because they germinate on plant cushions, which constitute a refuge ("nurse plants") against heaving by needle ice (Pérez 1987b).

My investigation has focused on rosettes. Cows do not actively seek these plants, but browse on their foliage while grazing on the low vegetation cover. Animals bite fresh rosette leaves but as only the upper part of each leaf is eaten, the leaf bases remain on the plant. Local *campesinos* have noticed that cows feed on rosettes mainly during the dry season because the milk and cheese become extremely bitter from the aromatic compounds and resins in the rosettes. Cuatrecasas (personal communication, 1990) noted that mules browsed the leaves of rosettes in the Colombian páramo, and reported that grazing by cattle can damage the apical meristem (Cuatrecasas 1979a). Schmidt and Verweij (1992) indicated that cattle also eat the inflorescence of *E. hartwegiana* in Colombia. Foliage browsing by native or introduced animals is a significant ecological factor for similar tropical-alpine rosettes in other areas like Hawaii (Yocom 1967) and Mount Kenya (Young 1994; Young and Smith 1994).

Cattle do not eat all rosette species present in Piedras Blancas. Animals

Fig. 6.3A: A 170-centimeter-tall pristine specimen of *C. timotensis* rosette with a well-preserved frill of dry marcescent leaves. December 1989.

Fig. 6.3B: A 195-centimeter-tall (excluding inflorescence) specimen of *C. timotensis* with a heavily damaged sheath of dry leaves. Note that the stems of two smaller rosettes on the lower right have also been damaged by cattle. December 1987.

restrict their feeding to *C. timotensis,* probably because of the unpalatability or high leaf pubescence of other species (Pérez 1992a). Rosette browsing shows a distinct spatial distribution. Plants on valley floors and near creeks are preferentially browsed; those on steep slopes usually escape predation. Some populations of *C. timotensis* along creeks have 30% to 55% of their plants browsed, with a resulting loss of plant cover of about 40%. Animals concentrate their feeding on plants between 60 centimeters and 120 centimeters in height. The upper limit is clearly imposed by cow stature, but it is not clear why cattle would not eat rosette seedlings. Perhaps the more pubescent seedlings are less palatable than adult plants.

The available data suggest that cattle feed on rosettes only when more palatable páramo plants become scarce, either permanently or on a seasonal basis. The fact that cattle eat a substantial amount of rosette foliage in Piedras Blancas during at least the dry season suggests that the ground cover there is overgrazed. Prolonged grazing in other Venezuelan páramos has resulted in changes in floristic composition and in an overall reduction of plant cover (Ewel et al. 1976). Dry páramos tend to be more adversely affected than moist ones. Overgrazing at high altitudes may help to create and maintain large bare patches (Smith 1974), as well as increase erosion on steep slopes (Flores 1979). It may also facilitate the extension of plant associations that are better able to withstand grazing (Cleef et al. 1983) and, in places with a long history of grazing, lead to plant extinction and the nearly complete disappearance of the native vegetation (Grubb 1970; Vargas and Rivera 1990).

Cattle cause other damage to tall rosettes. Stems of plants on locations not frequented by cattle retain a broad sheath of dry marcescent leaves (figure 6.3A); rosettes on areas heavily visited by cows show severely damaged trunks, which have lost this protective cover below a height of about 1 meter (figure 6.3B). This occurs because animals kick the plants and rub or scratch against them while feeding. I found a volumetric frill reduction of about 60% in a rosette population along a creek frequented by cows. This problem has also been reported in the Colombian páramos (Hofstede 1995; Verweij and Kok 1992). Trunk damage would diminish the beneficial effects of the sheaths (Smith 1979). The dead frill insulates the trunk against nightly freezing, reduces evaporative losses from stem pith tissue, collects dew and precipitation, and stores nutrients which are later released to the soil (Pérez 1992c). Field experiments have indeed shown that complete frill removal often results in rosette death (Smith 1979; Young and Smith 1994).

Cattle feeding might also have some indirect effects on the geomorphology and soils of the high páramo. A reduction of vegetation cover in grazed plots results in greater needle ice incidence and a consequent rise in soil erosion (Mahaney 1986; Pérez 1989, 1992d). Higher needle ice frequency also prevents the establishment of seedlings, which are easily heaved and killed by such frost activity (Pérez 1987b). In general, the vulnerability of páramo ecosystems increases with greater altitude because higher-elevation environments are less able to recover following animal disturbance.

Several direct effects of cattle on páramo soils are also of great concern. Cattle, after repeatedly kicking the sward, can rupture the ground cover. Animals break and dislodge chunks of turf and soil to eat plants or even soil (Mahaney et al. 1996). Soils exposed by cattle have high silt fractions and are especially prone to needle ice formation, which can enlarge the incipient depressions. Soil removal is also accomplished by runoff and wind. Animals may also speed up this process of soil erosion, known as "turf exfoliation" (Pérez 1992b; Troll 1973). Terrace edges offer easier access to plants already partially detached from the sward; cows kick the edge and lip of soil terracettes to dislodge plants. When standing on a terrace next to the edge, animal weight may also result in sudden collapse of segments of the soil/turf terrace up to 125 centimeters in length. Turf exfoliation may eventually remove all peaty layers above the coarse glaciofluvial C horizons, which then form a rubble pavement (Demangeot 1951; King 1971). Overall erosion by cattle in Piedras Blancas may be relatively small, but because it is concentrated in valley floor ecosystems, substantial damage can result. Similar cattle terraces and depressions have been found in other high Andean locations, including Peru (Hastenrath 1977), Argentina (Cabido and Acosta 1986), and Colombia (Vargas and Rivera 1990). Cattle overgrazing has also been linked to turf exfoliation in alpine areas of Africa (Hastenrath 1978; Hastenrath and Wilkinson 1973) and Asia (Mahaney and Linyuan 1991; Watanabe 1994).

Natural or human-caused fires are significant ecological factors in the páramos of Costa Rica (Janzen 1973; Horn 1989a, 1989b, this volume), Colombia (Fosberg 1944; Guhl 1968; Hofstede 1995; Lozano and Schnetter 1976; Vargas and Rivera 1990; Verweij and Kok 1992), and Ecuador (Lægaard 1992). Fire is also important in the comparable Afroalpine vegetation of Mount Kilimanjaro and Mount Kenya (Beck et al. 1983, 1986b; Hedberg 1964). Reports on fire in the Venezuelan páramo date back to 1891 when Goebel noted that some páramos used for grazing were burned late in the dry season. Drude (1890) also commented on the ease with which *Espeletia* rosettes could be set on fire due to their high resin content.

In Venezuela, however, fire is important mainly in the grassy páramo (*pajonal paramero*) common in the Sierra Nevada de Mérida, but essentially absent from the drier Sierra de La Culata (Monasterio 1980b). Vareschi (in Goebel 1891 [1975]: 385) mentioned a large fire in 1954 near Pico

Humboldt that devastated a sizable páramo patch at 4300 meters. When I visited this area in 1977 it had a dense grass cover and very few rosettes. Fire is also effective in the lower-altitude, densely vegetated páramos. Smith (1981) reported on a fire at 3200 meters that killed 55% of *E. schultzii* rosettes. Even higher rates of rosette mortality (> 80%) were found in *E. hartwegiana* after a fire at 4000 meters in Colombia (Verweij and Kok 1992). Cuatrecasas (1979b) believed that *Espeletia* should be adapted to periodic fires, but several reports imply that high fire frequencies destroy rosette communities and deflect páramo succession towards a grass-dominated landscape (Fosberg 1944: Lægaard 1992; Vargas and Rivera 1990; Verweij and Kok 1992).

I have never seen a single fire or signs of one in Piedras Blancas. Plant cover there is simply too open to convey fires for any distance, if at all. Hedberg (1964) and Coe (1967) also noted that the alpine vegetation of Mount Kenya and Mount Kilimanjaro is too sparse to spread any sizable fires. In 1982 I witnessed some young peasants set several tall rosettes on fire, seemingly for amusement. The dry leaf frills were very dry and burned easily. Although the plants were consumed by the fire, it did not spread to the adjacent vegetation. A similar instance of Indians setting gigantic *Puya raimondii* bromeliads on fire in the puna is cited by Budowski (1968). In conclusion, sizable fires seem to have been absent from Piedras Blancas, and it is likely that only the denser rosette/shrub associations along the lower fringe of this páramo will burn during medium-scale fire events.

Impact on Cryptogamic Crusts

The most significant threat to cryptogamic crusts in the páramo comes from off-road vehicles (ORVs). Vehicular access to Piedras Blancas was not possible until the late 1960s, when a dirt road stemming from the Panamerican Highway at Casa de Gobierno (map 6.1) was built. This road—apparently the highest in Venezuela—reaches up to 4440 meters (Schubert 1975). At first, this road was travelled by peasants commuting between high-altitude towns. Vehicular traffic, mainly for recreational purposes, increased rapidly during the early 1980s (Pérez 1991a). Most visitors head toward a group of small glacial tarns at 4450 meters. Vehicles leave the dirt road and cross the open páramo to reach other scenic spots. A myriad of vehicle tracks have visibly increased in recent years and criss-cross the landscape (figure 6.4). Some jeeps even venture up the steepest slopes; their tracks can be found up to the mountain crest at more than 4500 meters. The effects of ORVs have been studied in many tundra areas (Bellamy et al. 1971; Gersper and Challinor 1975), but research in mountains has focused on pedestrian or horse traffic along trails in temperate

Fig. 6.4. Off-road vehicle (jeep) tracks at 4380 meters. Tracks are occupied by light-colored pebbles and by white, vagrant lichens (*Thamnolia vermicularis*). Note the lack of vascular vegetation in this area. December 1985.

(Bell and Bliss 1973; Bryan 1977; Willard and Marr 1970) and tropical areas (Mahaney 1979, 1986).

ORVs damage páramo soils and cryptogams. Damage to substrate was studied in shallow lithic cryochrepts developed on weathered mica schist in an area where vehicles have produced tracks up to 16 centimeters deep (Pérez 1991a). Soils in the tracks had a three-centimeter-thick A11 (nubbin) horizon, which was three times thicker in the adjacent undisturbed soil. The A12 horizons were about 50% thicker in undisturbed areas. The lower AC and C horizons were similar in every respect, indicating that disturbance is restricted to upper horizons. The pristine soil had a total solum thickness of 25 centimeters, but the profile in the track area was only 16 centimeters deep. The drop in soil thickness represents a substantial loss of about 36% of the soil cover. The studied site was a gentle dry slope; negative effects of ORVs are even more pronounced in low-lying wet areas, such as tarn edges and near creeks, where tracks up to 50 centimeters in width and 45 centimeters in depth are common (Pérez 1991a). Moist soils are easily compressed and intensely affected by ORVs (Willard and Marr 1970). Vehicles cause soil damage by compression and subsequent erosion. The drop in thickness of A11 horizons is caused by both processes, but reduction of the A12 layer is caused only by compaction (Davidson and Fox 1974; Gersper and Challinor 1975; Pérez 1991a).

Tire tracks generate several runoff-related problems. They naturally channel water which erodes them even further. Extreme compaction restricts needle ice growth (Pérez 1986); less needle ice results in a lower infiltration rate and thus greater runoff. Pebble accumulation on tracks—effected by needle ice formation on the elevated areas outside—creates a largely impervious surface that generates even greater runoff (Yair and Klein 1973).

The impact of ORVs on páramo vascular vegetation is not known, but the ground cover and the low-stature plants appear moderately to severely damaged in areas with heavy vehicular traffic. Microbiotic crusts and the rare communities of erratic cryptogams in the periglacial desert are being rapidly destroyed by a recent influx of motorcycles, which can reach even higher altitudes than four-wheel-drive vehicles. It is not uncommon to see groups of 12 to 20 motorcycles racing across the páramo, which is used for unrestricted "motor-cross" activities. I have found bike tracks up to nearly 4600 meters where the fragile cryptogamic communities had, until recently, not been threatened.

Andean vagrant and cryptogamic crust communities have only lately been discovered and described (Pérez 1991b, 1994, 1996). Erratic plants are also found in the East African highlands (Beck et al. 1986a; Hedberg 1964). Although cryptogamic crusts are well known from many lowland deserts (West 1990), they had never been reported for high-mountain landscapes. Microbiotic páramo crusts are composed of a diverse array of lichens (*Catapyrenium lachneum, Stereocaulon congestum, Xanthoparmelia vagans*), hepatics (*Marsupella* species), mosses (*Grimmia longirostris*), and, apparently, algae (Pérez 1996). Crusts often form discontinuous, garland-shaped patches that cover about 50% of the ground surface. On some high-altitude plateaus they form small, rounded, evenly-spaced, patches 3 to 12 centimeters in diameter, associated with a pronounced soil microtopography of 5- to 6-centimeter-tall "buds" formed by ground freezing (Pérez 1996).

It is unfortunate that these crusts are being destroyed before we have fully researched their ecological roles in the Andes. The crusts produce several beneficial effects on soil (Pérez 1996, 1997). The tiny plants increase organic matter in the soil, and presumably nutrient content (Pérez 1992c), from an average of 3% in bare areas to 7.2% in crusted zones. Soils in crust areas have greater contents of silt and clay (31%) than adjacent naked soils (about 20%) because the uneven surface of cryptogamic plants traps wind-blown fine particles (Danin and Yaalon 1981). A crust also prevents removal of fine grains by aeolian activity. The above trends produce a greater water storage capacity (mean: 70%) than that of bare soils (31%). This might be significant for other plants, but is certainly important for runoff generation in the high páramo. Presence of a crust doubles infiltration rates from an average of 37.1 millimeters/minute to 69.8 millimeters/minute in *Marsupella* crusts, and from 53.2 to 119.5 millimeters/minute in *Grimmia*-covered soils (Pérez 1996). Because soils below crusts need more water to reach saturation, a given rain event or period will produce runoff earlier and at greater rates in bare soils than in crust areas. Presence of a cryptogamic crust also increases the ability of the soil to resist grain detachment by raindrop impact, wind, runoff, or needle ice.

Destruction of crusts should result in a reduction of all these beneficial effects (Booth 1941; Mücher et al. 1988). Whatever the actual environmental consequences of ORV traffic are, their use threatens to destroy the unique cryptogamic plant communities of the high páramo.

Conservation Recommendations

The Páramo de Piedras Blancas has undergone distinct stages of increasing human influence and shifting disturbance regimes in the past millennium. During this period, human impact has also gradually extended to the highest mountain landscapes. This study shows that, in contrast with other Andean highlands (Ellenberg 1979; Young, this volume), the high Venezuelan páramos had suffered little human-induced disturbance by the time of European contact. Barely 300 years have elapsed since the Spanish colonization frontier moved up altitudinally to impact, to a lesser or greater degree, all Andean elevational belts (Young, this volume). The last cultural phase undoubtedly has caused the most severe impact to the landscape of Piedras Blancas, yet this region remains unusual by all accounts. It is one of few expanses of high-desert páramo and periglacial desert in the northern Andes and contains a rich and diverse flora, and many interesting periglacial features (Pérez 1992d; Schubert 1975). Serious efforts to preserve this region should be taken, and its recent inclusion in the "Parque Nacional Sierra La Culata" is a promising step in that direction. It is hoped that the *Instituto Nacional de Parques* (INAP) will strive to develop basic conservation measures. Some activities should be curtailed and others should be controlled.

Plant gathering by local peasants still goes on in the park, but this is probably of minor concern for most plants. Some species, however, can be severely affected by harvesting, and collection for medicinal purposes may result in severe reduction of plant density (Alliende and Hoffmann 1983), or even in local extinction (Pérez 1992c; Smith 1981). Cutting live plants and collecting fallen stems for fuel should be prohibited. This can be very damaging to giant rosettes, and has proven particularly deleterious for the vegetation of Mount Kenya National Park (Hedberg 1964; Mahaney 1986).

Cattle herds pose a more serious threat to the páramo. Their effects on vegetation and soil erosion should be closely monitored with permanent exclosures at different elevations. Data on grazing/browsing effects and the number of cattle and horses using this páramo are needed. The frequently browsed vegetation should be periodically sampled, particularly during the dry season. Once some data become available, it may prove necessary to (1) eliminate grazing entirely from the páramo, (2) limit or reduce animal numbers, and/or (3) exclude cattle from certain critical landscape segments such as exfoliating swards, where only fencing will prevent this type of severe erosion.

Recent studies in the páramos of Colombia (Hofstede 1995; Pels and Verweij 1992; Schmidt and Verweij 1992; Verweij 1995) agree that the low productivity and nutritive value of páramo vegetation result in poor cattle quality (Roseveare 1948), and, in order to meet nutritional requirements, the animals have to make a considerable effort by grazing long periods over large areas. Given this, the soundness and logic of maintaining grazing in the higher páramos must be questioned, because even at high stocking densities they cannot provide an economically sustainable basis for Andean peasants. The inherent fragility of high-altitude ecosystems, and their limited ability to recover following the disturbances described here, argue against their continued use. Moderate grazing, however, might be acceptable for the low páramo (Hofstede 1995). The issue of grazing exclusion is important, and should be seriously examined by the INAP.

In contrast to cattle, which provide some sustenance to local peasants, ORVs have no positive effects whatsoever on the páramo landscape. ORVs pose the most serious and rapidly growing threat to the páramo ecosystem, but this could be easily reduced with a few simple measures. Access to this páramo is restricted to the road stemming from the highway in the narrow lower Mifafí Valley. The INAP could install an entrance checkpoint permanently manned by rangers. The public should be educated about the problems created by ORVs, and be discouraged from leaving the road at any time. If the road were extended to allow access to the lagoons, the need for crossing the open páramo would be largely removed. In addition, motorcycles have no place in a protected ecosystem. They should be allowed only on roads, or perhaps prohibited from the páramo altogether, since they cause most of the destruction of cryptogamic crusts. Patrolling park rangers would probably reduce ORV excesses.

Conclusion: Instability, Uncertainty, and Prospects for Páramo Conservation

Several chapters in this volume have emphasized the theme of "conservation-with-development" and argue that sustainable development need not conflict with conservation. Is this a truly viable option for the fragile ecosystems of the high páramo? An acceptable answer needs to include at least three elements: (1) a detailed evaluation of the nature and degree of the landscape modifications brought about by continued human use; (2) the recognition of the existence of obviously conflicting activities possibly harmful to the páramo; and (3) a consideration of the expectations of visitors to the páramo. This chapter offers some basic data regarding the first point. Hofstede (1995: 150) addresses the second issue in a recent essay on páramo conservation, in which he notes that

. . . It is not controversial to state that the ideal management regime for the páramo is one that does not have negative consequences for any function of the ecosystem. However, the demands that every single function places on the ecosystem might be conflicting and therefore this ideal management probably does not exist.

While these ideas may clearly be pertinent for *any* protected area, they seem particularly relevant for Piedras Blancas, where serious, unavoidable conflicts exist between long-term conservation of the biogeographical landscape and its unrestricted, uncontrolled human use.

Horn (this volume) cogently contrasts the pros and cons of different management plans for a Costa Rican páramo. To echo her ideas, when weighing conservation options for Piedras Blancas, we should "compare visions" of future páramos. If no action is taken, a very unattractive landscape crisscrossed by ORV tracks, affected by severe soil erosion, denuded of vegetation, populated by herds of imported cattle, and used by noisy motorized vehicles will result. A second alternative ("compromise" option) is to try to combine uses, restricting some and controlling others. It has yet to be proven if a compromise will work, but I offer the above conservation recommendations with the assumption and hope that *some* protection measures will eventually be adopted. This option would need to be based on possibly conflicting visitor expectations. The third possibility is to create a wilderness landscape where conditions still resemble those that the first Spaniards entering the páramo in 1559 might have encountered. This last option would be diametrically opposed to the first one, resulting in a páramo where people have the opportunity to enjoy the "wilderness experience" (Hofstede 1995) or recapture a "sense of wilderness" (Botkin 1990).

Which of the three end points is more acceptable? Management of some sort would prevent the undesirable effects of no action, which entails the highest degree of ecological risk. Managing the landscape for conservation undoubtedly involves significant uncertainty, but the uncertainty of managing for a balanced ecosystem seems vastly preferable to the certainty of otherwise witnessing an environmental disaster. Devising viable management plans poses a difficult challenge to government officials. Many uncertainties regarding management arise, mainly from lack of data on how the ecological processes of the páramo acted in pre-European times, how they evolved under anthropogenic pressure during the colonial period, and how they might operate in the future. Considering that the risks in conservation are minimal, and those of laissez-faire are unacceptably high, some type of action is certainly preferable to inaction. It is also obvious that managing Piedras Blancas to allow grazing and ORV access is bound to interfere with the goals of maintaining its landscape in a natural, pristine condition, and these activities may negatively impact the visit of hikers seeking a wilderness experience (Hofstede 1995). When confronted by the high risk and uncertainty that go with human use, it surely seems more prudent to err on the side of too much conservation and consider that overt human disturbance may not have a place in the high-Andean landscape.

Ultimately, any conservation measures that are chosen will have to include both the realization that some kind of human or animal impact will always be present (Botkin 1990; Sauer 1956), and that these páramo ecosystems have been already affected by humans and cannot be considered 'natural' anymore (Vale 1987). The minimum goal should be to control and minimize the negative effects of human disturbance. As in other páramos, conservation through utilization may be the only realistic solution to the problems that affect this unique Andean region (Hofstede 1995).

Acknowledgments

Since 1980, my Andean research has been funded by the Tinker Foundation via the Center for Latin American Studies and the Graduate Opportunity Fellowship program at the University of California at Berkeley. Support was also provided by the Andrew W. Mellon Foundation via the Institute of Latin American Studies and the Research Institute at the University of Texas at Austin. I thank my field assistants E. F. Pérez, J. M. Pérez, M. Schnitzer, O. Vera, H. Hoenicka, and A. Lucchetti for their valuable help. My parents provided logistical support and suitable vehicles to reach the páramo. I. L. Bergquist, R. Clark, J. L. Luteyn, and W. Wyckoff read earlier versions of this manuscript. I thank the Andean peasants for their kindness, and for tolerating my presence in their beautiful páramos for so many years.

References

Acosta, M. 1954. *Estudios de Etnología Antigua de Venezuela*. Caracas: Universidad Central de Venezuela, Instituto de Antropología y Geografía.

Alliende, M. C., and A. J. Hoffmann. 1983. *Laretia acaulis,* a Cushion Plant of the Andes: Ethnobotanical Aspects and Impact of Its Harvesting. *Mountain Research and Development* 3: 45–51.

Aristeguieta, L. 1964. *Flora de Venezuela, Volumen X, Compositae.* 2 volumes. Caracas: Instituto Botánico, Dirección de Recursos Naturales Renovables, MAC.

Balslev, H., and J. L. Luteyn. 1992. *Páramo: An Andean Ecosystem under Human Influence.* London: Academic Press.

Bauman, J., and L. Young. 1986. *Guia de Venezuela,* 2nd edition. Caracas: Editorial Armitano.

Beck, E., K. Mägdefrau, and M. Senser. 1986a. Globular Mosses. *Flora* 178: 73–83.

Beck, E., R. Scheibe, and E. D. Schulze. 1986b. Recovery from Fire: Observations in the Alpine Vegetation of Western Mt. Kilimanjaro (Tanzania). *Phytocoenologia* 14: 55–77.

Beck, E., R. Scheibe, and M. Senser. 1983. The Vegetation of the Shira Plateau and the Western Slopes of Kibo (Mt. Kilimanjaro, Tanzania). *Phytocoenologia* 11: 1–30.

Bell, K. L., and L. C. Bliss. 1973. Alpine Disturbance Studies: Olympic National Park, USA. *Biological Conservation* 5: 25–32.

Bellamy, D., J. Radforth, and N. W. Radforth. 1971. Terrain, Traffic and Tundra. *Nature* 231: 429–432.

Booth, W. E. 1941. Algae as Pioneers in Plant Succession and Their Importance in Erosion Control. *Ecology* 22: 38–46.

Botkin, D. B. 1990. *Discordant Harmonies: A New Ecology for the Twenty-first Century.* New York: Oxford University Press.

Brito, F. 1962. *Población y Economía en el Pasado Indígena Venezolano.* Caracas: Editorial Remar.

Brito, F. 1972. *Historia Económica y Social de Venezuela.* Havana: Instituto Cubano del Libro, Ediciones Ciencias Sociales.

Bryan, R. B. 1977. The Influence of Soil Properties on Degradation of Mountain Hiking Trails at Grövelsjön. *Geografiska Annaler* 59A: 49–65.

Budowski, G. 1968. La influencia humana en la vegetación natural de montañas tropicales americanas. *Colloquium Geographicum* 9: pp. 157–162.

Burguera, M. 1982. *Historia del Estado Mérida.* Caracas: Ediciones Presidencia de la Republica.

Cabido, M., and A. Acosta. 1986. Variabilidad florística a lo largo de un gradiente de degradación en céspedes de la Pampa de Achala, Sierras de Córdoba, Argentina. *Documents Phytosociologiques* 10: 289–304.

Cardozo, A. 1965. *Proceso de la Historia de los Andes.* Caracas: Biblioteca de Autores y Temas Tachirenses.

Cleef, A. M., O. Rangel, and S. Salamanca. 1983. Reconocimiento de la vegetación de la parte alta del transecto Parque Los Nevados. In *La Cordillera Central*

Colombiana, Transecto Parque Los Nevados. Introducción y Datos Iniciales, eds. T. Van der Hammen, A. Pérez, and P. Pinto, pp. 150–173. Vaduz: J. Cramer.

Coe, M. J. 1967. *Ecology of the Alpine Zone of Mount Kenya.* The Hague: W. Junk Publishers.

Crist, R. E. 1943. Wheat Raising in the Venezuelan Andes. *Scientific Monthly* 56: 332–338.

Cuatrecasas, J. 1968. Páramo Vegetation and Its Life Forms. *Colloquium Geographicum* 9: pp. 163–186.

Cuatrecasas, J. 1979a. Growth Forms of the Espeletiinae and Their Correlation to Vegetation Types of the High Tropical Andes. In *Tropical Botany,* eds. M. L. Larsen and L. B. Holm-Nielsen, pp. 397–410. London: Academic Press.

Cuatrecasas, J. 1979b. Comparación fitogeográfica de páramos entre varias cordilleras. In *El Medio Ambiente Páramo,* ed. M. L. Salgado-Labouriau, pp. 89–99. Caracas: Instituto Venezolano de Investigaciones Científicas.

Cuatrecasas, J., and A. Torres. 1988. *Páramos.* Bogotá: Villegas Editores.

Danin, A., and D. H. Yaalon. 1981. Trapping of Silt and Clay by Lichens and Bryophytes in the Desert of the Dead Sea Region. *International Conference on Aridic Soils, Jerusalem, Israel:* p. 29.

Davidson, E., and M. Fox. 1974. Effects of Off-road Motorcycle Activity on Mojave Desert Vegetation and Soil. *Madroño* 22: 381–412.

Demangeot, J. 1951. Observations sur les "sols en gradins" de L'Apennin central. *Revue de Géomorphologie Dynamique* 2: 110–119.

Drude, O. 1890. *Handbuch der Pflanzengeographie.* Stuttgart: J. Engelhorn.

Ellenberg, H. 1979. Man's Influence on Tropical Mountain Ecosystems in South America. *Journal of Ecology* 67: 401–416.

Ewel, J. J., A. Madriz, and J. A. Tosi. 1976. *Zonas de Vida de Venezuela,* 2nd edition. Caracas: Ministerio de Agricultura y Cria.

Flores, J. 1979. Desarrollo de las culturas humanas en las altas montañas tropicales (estrategias adaptativas). In *El Medio Ambiente Páramo,* ed. M. L. Salgado-Labouriau, pp. 225–234. Caracas: Instituto Venezolano de Investigaciones Científicas.

Fosberg, F. R. 1944. El Páramo de Sumapaz, Colombia. *Journal of the New York Botanical Garden* 45: 226–234.

Gade, D. W. 1975. Plants, Man and Land in the Vilcanota Valley of Peru. *Biogeographica* 6: 1–240.

Gersper, P. L., and J. L. Challinor. 1975. Vehicle Perturbation upon a Tundra Soil-Plant System: I. Effects on Morphological and Physical Environmental Properties of the Soils. *Soil Science Society of America Proceedings* 39: 737–744.

Goebel, K. 1891 [1975]. La vegetación de los páramos Venezolanos (trans. of *Die Vegetation der Venezolanischen Paramos,* Marburg, [1891]). *Acta Botánica Venezuelica* 10: 337–395.

Grubb, P. J. 1970. The Impact of Man on the Páramo of Cerro Antisana, Ecuador. *Journal of Applied Ecology* 7: 7p–8p.

Guhl, E. 1968. Los páramos circundantes de la Sabana de Bogotá, su ecología y su importancia para el régimen hidrológico de la misma. *Colloquium Geographicum* 9: pp. 195–212.

Gumilla, J. 1741. *El Orinoco Ilustrado: Historia Natural, Civil y Geográfica de Este Gran Río.* Madrid.

Hastenrath, S. 1977. Observations on Soil Frost Phenomena in the Peruvian Andes. *Zeitschrift für Geomorphologie, N. F.* 21: 357–362.

Hastenrath, S. 1978. On the Three-Dimensional Distribution of Subnival Soil Patterns in the High Mountains of East Africa. *Erdwissenschaftliche Forschung* 11: 458–481.

Hastenrath, S., and J. Wilkinson. 1973. A Contribution to the Periglacial Morphology of Lesotho, Southern Africa. *Biuletyn Peryglacjalny* 22: 157–167.

Hedberg, O. 1964. Features of Afro-alpine Vegetation Ecology. *Acta Phytogeographica Suecica* 49: 1–144.

Hodge, W. H. 1960. Yareta: Fuel Umbellifer of the Andean Puna. *Economic Botany* 14: 113–118.

Hofstede, R. 1995. *Effects of Burning and Grazing on a Colombian Páramo Ecosystem.* Amsterdam: University of Amsterdam, Hugo de Vries Laboratory.

Horn, S. P. 1989a. The Inter-American Highway and Human Disturbance of Páramo Vegetation in Costa Rica. *Yearbook, Conference of Latin Americanist Geographers* 15: 13–22.

Horn, S. P. 1989b. Postfire Vegetation Development in the Costa Rican Páramos. *Madroño* 36: 93–114.

Horn, S. P. 1998. Fire Management and Natural Landscapes in the Chirripó Páramo, Chirripó National Park, Costa Rica. In *Nature's Geography: New Lessons for Conservation in Developing Countries,* eds. K. S. Zimmerer and K. R. Young, pp. 125–46. Madison: University of Wisconsin Press.

Jahn, A. 1927. *Los Aborígenes del Occidente de Venezuela. Su Historia, Etnografía y Afinidades Lingüísticas.* Caracas: Litografía del Comercio.

Janzen, D. H. 1973. Rate of Regeneration After a Tropical High Elevation Fire. *Biotropica* 5: 117–122.

King, R. B. 1971. Vegetation Destruction in the Sub-alpine and Alpine Zones of the Cairngorn Mountains. *Scottish Geographical Magazine* 87: 103–115.

Lares, J. I. 1907. *Etnografía del Estado Mérida.* Mérida: Imprenta del Estado Mérida.

Lægaard, S. 1992. Influence of Fire in the Grass páramo Vegetation of Ecuador. In *Páramo: An Andean Ecosystem under Human Influence,* eds. H. Balslev and J. L. Luteyn, pp. 151–170. London: Academic Press.

Levi-Strauss, C. 1950. The Use of Wild Plants in Tropical South America. In *Handbook of South American Indians,* volume 6, pp. 465–486. Washington, D. C.: Smithsonian Institution.

Lozano, G., and R. Schnetter. 1976. Estudios ecológicos en el Páramo de Cruz Verde, Colombia. II. Las comunidades vegetales. *Caldasia* 11: 53–68.

Luteyn, J. L. 1992. Páramos: Why Study Them? In *Páramo: An Andean Ecosystem under Human Influence,* eds. H. Balslev and J. L. Luteyn, pp. 1–14. London: Academic Press.

Mahaney, W. C. 1979. Soil Erosion Along the Naro Moru Track, a Western Approach to Mount Kenya. *East African Agriculture and Forestry Journal* 45: 158–166.

Mahaney, W. C. 1986. Environmental Impact in the Afroalpine and Subalpine Belts of Mount Kenya, East Africa. *Mountain Research and Development* 6: 247–260.

Mahaney, W. C., M. Bezada, R. G. V. Hancock, S. Aufreiter, and F. L. Pérez. 1996. Geophagy of Holstein Hybrid Cattle in the Northern Andes, Venezuela. *Mountain Research and Development* 16: 177–180.

Mahaney, W. C., and Z. Linyuan. 1991. Removal of Local Alpine Vegetation and Overgrazing in the Dalijia Mountains, Northwestern China. *Mountain Research and Development* 11: 165–167.

Metraux, A., and P. Kirchhoff. 1948. The Northeastern Extension of Andean Culture. In *Handbook of South American Indians,* volume 4, pp. 349–368. Washington, D. C.: Smithsonian Institution.

Monasterio, M. 1980a. Poblamiento humano y uso de la tierra en los altos Andes de Venezuela. In *Estudios Ecológicos en los Páramos Andinos,* ed. M. Monasterio, pp. 170–198. Mérida: Universidad de Los Andes.

Monasterio, M. 1980b. Las formaciones vegetales de los páramos de Venezuela. In *Estudios Ecológicos en los Páramos Andinos,* ed. M. Monasterio, pp. 93–158. Mérida: Universidad de Los Andes.

Morón G. 1961. *Historia de Venezuela.* 3rd edition. Caracas: Editorial Guadarrama.

Mücher, H. J., C. J. Chartres, D. J. Tongway, and R. S. B. Greene. 1988. Micromorphology and Significance of the Surface Crusts of Soils in Rangelands Near Cobar, Australia. *Geoderma* 42: 227–244.

Parsons, J. J. 1982. The Northern Andean Environment. *Mountain Research and Development* 3: 253–262.

Pels, B., and P. A. Verweij. 1992. Burning and Grazing in a Bunchgrass Páramo Ecosystem: Vegetation Dynamics Described by a Transition model. In *Páramo: An Andean Ecosystem under Human Influence,* eds. H. Balslev and J. L. Luteyn, pp. 243–263. London: Academic Press.

Pérez, F. L. 1986. The Effect of Compaction on Soil Disturbance by Needle Ice Growth. *Acta Geocriogénica* 4: 111–119.

Pérez, F. L. 1987a. Downslope Stone Transport by Needle Ice in a High Andean Area (Venezuela). *Revue de Géomorphologie Dynamique:* 36: 33–51.

Pérez, F. L. 1987b. Needle-Ice Activity and the Distribution of Stem-Rosette Species in a Venezuelan Páramo. *Arctic and Alpine Research* 19: 135–153.

Pérez, F. L. 1989. Some Effects of Giant Andean Stem-Rosettes on Ground Microclimate, and Their Ecological Significance. *International Journal of Biometeorology* 33: 131–135.

Pérez, F. L. 1991a. Particle Sorting Due to Off-Road Vehicle Traffic in a High Andean Páramo.*Catena* 18: 239–254.

Pérez, F. L. 1991b. Ecology and Morphology of Globular Mosses of *Grimmia longirostris* in the Páramo de Piedras Blancas, Venezuelan Andes. *Arctic and Alpine Research* 23: 133–148.

Pérez, F. L. 1992a. The Ecological Impact of Cattle on Caulescent Andean Rosettes in a High Venezuelan Páramo. *Mountain Research and Development* 12: 29–46.

Pérez, F. L. 1992b. Processes of Turf Exfoliation (Rasenabschälung) in the High Venezuelan Andes. *Zeitschrift für Geomorphologie, N. F.* 36: 81–106.

Pérez, F. L. 1992c. The Influence of Organic Matter Addition by Caulescent Andean Rosettes on Surficial Soil Properties. *Geoderma* 54: 151–171.

Pérez, F. L. 1992d. Miniature Sorted Stripes in the Páramo de Piedras Blancas (Venezuelan Andes). In *Periglacial Geomorphology,* eds. J. C. Dixon and A. D. Abrahams, pp. 125–157. London: Wiley and Sons.

Pérez, F. L. 1993. Turf Destruction by Cattle in the High Equatorial Andes. *Mountain Research and Development* 13: 107–110.

Pérez, F. L. 1994. Vagant Cryptogams in a Páramo of the High Venezuelan Andes. *Flora* 189: 263–276.

Pérez, F. L. 1996. Cryptogamic Soil Buds in the Equatorial Andes of Venezuela. *Permafrost and Periglacial Processes* 7: 229–255.

Pérez, F. L. 1997. Microbiotic Crusts in the High Equatorial Andes, and Their Influence on Páramo Soils. *Catena* 31: 173–198.

Pittier, H. 1926. *Manual de las Plantas Usuales de Venezuela.* Caracas: Litografía del Comercio.

Pittier, H. 1939. *Suplemento a las Plantas Usuales de Venezuela.* Caracas: Editorial Elite.

Ralph, C. P. 1978. Observations on *Azorella compacta* (Umbelliferae), a Tropical Andean Cushion Plant. *Biotropica* 10: 62–67.

Ramón y Rivera, L. F., and I. Aretz. 1963. *Folklore Tachirense, Tomo II, Volumen 3.* Caracas: Biblioteca de Autores y Temas Tachirenses.

Roseveare, G. M. 1948. The Grasslands of Latin America. *Imperial Bureau of Pastures and Field Crops Bulletin, Aberystwyth* 36: 1–291.

Rouse, J. E. 1977. *The Criollo: Spanish Cattle in the Americas.* Norman, Oklahoma: University of Oklahoma Press.

Salas, J. C. 1956. *Etnografía de Venezuela (Estados Mérida, Trujillo y Táchira). Los Aborígenes de la Cordillera de Los Andes.* Mérida: Universidad de Los Andes.

Salas, J. C. 1971. *Tierra Firme (Venezuela y Colombia): Estudios sobre Etnología e Historia.* Mérida: Editorial Paz y Progreso.

Salgado-Labouriau, M. L. 1975. Compositae versus Gramineae in Pollen Analysis. *The Palaeobotanist* 25: 439–447.

Salgado-Labouriau, M. L. 1979. Modern Pollen Deposition in the Venezuelan Andes. *Grana* 18: 53–68.

Salgado-Labouriau, M. L., and C. Schubert. 1977. Pollen Analysis of a Peat Bog from Laguna Victoria (Venezuelan Andes). *Acta Científica Venezolana* 28: 328–332.

Sauer, C. O. 1956. The Agency of Man on the Earth. In *Man's Role in Changing the Face of the Earth,* volume 1, ed. W. L. Thomas, pp. 49–69. Chicago: University of Chicago Press.

Sauer, J. D. 1988. *Plant Migration: The Dynamics of Geographic Patterning in Seed Plant Species.* Berkeley: University of California Press.

Schmidt, A. M., and P. A. Verweij. 1992. Forage Intake and Secondary Production in Extensive Livestock Systems in Páramo. In *Páramo: An Andean Ecosystem under Human Influence,* eds. H. Balslév and J. L. Luteyn, pp. 197–210. London: Academic Press.

Schnee, L. 1973. *Plantas Comunes de Venezuela.* 2nd edition. Maracay: Facultad de Agronomía, Universidad Central de Venezuela.

Schubert, C. 1975. Glaciation and Periglacial Morphology in the Northwestern Venezuelan Andes. *Eiszeitalter und Gegenwart* 26: 196–211.

Smith, A. P. 1974. Population Dynamics and Life Form of *Espeletia* in the Venezuelan Andes. Ph. D. dissertation, Duke University.

Smith, A. P. 1979. Function of Dead Leaves in *Espeletia schultzii* (Compositae), an Andean Caulescent Rosette Species. *Biotropica* 11: 43–47.

Smith, A. P. 1981. Growth and Population Dynamics of *Espeletia* (Compositae) in the Venezuelan Andes. *Smithsonian Contributions to Botany* 48: 1–45.

Smith, A. P., and T. P. Young. 1987. Tropical Alpine Plant Ecology. *Annual Review of Ecology and Systematics* 18: 137–158.

Soil Survey Staff. 1994. *Keys to Soil Taxonomy.* 6th edition. Washington, D. C.: U.S. Department of Agriculture, Soil Conservation Service.

Troconis, L. 1962. *La Cuestión Agraria en la Historia Nacional.* Caracas: Biblioteca Nacional de Autores y Temas Tachirenses.

Troll, K. 1973. Rasenabschälung (Turf Exfoliation) als periglaziales Phänomen der subpolaren Zonen und der Hochgebirge. *Zeitschrift für Geomorphologie, Supplement* 17: 1–32.

Vale, T. R. 1987. Vegetation Change and Park Purposes in the High Elevations of Yosemite National Park, California. *Annals of the Association of American Geographers* 77: 1–18.

Vareschi, V. 1970. *Flora de los Páramos de Venezuela.* Mérida: Ediciones del Rectorado, Universidad de Los Andes.

Vargas, O., and D. Rivera. 1990. El páramo, un ecosistema fragil. *Cuadernos de Agroindustria y Economía Rural, Universidad Javeriana, Bogotá* 25: 145–163.

Verweij, P. A. 1995. *Spatial and Temporal Modelling of Vegetation Patterns. Burning and Grazing in the Paramo of Los Nevados National Park, Colombia.* Enschede: ITC, International Institute for Aerospace Survey and Earth Sciences.

Verweij, P. A., and K. Kok. 1992. Effects of Fire and Grazing on *Espeletia hartwegiana* populations. In *Páramo: An Andean Ecosystem under Human Influence,* eds. H. Balslev and J. L. Luteyn, pp. 215–229. London: Academic Press.

Vila, M. A. 1962. *Geografía de Venezuela.* Caracas: Fundación Eugenio Mendoza.

Vila, M. A. 1976. *Notas Sobre Geoeconomía Prehispánica de Venezuela.* Caracas: Ediciones Facultad de Humanidades y Educación, Universidad Central de Venezuela.

Vila, M. A. 1978. *Antecedentes Coloniales de Centros Poblados de Venezuela.* Caracas: Universidad Central de Venezuela.

Vila, M. A. 1980. *Síntesis Geohistórica de la Economía Colonial de Venezuela.* Caracas: Banco Central de Venezuela.

Vila, M. A. N.d. *Geoeconomía de Venezuela.* Caracas: Corporación Venezolana de Fomento, Editorial Vargas.

Wagner, E. 1967. The Prehistory and Ethnohistory of the Carache Area in Western Venezuela. *Yale University Publications in Anthropology* 71: 1–138.

Wagner, E. 1973. The Mucuchíes Phase: An Extension of the Andean Cultural Pattern into Western Venezuela. *American Anthropologist* 75: 195–213.

Wagner, E. 1978. Los Andes Venezolanos: Arqueología y ecología cultural. *Ibero-Amerikanisches Archiv, N. F.* 4: 81–91.

Wagner, E. 1979. Arqueología de los Andes Venezolanos. In *El Medio Ambiente Páramo*, ed. M. L. Salgado-Labouriau, pp. 207–218. Caracas: Instituto Venezolano de Investigaciones Científicas.

Washburn, A. L. 1980. *Geocryology: A Survey of Periglacial Processes and Environments.* New York: Wiley and Sons.

Watanabe, T. 1994. Soil Erosion on Yak-Grazing Steps in the Langtang Himal, Nepal. *Mountain Research and Development* 14: 171–179.

West, N. E. 1990. Structure and Function of Microphytic Soil Crusts in Wildland Ecosystems of Arid to Semi-arid Regions. *Advances in Ecological Research* 20: 179–223.

Willard, B. E., and J. W. Marr. 1970. Effects of Human Activities on Alpine Tundra Ecosystems in Rocky Mountain National Park, Colorado. *Biological Conservation* 2: 257–265.

Winterhalder, B. P., and R. B. Thomas. 1978. *Geoecology of Southern Highland Peru: A Human Adaptation Perspective.* Occasional Paper 27. Boulder: Institute of Arctic and Alpine Research, University of Colorado.

Yair, A., and M. Klein. 1973. The Influence of Surface Properties on Flow and Erosion Processes on Debris-Covered Slopes in an Arid Area. *Catena* 1: 1–18.

Yocom, C. F. 1967. Ecology of Feral Goats in Haleakala National Park, Maui, Hawaii. *American Midland Naturalist* 77: 418–451.

Young, K. R. 1998. Deforestation in Landscapes with Humid Forests in the Central Andes: Patterns and Processes. In *Nature's Geography: New Lessons for Conservation in Developing Countries,* eds. K. S. Zimmerer and K. R. Young, pp. 75–99. Madison: University of Wisconsin Press.

Young, T. P. 1994. Population Biology of Mount Kenya *Lobelias.* In *Tropical Alpine Environments: Plant Form and Function,* eds. P. W. Rundel, A. P. Smith, and F. C. Meinzer, pp. 251–272. Cambridge: Cambridge University Press.

Young, T. P., and A. P. Smith. 1994. Alpine Herbivory on Mount Kenya. In *Tropical Alpine Environments: Plant Form and Function,* eds. P. W. Rundel, A. P. Smith, and F. C. Meinzer, pp. 319–335. Cambridge: Cambridge University Press.

7

Grazing the Forest, Shaping the Landscape?
Continuing the Debate about Forest Dynamics
in Sagarmatha National Park

Barbara Brower and Ann Dennis

Sagarmatha, the International Park

In 1976 His Majesty's Government of Nepal responded to considerable international pressure and its own national agenda by creating Sagarmatha National Park (SNP), a reserve on Mount Everest's southern flanks. The park encompasses not only the world's highest mountain and its spectacular sister peaks, but also the valleys and interfluves of Khumbu, homeland of the Sherpa people (map 7.1). The agropastoral Sherpa, famous for exploits on Everest, were an integral part of the park design (Blower 1971; Mishra 1973; L. N. Sherpa 1979). They retained claim to private dwellings and fields within its boundaries, but found themselves subject to a variety of regulations brought about by the expectations of the park's new international constituency. The region's forest landscape in particular was considered to be at risk from human use, and activities believed to jeopardize the survival of woodlands, including wood-cutting and livestock grazing (Blower 1971; Lucas et al. 1974; Mather 1973; Mishra 1973; Naylor 1970; Speechly 1976, among others), were curtailed (Garrett 1981).

Management of the park's resources in the years since has continued to be preoccupied with trees (Drew and Sharma 1977; Halkett 1981; Hardie et al. 1987; Ledgard and Baker 1990). Rules promulgated by the park spell

Map 7.1. Sagarmatha National Park in Khumbu, Nepal.

out where village fuel and construction wood may be gathered, restrict the kinds and movements of livestock, and promote the establishment of nurseries and plantations. This arbor-centric management persists despite a growing uncertainty about the processes that shape the area's forest landscape (Brower 1991a; Byers 1987a, 1987b, 1987c; Houston 1982, Ives and Messerli 1989; Thompson and Warburton 1985; Stevens 1993), and in the face of continuing local resentment about the infringement of accustomed access to resources (Brower 1983, 1993; Halkett 1981; Stevens 1993).

Khumbu's forests were initially assumed to be degraded remnants of a

formerly uninterrupted fir-dominated forest landscape covering all slopes
and aspects below about 4100 meters that had been severely diminished
in recent times by the collective assault of local people, their livestock,
and swelling ranks of tourists (Blower 1971; Hardie et al. 1987; Lucas et
al. 1974; MacDonald 1984; Mishra 1973; Naylor 1970). But increasingly
there are questions about the extent of the former forest landscape, re-
cency of deforestation, and role of humans and livestock in the present
patterns of forest cover (Brower 1987, 1991a; Byers 1987a; Dennis and
Brower 1995; Stevens 1993).

In this paper we investigate the ideas and evidence that have accumulated about the dynamics of the forest landscape in Sagarmatha National
Park, and offer additional information from our biogeographical analysis
of stand dynamics in the neighborhood of Khumbu's main settlements.
Our study refutes the conventional wisdom about a recently accelerated
retreat of forest in Sagarmatha National Park, challenges assumptions
about the impact of grazing, and provokes a set of questions about forest
interpretation and management. Our object is to augment the emerging
picture of people/livestock/forest interactions in Sagarmatha National
Park, warn against the use of hasty and unsupported intuition as a guide
to resource management, advocate research approaches grounded in biogeographical concepts, and argue for conservation policy that, in real collaboration with local people, draws on both social and natural science,
privileging neither and recognizing the limitations of both.

The Changing Face of Conservation

The year 1976, when Sagarmatha National Park was established, might
be one of the last moments in recent history during which conservationists
were confident of their goals and means: The function of national parks
was clear, resource managers were comfortable in their expertise, and the
natural world still worked in understood and predictable ways. The last
two decades have seriously undermined that confidence. Nature turns out
to be more complicated, varying, mysterious, and unpredictable than the
science of ecology once taught (Botkin 1990; Worster 1990; Zimmerer
1994). The tools of resource managers don't answer the needs of many of
the world's resource users (Brower 1991b, 1993, Fortmann and Fairfax
1989; Guha 1989). And national parks and reserves, initially derived from
the Yellowstone precedent, finer-tuned through biosphere reserve models,
still fall short as areas to protect the wonders and diversity of nature in a
world full of people.

Until perhaps fifteen years ago the primary concern voiced by international institutions advocating the conservation of nature was with nature

itself. (This generalization obscures an important countercurrent within the environmental movement: A concern for cultures as well as nature was evidenced at least 30 years ago in the Sierra Club's *Almost Ancestors* [Kroeber and Heizer 1968], for example.) As a legacy of that period and preoccupation we have the World Conservation Strategy, a biosphere reserve system, and a proliferation of national parks and protected areas. Concern with biotic resources is still high. But conservation is increasingly recognized as being a more complicated business. Today it is as much a cultural as an ecological issue. Ensuring the protection and wise use of nature has often come to mean acknowledging and somehow accommodating the demands of the people dependent on those resources.

This of course complicates the task of conservation enormously. A national park created by throwing out the natives, as Yosemite was, is a far simpler management proposition than that faced by park-makers in most of the developing world. In 1969, the International Union for the Conservation of Nature, establishing criteria for national parks that could legitimately hold the title, could say: "A National Park is a relatively large area . . . where . . . ecosystems are not materially altered by human exploitation and occupation . . . [and] where the highest competent authority . . . has taken steps to prevent or eliminate as soon as possible exploitation or occupation in the whole area." (IUCN Tenth General Assembly, New Delhi, cited in Allin [1990]). Yet in most parts of the world, areas marked for protection are somebody's homeland. Today, sensitive to the charge of green imperialism, worried about the disappearance of peoples as well as other species, and increasingly conscious of the futility of protecting places without the support of local people, international conservation efforts are slower to cast out resident humans. Inclusion rather than exclusion is increasingly advocated.

We have witnessed something of a sea change: The nature-centered perspective of a generation ago has given ground before a preoccupation with local people today. In the publications (see for example Davidson [1993]), and even the advertisements of the big conservation groups, indigenous people that were once cast as villains have new roles as stewards, environmental sages, and collaborators in conservation. Once to be excluded from parks, local people, particularly the more exotic among them, are targeted as recruits in the new conservation wars. Their participation is held to be a pragmatic and moral necessity, given not only their ineluctable presence and need for resources, but also the uncertain boundary that has developed between endangered species and endangered cultures.

As a changing perspective on the place of people in parks has added complexity to conservation, so has the evolution in ideas about the way the environment works. Models predicated on equilibrium and stasis give ground before increasing evidence for dynamism and chaos (Botkin 1990; Worster 1990). Efforts to reach broadly applicable generalizations lose favor, and the emphasis shifts to case studies and the specific. Yet the changing paradigms of new ecology are slow to dislodge older ideas and ways of doing things (Zimmerer 1994); resource managers, although perhaps uneasy with the shift of ground under their feet, still hang on to the world as they learned it (Brower 1991b, 1993; Fortmann and Fairfax 1989).

The Case of Sagarmatha

The history of environmental research, park planning, and management in Sagarmatha National Park reflects the transition from complacency through uncertainty about both people and environment that has taken place over the last 20 years. Khumbu/Sagarmatha National Park is especially intriguing for a number of reasons. Sagarmatha has been a crucible of experimentation on ways to incorporate local people and institutions into park planning and management. Khumbu is one of several study areas in the Himalayan region where misgivings about conventional explanations for ecological processes first took root among researchers whose findings were at odds with expectations. Uncertainty on this Himalayan scale (Thompson et al. 1986) was of sufficiently high profile that researchers elsewhere took note, beginning to explore anew what had seemed to be settled questions about man's role in changing the environment. And the intensity of attention that has been paid to the place since Nepal opened its borders to the West 55 years ago is perhaps unprecedented. When the Survey of India's technicians identified the peak we call Everest as the earth's highest in 1854, they ensured the perennial interest and attention of hosts of adventurers from all over the world.

Tourists have visited in the tens of thousands, drawn both by mountain spectacle and challenge and by exotic culture. Development specialists have been busy with projects ranging from public health and education through hydroelectricity generation to cultural survival. Scholars, too, have been active in Sagarmatha National Park/Khumbu. Geographers, anthropologists, geologists, foresters, ecologists, biologists, and others have studied and written about the region, producing everything from magazine pieces based on a two-week trek to substantial books grounded in many years' residence. The long-term interest and deepening familiarity of a number of commentators provide insight, too, into the evolution of understanding about the workings of people and environment here (Adams 1996; Adams and Pawson 1984; Brower 1982, 1996; Fisher 1990; Haimendorf 1964, 1975, 1984; Hillary 1964, 1982; Ives 1979; Ives and Messerli 1989; Stevens 1983, 1993). In addition to these visitors, there is a significant voice of Khumbu-born Sherpas themselves, writing either in their role as young scholars trained in the West (L. N. Sherpa 1979, 1987, 1988; M. N. Sherpa 1993; N. W. Sherpa 1979) or as representatives of older Sherpa traditions (Rimpoche and Klatzell 1986). All these eyes and voices mean a wide array of interpretations of the landscape of Khumbu/ Sagarmatha National Park.

Grazing and Forests

This is perhaps nowhere better illustrated than by outsiders' perceptions about dynamics of the forest landscape and a presumed role of livestock in deforestation. People who have commented in print about the state of Khumbu's forests include pleasure-seekers, adventurers, journalists, scientists, and government agents. Some are passing through; others stop to study—each brings the preoccupations and assumptions of a particular point of view. Their interpretations of what they witness, reported in casual accounts, news reports, scientific journals, and government studies, have shaped a public perception of the interaction between local people and wildland resources that has determined national and international policy regulating the use of lands and resources in Mount Everest National Park.

One such perception, widely held and popularly promulgated, sees Khumbu's landscape as sharing in the pattern of worsening human-induced deforestation hypothesized for the Himalaya as a region:

A number of widespread practices—such as over-collection of fuel and fodder, over-grazing, shifting agriculture, and regeneration of fodder grasses through annual burning—are well known. These ecologically unsound activities are not new; but in recent years they have caused an unprecedented amount of environmental damage and now threaten to virtually destroy the Himalaya's natural resources. (Campbell, 1979, cited in Messerschmidt [1987])

This regional scenario, with its sense of crisis and implicit call to action, has been widely reviewed and refuted (Guthman 1995; Ives and Messerli 1989; Metz 1991; Thompson and Warburton 1985). It nevertheless persists as the guiding vision for the Himalayan environment.

The Khumbu variant of this crisis scenario compounds the resource demands by the local population with the added impacts of tourists, who demand fuelwood and other products (Bjønness 1979; Blower 1971; Lucas et al. 1974; Mather 1973; Mishra 1973):

Increasing numbers of visitors of all kinds . . . place heavy demands on the limited natural resources of timber and fuel-wood in an area of high altitude and low productivity. This, together with the impact of visitors on tourist routes with virtu-

ally no sanitation and rubbish facilities, sows the seed of environmental and economic disaster. (Lucas et al. [1974])

It has long been assumed that the arrival of the first Sherpa settlers,
who came to Khumbu perhaps 25 generations ago, initiated a process of
landscape transformation that began to accelerate in the last few decades
under pressure from increasing numbers of visiting tourists. The most frequently decried manifestation of human presence is the reduction of the
forest landscape, ascribed to a spectrum of factors usually including tree-
felling and livestock grazing:

The forests and alpine scrub-grasslands of the park have been extensively modified
and severely depleted. Apart from inaccessible country or country of colder aspect,
elsewhere the forest is under pressure from human exploitation and stock-grazing,
causing both a deterioration and diminution of area. (Halkett [1981] section b)

For Khumbu, as for the whole Himalayan region, work in the last decade has forced a reevaluation of the crisis scenario. Repeat photography
(Byers 1987a; Houston 1982), forest analysis (Byers 1987c; Jordan 1993;
MacDonald 1984), oral history (Stevens 1993), and range analysis
(Brower 1987) all suggest the condition of Sagarmatha National Park forests has been misinterpreted. Evidence for a significant *recent* reduction
in forest cover is absent (except in the case of highest-altitude juniper
woodland; the important issue of changes in forest *composition* remains
inadequately studied, although work in progress by L. N. Sherpa should
address this). But researchers have been quite comfortable with the still-
prevailing view that the Sherpa's arrival coincided with—and probably
initiated—a process of deforestation. And the Sherpa's herds of yak and
other livestock are consistently targeted as agents of forest destruction.

Livestock and Landscape Change

Sherpa livestock enjoy pride of place in most explanations of the initial
removal of forest and the subsequent suppression of regeneration. "I think
there is something poisonous in goat saliva." So one of us was told by the
first Sherpa warden of Sagarmatha National Park, newly returned from
New Zealand with a diploma in park management from Lincoln College,
who was explaining the Himalayan Trust's efforts to rid the park of
goats—those quintessentially destructive domestic beasts. The warden
may have spoken in metaphor, but he clearly grasped the essence of the
message taught at Lincoln College. It has been a common-enough message in forestry schools, environmental studies programs, and among con-

servationists: Livestock grazing and forests don't mix (Dambach 1944; Day 1930; DeWitt 1989; Patric and Helvey 1986).

That is the message, certainly, in nearly all attempts to come to grips with the dynamic interaction of forests and Sherpas in Sagarmatha National Park. In almost every account promoting SNP, in plans for management of the park, and in reports by trained and lay visitors alike who comment upon the status of forests, livestock rank high as culprits in the retreat and degradation of woodland (Bjønness 1980; Blower 1971; Byers 1989c; Garret 1981; Haimendorf 1984; Jeffries 1982; Ledgard 1989; Lucas et al. 1974; Mather 1973; Miehe 1987; Mishra 1973; Naylor 1970; Speechly 1976). Hardie et al. (1987) arrived at the following conclusion:

The remaining forest . . . is fast disappearing and will have probably vanished altogether in a few years unless properly protected and managed. . . . The main problem . . . is that heavy grazing by yaks and other livestock destroys all young tree growth through browsing and trampling and thereby prevents natural regeneration of the forest.

Case Study: Testing Assumptions about Grazing and the Forest Landscape

Overgrazing is commonly associated with environmental deterioration, particularly in vulnerable high-elevation environments (see chapters 6 and 11). But we were uneasy at this readiness to lay responsibility for forest damage on Sherpa herds in the absence of any corroborating study, especially in light of the array of erroneous assumptions commonly made about ecological process in the Himalaya, and in the face of conflicting preliminary work (Brower 1987). We sought to clarify the implications of livestock grazing for Khumbu's forest landscape, and set out to explore two hypotheses: (1) that forests are declining, and (2) that the scrub vegetation dominating most slopes near villages is degraded remnant forest. Evidence of declining forest would include lack of regeneration within stands, in canopy gaps, and at edges of stands; regeneration within stands and at stand edges and a pyramidal age structure would refute this hypothesis. Scrub vegetation composed of a subset of forest understory species plus species with "invader" biology would be expected in shrub-grassland produced from degraded forest; floristically rich and distinctive shrubland argues against such origins. Dennis designed a small-scale, exploratory study of forest plots located above the confluence of the Bhote Kosi and Imja Khola Rivers, near the Sherpa settlements of Namche Bazar, Kunde, Khumjung, and Tashinga (map 7.2) in order to test these hypotheses. We conducted the study in monsoon of 1990, with generous support from the Illinois Natural History Society (Dennis) and Fulbright-Hays Faculty Research Abroad Program (Brower).

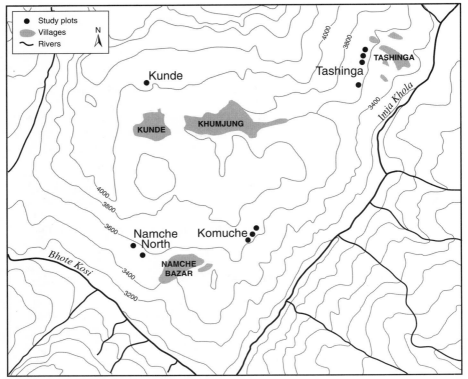

Map 7.2. Fir woodland study plots near the Sherpa settlements of Namche Bazar, Kunde, Khumjung, and Tashinga.

The Location

The triangular "plateau" defined by these joined rivers is the area of densest population and most concentrated land use within Khumbu. It is also the place most likely to be singled out as evidence that people and livestock destroy forest landscapes. The brush-dominated slopes around these south-facing settlements are frequently characterized as being former fir forest degraded to scrub through some combination of Sherpa burning, grazing, and forest felling (Halkett 1981; Hardie et al. 1987; Mather 1973; many others). The area is underlain by migmatitic ortho-gneisses; surface topography is the manifestation of preglacial V-shaped troughs and platforms overprinted by glaciation and subsequent mass wasting, and relief from valley bottom to peak ranges from 2000 to 4000 meters (Vuichard 1986). Monsoon delivers most yearly rainfall between

June and October, some rain falls in April and May as a result of convective heating in the plains, and winter storms, while sporadic, can produce substantial and destructive snowfalls that are described in Sherpa oral history (Brower 1983) as well as experienced by unfortunate travelers, such as those hit by record snows in December of 1995. Specific weather data are scarce, but a station maintained at Namche Bazar (3440 meters) reported a January mean of $-0.4°C$ (the coldest month), a July mean of $12°C$, and average annual precipitation 1048 millimeters with a range from 708 to 1710 millimeters (Joshi 1982); according to both researchers (Byers 1986) and Sherpa informants, *Dawa duinba*—the sixth month, falling somewhere in July/August—is the wettest month.

The study area is subject to grazing by a total population of about 1200 animals, maintained by 300 households and a government yak-breeding facility. Yak (*Bos grunniens*), "hilly cattle," (*B. taurus*), and yak-cattle hybrids (female: *dzum;* male: *zopkio*) are stabled at night in the villages and driven out each morning to forage for themselves. In winter, limited feeding of hay and household scraps may supplement the diet of particularly valuable animals, but most depend primarily on grazing and browsing. According to previous studies (Bjønness 1980; Brower 1987) and our observations, grazing pressure in the vicinity of these villages varies from moderate to severe. Six of the sample sites (Namche North and Komuche) are subject to grazing year-round; the others—at Tashinga and Kunde— are protected, at least in principal, by restrictions on grazing during part of the summer (Brower 1990). Each sample plot is well within the daily wandering of all types of livestock maintained by Sherpas, although pure cattle are a larger proportion with access to some sites (Namche North), crossbreeds dominate others (Komuche), and pure yak are a larger percentage of the total stock likely to be grazing those plots nearer Tashinga. (These differences may be reflected in grazing effects, for the feeding behaviors and preferences of Sherpa cattle vary according to breed [Brower 1996; Tessier and Bonnemaire 1976]). In addition to impacts of wandering livestock, sample sites are also subject to collection of forest litter and deadfall, occasional illicit wood-cutting, and browsing by musk deer and tahr (a wild goat).

Within this area we chose a dispersed set of sample sites in fir (*Abies spectabilis*) woodlands within a few minutes walk of settlements (map 7.2). Sites represent a spectrum in both observed and inferred grazing conditions and levels of stand closure. All but one of our sites lie between 3500 and 3680 meters in three locations representing different combinations of slope, aspect, and level of grazing use; we designate these areas Tashinga, Komuche, and Namche North. We chose an additional site at 4000 meters within a stand near Kunde specifically identified as manifesting stand de-

Table 7.1. *Abies* population structure.

		Number of individuals/1000m²				
	Tashinga fir sites	Tashinga birch site	Komuche	Namche North fir sites		Kunde fir sites
AGE						
Year established						
Since 1976	1829	403	94	12,338	7213	77
1961–75	72	68	52	78	48	8
1946–60	10	30	10	60	76	3
1931–45	12	6	0	14	10	0
1930 and before	0	3	0	6	0	8
Oldest tree	*1935*	*1941*	*1947*	*1915*	*1937*	*1891*
SIZE						
Seedlings	1742	225	33	12,288	7150	50 (+ 43 dead)
Saplings (to 5 cm dbh)	115	254	100	95	88	28 (+ 60 dead)
Small trees (to 20 cm dbh)	60	30	20	100	75	8
Large trees (>20 cm dbh)	6	0	3	13	35	10
Largest tree (dbh cm)	*24*	*17*	*21*	*36*	*29*	*43*

cline. At these elevations within Khumbu, more southerly aspects support shrub/grassland vegetation dominated by low-growing rhododendron (*Rhododendron lepidotum* Wallich ex G. Don) and cotoneaster (*Cotoneaster microphyllus* Wallich ex Lindley). Tall shrub vegetation with birch (*Betula utilis* D. Don), fir (*Abies spectabilis* (D. Don) Mirbel), and tree rhododendron (*Rhododendron campanulatum* D. Don, *R. campylocarpum* Hook f., and others) dominate the more northerly aspects. East and southeast aspects support fir forest of various densities mixed with shrub vegetation including tall species of *Rhododendron* and *Cotoneaster, Salix,* and *Rosa,* as well as many species of the rhododendron-cotoneaster shrub/grassland assemblage.

Our findings, summarized here and reported elsewhere (Dennis and Brower 1995), suggest that an interplay of site factors is shaping the present configuration of the forest landscape of Sagarmatha National Park. Although each of the four study locations tells a somewhat different story, together the sites suggest that altitude, aspect, and slope steepness are the most important elements determining whether forest becomes established and subsequently thrives. Our study sites show that contemporary grazing has at most a limited effect on the regeneration of forest; it is significant— if at all—only when acting in conjunction with other site stresses.

Tashinga

Tashinga is a *gunsa* (winter settlement) of small houses and associated fields, occupying an area of moderate slopes facing dominantly east-southeast. All four of the sites at Tashinga show light to moderate grazing; the only evidence of wood-cutting were a few cut stumps of small trees on the forest sites. The two plots on east-southeast aspects support fir forest, the oldest trees established in the 1930s. Plot 1's trees are larger, the peak period of establishment was between 1960 and 1964, and it is still a fairly open stand. Plot 2 is denser, and its trees smaller and younger with high densities of seedlings and saplings mostly established since 1970. Height growth of large saplings on both sites exceeds the rates on all other sites we examined. About 5% of plot 2's saplings less than three meters tall indicate damage from browsing, but leader growth is equal in both browsed and unbrowsed trees, suggesting minimum impact. These can be characterized as actively growing *Abies* stands, within which human activities are having small effect.

Tashinga's north-northeast aspect supports tree cover comparable to these sites, but a third of the trees over one meter in height are birch, rather than fir. *Abies* seedlings and small saplings are fairly abundant, but growth apparently stagnates: trees up to 60 years old have diameters less than 10 centimeters; the oldest fir tree on this plot became established about 1926 and is only 17 centimeters dbh. The more mesic conditions of this site, as for all north-oriented slopes in this elevation zone, appear to favor birch. Such sites permit the establishment of fir seedlings but forestall their continued growth.

The Tashinga plot with the most southerly aspect is a forest-free site on a south-southeast slope, and supports rhododendron-cotoneaster shrub/grassland vegetation; thorough search revealed a number of small juniper (*Juniperus recurva*) plants but no fir or birch.

Komuche

Komuche lies about one kilometer northeast of the village of Namche Bazar, five kilometers south of the Tashinga site. The general aspect here is more southerly and the slope steeper than at Tashinga; at a smaller scale there is again a range of aspects from southwest to north-northeast. The Komuche area shows high current grazing use (a majority of grass clumps had been grazed short by livestock [particularly hybrids], musk deer, and an expanding herd of tahr). Grazing pressures are concentrated on reduced range resources in the area near Namche Bazar, because approximately 35 of the national park's 45 hectares of fenced plantation (Ledgard 1989) are located here, and limit access to forage formerly available. At Komuche we sampled one forest plot (plot 5) and two shrub/grassland plots (plots 6 and 7).

Forest stands in this area occur in isolated positions above rocky outcrops or in ravines diagonal to the generally steep slope of the hillside. This is the setting for plot 5, which has the same east-southeast aspect and approximate elevation as the fir forest plots at Tashinga. The oldest trees here were established in about 1946—more recently than those at Tashinga. Establishment has apparently been intermittent since, with clear peaks between 1960 and 1965 and between 1970 and 1975. Long-term growth rates are less than at other sites in this elevation zone, and there are few seedlings or small saplings; sapling growth is much slower than on the plot with similar canopy closure at Tashinga. Although about 45% of saplings and small trees show browsing damage, as at Tashinga browsed saplings show no reduced growth relative to unbrowsed saplings. There was no evidence of wood cutting—no surprise, perhaps, since the site is close to Sagarmatha National park headquarters and the patrolling park rangers billeted there. Conditions on this rocky, steep, and exposed site appear to be more hostile for seedlings and less nurturing for growing fir than Tashinga's fir plots.

One shrub/grassland plot here is very like that at Tashinga: predominantly *Rhododendron lepidotum* and *Cotoneaster microphyllus* less than one-half meter in height. The other occupies a more northerly aspect and more closely resembles the Tashinga fir/birch site, although only a few birch, most with multiple stems less than three meters in height, occur on the plot. These show signs of having been repeatedly browsed, although there is little evidence of browsing on current growth. Willows apparently have not been browsed, and neither willow nor birch appear to have been cut.

Namche North
The sites occupying the southwest-facing slope of the Bhote Kosi Valley about one-half kilometers northwest of Namche contain more well-developed forest cover than Tashinga and Komuche. The site nearer Namche (plot 8) shows somewhat heavy grazing use; the farther one (plot 9) light–moderate use. We saw crossbreeds at these sites but found no sign of musk deer or tahr.

Plot 8 has the largest and oldest trees of all the plots in this elevation zone: up to 77 rings and 36 centimeters dbh, indicating the trees were established about 1914. This site shows a peak of establishment in 1956–1958, but also substantial recruitment since. Plot 9 has more complete canopy closure and more trees in larger-size classes. Maximum age and size, however, are somewhat less than at plot 8. The oldest tree on plot 9 was established in about 1936 and measures 29 centimeters dbh. Long-term growth rates are high—similar to those at the Tashinga fir sites and considerably higher than at Komuche. Sapling growth, on the other hand, is quite slow. Both plots contain low numbers of small saplings but very high numbers of seedlings. Low recruitment of seedlings into the sapling class and slow current sapling growth may be a response to shading by the fairly closed canopy of these sites.

A Separate Case: Kunde
Above Kunde at 4020 meters on a southwest-facing slope, the highest elevation fir plot occupies a gently sloping, smooth herbaceous sward, a contrast to the steep terracette topography of all other sites in the study. The Kunde site has the largest and oldest trees (up to 43 centimeters dbh, established as long ago as 1890), but the stand is very sparse: only 16% canopy closure. There are few seedlings. Over half of the individuals in the small sapling class are dead, and there is abundant evidence of severe browsing damage. Sapling growth estimates, based only on the few live individuals in this size class, are certainly very low. Long-term growth rates of the larger trees have also been very slow. In contrast to other sites, cut stumps are obvious here—far from national park headquarters, although within the Sherpa's traditionally designated protected forest. Most of these are less than five centimeters in diameter, but there are some stumps

of larger trees as well as holes where large stumps have been removed in the vicinity of the plot. There is little shrub cover at this site, and species composition has little resemblance to other sites.

Summarizing Forest Dynamics

Variability within and between sample sites provides intriguing and enigmatic evidence about the life history of Khumbu's forest landscape. With the single exception of the highest-altitude plot at Kunde, all fir-dominated stands are relatively young, with a range of age and size classes represented and ongoing recruitment indicated by seedlings and saplings. Evidence of grazing and browsing can be seen in most plots, yet again with the exception of the Kunde site (possibly Komuche), there is no evidence that the impacts of livestock are limiting the survival and regeneration of these forests. Neither is there evidence that either soil erosion or floristic diversity is significantly influenced by livestock.

Soil Erosion Potential

All sites, both forest and shrub/grassland, have similarly low levels of exposed soil. Estimates range from 0% at the gently-sloping Kunde site to 6% at the Komuche shrub/grassland site (plot 6), one of the steepest sites sampled. These two sites are the most heavily grazed in our sample set.

Floristic Analysis

We made complete plant species lists for five of the study plots: the Komuche rhododendron-cotoneaster shrub/grassland site (plot 6) and four forest sites (plots 5, 8, 9, and 10). Species richness was remarkably constant: 83–85 species per 100 square meters at four of the five sites. Only plot 9 in the Namche North set stands out with 72 species. The three species that we would classify as ruderal or introduced were only found on plot 8, the site nearest Namche Bazar, Sagarmatha National Park's urban hub. The rest of the species we encountered are native species typical of intact vegetation.

Kunde stands out as floristically distinct from the other sites. It has by far the most singleton species, and was missing almost half the common species. At the other extreme the Komuche scrub/grassland site was missing only six of the common species and had relatively few singleton species. In pairwise comparisons, all sites had least in common with Kunde. Except for the fact that the two Namche North sites had most in common with each other, each forest site had more in common with the scrub/

grassland than with any of the other forest sites. Plot 9, the most heavily forested site, was missing almost as many common species as Kunde, but had a relatively low number of singleton species. This, together with the high degree of floristic similarity to the other Namche North site, suggests that the low species richness at plot 9 is better described as a depletion in species richness rather than an indication of a distinct community.

Discussion

The 4100-meter site at Kunde clearly shows evidence of forest decline: few seedlings, high mortality of saplings, and few trees in intermediate size and age classes. Low growth rates over the lives of the few large trees show that conditions for growth at this site have been poor even for individuals in size classes that are little affected by current browsing and trampling, although browsing and tree-cutting have every appearance of being major factors in mortality of small trees. Clearly, unless conditions improve for survival and growth of small trees, the large trees on this site are unlikely to be replaced and the stand will thin even further or disappear. Yet despite apparently heavy grazing use, vegetation cover at this site is complete and there is clearly no accelerated soil loss taking place. This, together with observations by Byers (1987a) at similar sites, suggest that livestock impacts on soil productivity are unlikely to be a major factor in the decline of the forest at this site.

The understory community at Kunde, although equally diverse, shares relatively few species with any of the sites at lower elevations. All of the shrub species that form major components in both the lower-elevation forest understory and shrub/grassland sites are absent here, as are all the grasses and herbs that are major vegetation components on those sites. Clearly, many plant species typical of fir forests and associated communities reach their environmental limits below the elevation of this site, which lies at the uppermost limit for fir (Stainton 1976). Timberlines fluctuate, and older trees on this site are likely to have been established under environmental conditions more favorable than those that exist today. Grazing and wood-cutting are not the only limiting factors operating here.

Fir forests at Tashinga and Namche North do not have characteristics we expect in declining stands. What we see at these sites is consistent with young stands at various stages of filling in. In contrast to Kunde, at these sites we did not find any evidence to suggest that tree cutting in recent years has reduced the number of large trees. Rather, the oldest trees here appear to have been pioneers on a previously unforested site. Composition and floristics of the understory vegetation support the interpretation of these as sites that were shrub/grassland with scattered trees rather than dense forest in the recent past. On all these sites, the understories bear strong resemblance to nearby shrub/grassland. The herbaceous understory differs from the shrub/grassland by absence of individual common species (a different set in each case, rather than by presence or absence of a common set of species, or, alternatively, presence of species characteristic of more fully-developed fir forests elsewhere in these valleys [see Byers (1987c) or Miehe (1987) for lists of these species]). The understory shrub component, on the other hand, does show signs of acquiring a distinctive set of species not characteristic of shrub/grassland—the tall shrub/small tree species of *Rhododendron, Salix, Cotoneaster,* and *Rosa.* At present, these species are minor components of cover on these sites.

The Komuche fir site presents some ambiguity. Looking only at trees over one meter in height, we see a young, expanding stand, the pioneers of which became established in the late 1940s. However, populations of seedlings and small saplings are relatively small. Sapling numbers and growth rates resemble the rates in closed canopy at Namche North rather than the rapid rates of open stands at Tashinga. Growth rates of large trees are lower than at either Tashinga or Namche North. Direct or indirect effects of livestock and tahr, both of which use this area heavily, could be factors in low seedling numbers. However, contrary to the Kunde case, no evidence suggests that browsing is directly reducing growth or increasing mortality for saplings. As with Tashinga and Namche North, understory characteristics support an interpretation of this as a young stand, not a depleted remnant of a formerly more dense forest. Intermittent establishment of *Abies* appears to have been the pattern over the history of the stand. Animal impacts may interact with environmental factors to limit regeneration except in favorable years. It is possible that recent changes in livestock use, such as increasing numbers of poorly controlled zopkio, could reduce the frequency at which those favorable circumstances occur, as could the impact of a rapidly expanding herd of tahr whose range centers on this area. Trees now in the sapling stage can reasonably be expected to survive and produce some filling in of the stand, but it is not clear that the stand will develop further.

The north-facing birch-fir stand near Tashinga shows that poor conditions for growth, not grazing, is probably the factor limiting expansion of fir forest onto northerly aspects. Although the mesic environment of these sites apparently favors seedling establishment and initial growth, it clearly does not favor continued growth. Livestock trampling or browsing would be unlikely to limit growth in this size class without also affecting smaller trees. The shift in understory species composition between these sites and adjacent, more southerly aspects, shows that other species also reach environmental limits on these shady slopes.

It is striking that no fir seedlings were found in shrub/grassland plots, although plot vegetation is similar in structure and composition to the fir stand understories, and there were no apparent differences in level of livestock use, density, or condition of trails and terracettes, or level of herbage removal between adjacent scrub/grassland and forested sites. We do not have specific information on seedling physiology at such sites for *Abies spectabilis,* but all the North American *Abies* species are shade-tolerant and sensitive to heat and desiccation at seedling stage. For these *Abies* species, seedling survival is greatly enhanced by shading, although established trees tolerate and grow more rapidly in full sun (Burns and Honkala 1990). If *Abies spectabilis* regeneration at these altitudes is similar, high seedling mortality in full sunlight may largely account for the scarcity of seedlings on the scrub/grassland sites. This would fit with several other patterns we have observed: the delay between establishment of the first cohort on a site and subsequent population expansion, the abundance of seedlings on the Tashinga birch/fir site and the Namche North sites where conditions for subsequent growth are poor at the present time, and the scarcity of seedlings on the steep, exposed Komuche site.

Given the evidence for stand dynamics at these sites, which represent a range in altitude, aspect, slope, and livestock-use variables, we suggest the following process:

1. Establishment of fir seedlings in shrub/grassland is a rare event owing to the vulnerability of seedlings to desiccation. Establishment events are likely to be more frequent on shadier aspects, on sites protected from wind, and near seed sources.
2. Once past the critical establishment phase, shrub/grassland sites are suitable for fir growth.
3. Because of the relative unpalatibility of fir and its ability to rapidly replace damaged leaders, browsing by livestock and other herbivores does not result in major foliage losses or retard growth of trees that are otherwise healthy.
4. Once pioneer trees grow big enough to produce a shaded, protected area, seedling establishment increases rapidly.

Conclusions

Our results show that concern about livestock impacts on fir regeneration in Khumbu may be justified in some situations, but on the whole fir forests in the Namche-Kunde-Khumjung region are expanding and livestock grazing has little direct adverse effect on tree regeneration and growth. Regeneration problems can occur where environmental limits to growth

interact with browsing pressure near timberline. However, we found that
there was ample regeneration on sites representing a range of grazing conditions, from relatively moderate, controlled use near Tashinga to fairly
heavy and less controlled use within half a kilometer of Namche Bazar.
Observations of vigorous fir regeneration on very heavily grazed sites at
3800 meters just south of Kunde (in an area of much recent tree-cutting)
further support our conclusion that grazing alone would rarely prevent fir
regeneration. Our findings on age structure, vegetation composition, and
floristic characteristics all support this conclusion. We see the relationships between human impacts and forest stand dynamics hinging on the
particular biology of *Abies spectabilis* at this location, especially its inferred sensitivity to desiccation at seedling stage, its relative unpalatability
to large herbivores, and its tolerance of moderate browsing when otherwise healthy. We would not necessarily expect to see similar relationships
with species that differ from *Abies spectabilis* in these characteristics, or
with *Abies* in other environmental contexts.

Our conclusions agree with those drawn by Houston (1982) and Byers
(1987a): Forest cover is increasing in the 3400–3800-meter zone in the
Namche triangle. Sparse stands and lone trees in this zone are pioneers,
not remnants of formerly more dense forests as Halkett, Hardie, Miehe,
Bjønness, Speechly, and others have speculated.

We would like to make a further point regarding the shrub/grassland
vegetation, characterized as "degraded" by these authors: *Rhododendron
lepidotum–Cotoneaster microphyllus* shrub/grassland is a rich and diverse
plant community in its own right. It is in no way a collection of forest
species that have clung to life despite overstory removal and grazing, nor
is it composed of pioneering species. It is clear that boundaries between
shrub/grassland and forest have shifted over time on the sunny slopes of
Khumbu. Today's shrub/forest landscape mosaic represents one moment
in a dynamic process in which the role of humans and their livestock is by
no means clear. One might, for instance, speculate that Sherpa livestock
are factors in *increasing* the extent of forest: Grazing that reduced the fuel
load of the shrub/grassland might have reduced the frequency and severity
of wildfires on these warm-aspect slopes, permitting the establishment of
fir stands—a hypothesis suggested by our results that requires further test-
ing. The fact that human activities may play a major role in some of these
forest-to-shrub, shrub-to-forest shifts does not inherently make the community type favored by human activity degraded. We can in fact recognize
degraded conditions by accelerated soil loss or the absence of species characteristic of the shrub/grassland or forest community, and can find examples of both in Khumbu (the steeply eroding trails and heavily trampled
areas above Namche, for example). In contrast, intact shrub/grasslands

like the ones we examined in this study and observed elsewhere in Khumbu are not undergoing soil loss, and do provide habitats for a wide range of native species that do not occur in forests. Attempts to "restore" intact shrub/grasslands to forest cover are counterproductive to maintaining the diversity of habitats of Khumbu and the full array of native biota these habitats support.

Implications for Conservation

Our study suggests that ecology, to some extent, may be in the eye of the beholder. Foresters, park planners, and others convinced of the dangers of Himalayan deforestation and unkindly disposed toward livestock saw decline in the patterning of Khumbu forests, and agents of degradation in Sherpa herds. Imbued with notions of stasis and visions of a fir-forest climax, they perhaps neglected to consider the dynamic biogeographical underpinnings of landscape. They somehow overlooked what would seem to be clear evidence of forest vitality—conspicuous pyramidal form and abundant seedlings and saplings—distracted, perhaps, by other evidence, such as ubiquitous stock-terrecettes, that livestock clearly have been agents of other change in this landscape.

More partial to yaks, leery of conventional explanations for ecological processes in the Himalaya, with experience of unexamined prejudices against forest grazing, and tutored in geography's ideas about landscape change, ours was a different perspective; our study was framed specifically to address the livestock/deforestation hypothesis, and we offer a different conclusion. It is clearly an exploratory study, and raises no end of further questions about environmental processes, cultural practices, and park policy objectives. We await with interest the results of future work on some of these questions, and brace ourselves for the critique we hope will follow on our own attempt to make sense of the forest landscape of Khumbu. The outcome we would hope for is a conservation strategy for Sagarmatha National Park that reflects a thorough—and thoroughly neutral—understanding of both ecological processes and the changing role of local people through time, one that recognizes the mutability of scientific explanations and the legitimacy of local needs and values while safeguarding the critical values of biodiversity.

Sagarmatha, of course, is not unique in requiring understanding of both natural and cultural processes as the foundation for conservation planning and management. The landscape valued here—a high-altitude, peopled foreground to the world's biggest mountains—has its own distinctive history, and reflects responses to a very particular set of conditions. The inherent instability of high-mountain environments presents an ever-changing interaction of climate, slopes, and biota. People introduce another set of influences on landscape processes that change along with the lifeways of Khumbu's residents, and interventions of Sagarmatha's managers. The processes and patterns that characterize this particular Himalayan landscape are not, after all, part of some uniform regional scheme. Yet for all its distinctiveness, Sagarmatha/Khumbu's forest landscape is a product of the same broad dynamics of establishment and decline that shape landscapes everywhere. Understanding the particular expression of those dynamics, here as anywhere, requires a suspension of presumptions, a careful analysis of a complex spectrum of factors, and a willingness to re-examine interpretations of landscape-making derived from inadequate understanding of dynamism, complexity, and untested assumptions.

References

Adams, V. 1996. *Tigers of the Snow and Other Virtual Sherpas: An Ethnography of Himalayan Encounters.* Princeton: Princeton University Press.

Allin, C., ed. 1990. *International Handbook of National Parks and Nature Reserves.* New York: Greenwood Press.

Anon. 1976. *Memorandum of Understanding between the Government of New Zealand and His Majesty's Government of Nepal Concerning a Project for the Establishment of Sagarmatha (Mt. Everest) National Park, Nepal.*

Bajracharya, D. 1983. Deforestation in the Food/Fuel Context: Historical and Political Perspectives from Nepal. *Mountain Research and Development* 3(3): 227–240.

Bjønness, I. M. 1979. *Impacts on a High-Mountain Ecosystem: Recommendations for Action in Sagarmatha (Mt. Everest) National Park.* Unpublished report for Sagarmatha National Park.

Bjønness, I. M. 1980. Animal Husbandry and Grazing, a Conservation and Management Problem in Sagarmatha (Mt. Everest) National Park, Nepal. *Norsk Geografisk Tiddscrift* 34: 59–76.

Bjønness, I. M. 1983. External Economic Dependency and Changing Human Adjustment to Marginal Environment in the High Himalaya, Nepal. *Mountain Research and Development* 3(3): 263–272.

Blower, J. H. 1971. *Proposed National Park in Khumbu District.* Memorandum, Wildlife Conservation Officer, F. A. O., to His Majesty's Government's Secretary of Forests, March 21, 1971.

Botkin, D. B. 1990. *Discordant Harmonies: A New Ecology for the Twenty-first Century.* New York: Oxford University.

Brower, B. 1982. *The Problem of People in Natural Parks: Chitwan and Sagarmatha National Parks, Nepal.* Unpublished paper, University of California–Berkeley.

Brower, B. 1983. *Mountain Hazards and the People of Khumbu-Pharag.* Report, Nepal Mountain Hazards Mapping Project. Bern, Switzerland: UNU MHMP.

Brower, B. 1987. *Livestock and Landscape: The Sherpa Pastoral System in Sagarmatha (Mt. Everest) National Park, Nepal.* Ph. D. dissertation, University of California–Berkeley.

Brower, B. 1990. Range Conservation and Sherpa Livestock Management in Khumbu, Nepal. *Mountain Research and Development* 10(1): 34–42.

Brower, B. 1991a. *Sherpa of Khumbu: People, Livestock, and Landscape.* Delhi: Oxford University Press.

Brower, B. 1991b. Crisis and Conservation in Sagarmatha National Park. *Society and Natural Resources* 4(2): 151–163.

Brower, B. 1993. Co-Management versus Co-Option: Reconciling Scientific Management with Local Needs, Values and Expertise. In *Conservation and Development: Bottom-up Strategies for the Roof of the World,* eds. B. Brower, S. J. Loomus, and M. J. Enger, pp. 34–52. New Haven: Yale University.

Brower, B. 1996. Geography and History in the Solukhumbu Landscape. *Mountain Research and Development* 16(3): 249–256.

Burns, R. M., and B. H. Honkala. 1990. *Silvics of North America Conifers* (volume 1). Agriculture Handbook 271. Washington, D.C.: US Department of Agriculture.

Byers, A. 1986. A Geomorphic Study of Man-Induced Soil Erosion in Sagarmatha (Mt. Everest) National Park, Nepal. *Mountain Research and Development* 6(1): 83–87.

Byers, A. 1987a. An Assessment of Landscape Change in the Khumbu Region of Nepal Using Repeat Photography. *Mountain Research and Development* 7(1): 77–81.

Byers, A. 1987b. Landscape Change and Man-Accelerated Soil Loss: The Case of Sagarmatha (Mt. Everest) National Park, Nepal. *Mountain Research and Development* 7(3): 209–216.

Byers, A. 1987c. *A Geoecological Study of Landscape Change and Man-Accelerated Soil Loss: The Case of Sagarmatha (Mt. Everest) National Park, Nepal.* Ph. D. dissertation, University of Colorado–Boulder.

Campbell, G. 1979. *Community Involvement in Conservation: Social and Organizational Aspects of the Proposed Resource Conservation and Utilization Project in Nepal.* Kathmandu: Agricultural Projects Services Centre (APROSC).

Dambach, C. A. 1944. A Ten-Year Ecological Study of Adjoining Grazed and Ungrazed Woodlots in North-Eastern Ohio. *Ecological Monographs* 14: 257–270.

Davidson, A. 1993. *Endangered Peoples.* San Francisco: Sierra Club.

Day, R. K. 1930. Grazing Out the Birds. *American Forests* 36: 555–557, 594.

Day, R. K., and D. DenUyl. 1932. *The Natural Regeneration of Farm Woods Following the Exclusion of Livestock* (Purdue University Agricultural Experiment Station Bulletin No. 368). Purdue: Purdue University.

Dennis, A., in preparation. *Fir Reproduction in Subalpine Forests of Sagarmatha National Park.*

Dennis, A., and B. Brower. 1995. *Forest Reproduction in Subalpine Forests of Khumbu.* Paper presented at the 24th Annual Conference on South Asia, Madison, Wisconsin.

DeWitt, B. 1989. Forest Grazing Hurts. *Missouri Conservationist* 50: 18–19.

Drew, I. K., and U. Sharma. 1977. *A Proposal for a Reforestation Programme in the Sagarmatha (Mt. Everest) National Park.* Report, Sagarmatha National Park.

Eckholm, E. 1976. The Deterioration of Mountain Environments. *Science* 169: 764–766.

Fisher, J. 1990. *Sherpas: Reflections on Change in Himalayan Nepal.* Berkeley: University of California Press.

Fortmann, L., and S. K. Fairfax. 1989. American Forestry Professionalism in the Third World: Some Preliminary Observations. *Economic and Political Weekly* 24(32): 1839–1844.

Garret, K. 1981. *Sagarmatha National Park Management Plan.* Wellington: Department of Lands and Surveys.

Gilmour, D. 1989. *Forest Resources and Indigenous Management in Nepal* (Working Paper 17). Honolulu: East-West Environment and Policy Institute.

Guha, R. 1990. *The Unquiet Woods.* Berkeley: University of California Press.

Guise, C. H. 1950. *The Management of Farm Woodlands.* New York: McGraw-Hill.

Haimendorf, C. von F. 1964. *The Sherpas of Nepal.* Berkeley: University of California Press.

Haimendorf, C. von F. 1975. *Himalayan Traders: Life in Highland Nepal.* London: John Murray.

Haimendorf, C. von F. 1984. *The Sherpas Transformed: Social Change in a Buddhist Society of Nepal.* New Delhi: Sterling Publishers.

Halkett, L. M. 1981. *Forest Management Sagarmatha National Park.* Report by New Zealand Project Forester. *Sagarmatha National Park Library*

Hardie, N. 1977. *The Sherpa People.* Manuscript (New Zealand Mission Report).

Hardie, N., U. Benecke, P. Gorman, and P. Gorman. 1987. *Nepal-New Zealand Project on Forest Management in Khumbu-Pharak.* Christchurch: Forestry Research Center.

Hillary, E. 1964. *Schoolhouse in the Clouds.* London: Hodder and Stoughton.

Hillary, E. 1982. Preserving a Mountain Heritage. *National Geographic* 161(6): 696–702.

Houston, C. 1982. Return to Everest: A Sentimental Journey. *Summit* 28: 14–17.

Ives, J. 1979. Applied Mountain Geoecology: Can the Scientist Assist in the Preservation of Mountains? In *High Mountain Geoecology,* ed. P. Webber. Boulder: Westview Press.

Ives, J., and B. Messerli. 1989. *The Himalayan Dilemma: Reconciling Development and Conservation.* London: Routledge.

Jeffries, B. E. 1982. Sagarmatha National Park: The Impact of Tourism in the Himalayas. *Ambio* 11(5): 274–281.

Jordan, G. 1993. *GIS Modeling and Model Variation of Erosion and Deforestation Risks, Nepal.* M. Sc. dissertation, University of Wales.

Joshi, D. P. 1982. The Climate of Namche Bazar. *Mountain Research and Development* 2(4): 399–402.

Kroeber, T., and R. Heizer. 1968. *Almost Ancestors: The First Californians,* ed. D. Hales. San Francisco: Sierra Club.

Lall, J. S. 1981. *The Himalaya.* Delhi: Oxford University Press.

Ledgard, N. 1989. *Nepal (Khumbu/Pharak) Visit-April 1989: Forestry Notes for the Himalayan Trust.* Christchurch: Forest Research Institute.

Ledgard, N., and G. Baker. 1990. *Nepal (Solu-Khumbu) Visit-June/July 1990: Forestry Notes for the Himalayan Trust* (Contract Report FWE 90/OSI). Christchurch: Forest Research Institute.

Lucas, P. H. C., N. D. Hardie, and R. A. C. Hodder. 1974. *Report of New Zealand Mission on Sagarmatha (Mt. Everest) National Park, Nepal.* Wellington, New Zealand: Ministry of Foreign Affairs.

MacDonald L. 1984. Field notes.

March, K. 1977. Of People and Nak: The Management and Meaning of High Altitude Herding among Contemporary Solu Sherpa. *Contributions to Nepalese Studies* 4(2): 83–97.

Mather, A. D. 1973. *Reforestation in the Proposed Mt. Everest National Park.* Sagarmatha National Park.

Messerschmidt, D. 1987. Conservation and Society in Nepal: Traditional Forest Management and Innovative Development. In *Lands at Risk in the Third World: Local Level Perspectives,* eds. P. D. Little and M. M. Horowitz, pp. 373–397. Boulder: Westview Press.

Metz, J. 1991. A Reassessment of the Causes and Severity of Nepal's Environmental Crisis. *World Development* 19(7): 805–820.

Miehe, G. 1987. An Annotated List of Vascular Plants Collected in the Valleys South of Mt Everest. *Bulletin British Museum Natural History* 16(3): 225–268.

Mishra, H. 1973. *Conservation in Khumbu: The Proposed Mt. Everest National Park* (confidential report). Department of Wildlife and National Parks.

Naylor, R. 1970. *Colombo Plan Assignment in Nepal.* Wellington: New Zealand Forest Service.

Patric, J. H., and J. D. Helvey. 1986. *Some Effects of Grazing on Soil and Water in the Eastern Forest* (General Technical Report NE-115). Washington, D.C.: USDA Forest Service.

Pawson, G., D. Stanford, and V. Adams. 1984. Growth of Tourism on Nepal's Everest Area: Impact on the Physical Environment and Structure of Human Settlements. *Mountain Research and Development* 4(3): 237–246.

Sakya, K. 1985. *The Need for Environmental Education.* Paper presented to the International Workshop on the Management of National Parks and Protected Areas of the Hindu Kush-Himalaya, Kathmandu, Nepal.

Sherpa, L. N. 1979. *Considerations for Management Planning of Sagarmatha National Park.* Diploma dissertation, University of Canterbury, New Zealand.

Sherpa, L. N. 1987. *Social Functions of Rara National Park: A Case Study of Park and People Interaction at Rara National Park, Nepal.* Unpublished manuscript.

Sherpa, L. N. 1988. *Conserving and Managing Biological Resources in Sagarmatha (Mt. Everest) National Park, Nepal.* Working Paper No. 8. Honolulu: East-West Center Environment and Policy Institute.

Sherpa, M. N. 1993. Grass Roots in a Himalayan Kingdom. *Indigenous Peoples and Protected Areas.* San Francisco: Sierra Club.

Sherpa, N. W. 1979. *A Report on Firewood Use in Sagarmatha National Park,*

Khumbu Region, Nepal. Report, HMG Department of Wildlife and National Parks.

Speechly, H. 1976. *Proposal for Forest Management in Sagarmatha National Park.* Report, HMG Department of National Parks and Wildlife Conservation, Nepal.

Stainton, J. D. A. 1972. *Forests of Nepal.* New York: Hafner Publishing Company.

Stettler, A., and L. MacDonald. 1982. Unpublished data.

Stevens, S. F. 1983. *Tourism and Change in Khumbu.* B.A. thesis, University of California–Berkeley.

Stevens, S. F. 1989. *Sherpa Settlement and Subsistence: Cultural Ecology and History in Highland Nepal.* Ph. D. dissertation, University of California–Berkeley.

Stevens, S. F. 1993. *Claiming The High Ground.* Berkeley: University of California Press.

Tengboche Reincarnate Lama, F. Klatzel, and T. Sherpa. n.d. *The Stories and Customs of the Sherpas.* No publisher listed.

Thompson, M., and M. Warburton. 1985. Decision-Making under Contradictory Certainties: How to Save the Himalayas When You Can't Find Out What's Wrong with Them. *Journal of Applied Systems Analysis* 12: 3–34.

Thompson, M., M. Warburton, and T. Hatley. 1986. *Uncertainty on a Himalayan Scale.* London: Milton Ash Editions Ethnographia.

Vuichard, D. 1986. Geological and Petrographical Investigations for the Mountain Hazards Mapping Project, Khumbu Himal, Nepal. *Mountain Research and Development* 6(1): 41–51.

Worster, D. 1990. The Ecology of Order and Chaos. *Environmental History Review* 14 (1/2): 1–18.

Zimmerer, K. S. 1994. Human Geography and the "New Ecology": The Prospect and Promise of Integration. *Annals of the Association of American Geographers* 84(1): 108–125.

PART 3

LAND USE IN SETTLED AREAS AND
THE PROSPECTS FOR CONSERVATION-
WITH-DEVELOPMENT

Introduction

Areas of land use associated with primary economic activities make up a distinct set of challenges for conservation in developing countries. The five chapters in this section are focused on a variety of biogeographical landscapes that are strongly shaped by agriculture, livestock-raising, and the extractive use of forest products. The case studies cover a wide range of environments, resource systems, and regions. The first study (chapter 8, Blumler) presents an historical and ecological biogeographical analysis of the diverse landscapes of forests, shrublands, and grasslands in the Middle East that for several thousand years have been impacted by the combination of agriculture, pastoralism, and forest use, which together may be thought of as agro-sylvo-pastoralism. The studies that follow examine the topics of pastoralism in the Sahel of Africa (chapter 9, Turner), agriculture in the highlands of South America (chapter 10, Zimmerer), and the extraction of plant products from the forests of the Himalaya (chapter 11, Metz) and Southeast Asia (chapter 12, Voeks).

These case studies contribute to a new perspective on the nature of environmental change in the land use of developing countries. As a way of introducing the varied land-use systems, the following descriptions provide a brief overview of the biological diversity, livelihood value, landscape function, and environmental change that are addressed by the case studies. It may be helpful to recall that the central concern of our volume is the nature of change in the biogeographical landscapes of developing countries. We recognize that a large number of important landscapes and resources are threatened by panoplies of change that are social, political, cultural, and environmental. The chapters in this section evaluate the biogeographical properties of certain major change processes and ascertain the prospects for conservation by delving into particular facets of utilized landscapes.

The *biological diversity* of agriculture, rangeland, and forests are amply represented in the case studies. The history of land use and landscape change in the diverse environments of the Middle East is explored where, to an extent, there persists a wide repertoire of crop and forage species as well as forest taxa. The livestock-raising of peasant pastoralists in the highly variable, semiarid climate of the Sahel area of sub-Saharan Africa is seen as making strategic use of a floristically diverse rangeland. Biodiversity is notable at the subspecies level as well. Andean farmers' world-renowned agrodiversity of the potato crop provides part of a strategy to deal with the highly varied disturbance regimes of tropical mountains. The dynamics of the temperate and subalpine forests found above 2500 meters in the central Himalaya have produced unique assemblages of rhododen-

dron, fir, oak, maple, and birch species. Finally, the biologically-rich dipterocarp rain forest of Borneo, which to an extent is human-managed, is
found to contain an almost unrivalled diversity of 303 tree species within
an area that measures slightly less than one hectare.

The biological richness of these arrangements of land use offer much
livelihood value to local inhabitants in the developing countries. Forest
landscapes are particularly notable for providing wild and semidomesti-
cated foods, fuelwood, building material, fodder, fiber, and medicinal
plants. Rangelands and agriculture are of course primary economic activities that furnish grazing resources and crop plants. In the African Sahel
and the Andean highlands, the immense diversity of useful range and farm
plants is a locally valued resource for land-users that must accommodate
a variety of landscape changes. The studies describe the value of these
resources to the livelihoods of diverse peoples, and a few of the social and
cultural conditions of usefulness as well. Livelihood values and knowledge
of the Borneo rain forest and the Andean food plants, for example, involve
local personages and rationales such as spiritual healers and customs related to cultural identity formation.

The *landscape functions* of land and resource use that are described in
the case studies present a crucial aspect of nature-society interrelations.
Patchwork mosaics of forests, agriculture, and rangeland are found to
characterize each of the study regions. In fact a combination of all three
resource types is typically relied on by many inhabitants of developing
countries, especially the small-scale land-users frequently referred to as
peasants. Of chief interest to conservation-with-development is that certain types and levels of resource use in these farming, grazing, and forest
extraction activities, as well as the right mosaic-like combination on the
landscape, can continue to furnish important environmental functions. In
a particularly important example, many peasant land-users in Asia, Africa, and Latin America have utilized the vegetation component of landscapes while, under the right conditions, managing to control soil loss and
protecting watersheds.

A main goal of this section is to evaluate how *environmental changes*
under certain social conditions enable conservation to take place, while
under other conditions the outcome is resource degradation. Each study
considers the scale, magnitude, and character of environmental change
stemming from the "natural" and human-induced disturbances of landscapes. The short-term variation of climate, typically interannual, plays
a key role in Sahelian pastoralism and Andean agriculture, for example.
Rainfall and temperature fluctuations clearly disrupt these production systems. Yet it is also apparent how the pastoralists and farmers of these
regions, under certain favorable conditions, can retain a flexibility of land

use that allows them to accommodate such landscape change. Under other social conditions less favorable for the land-users, however, the same sort of change may be devastating.

The analysis of environmental change in this section also draws comparisons between the regimes of "natural" and human-induced disturbances. Disturbances are compared in the analysis of the ecological and historical biogeography of diverse Middle Eastern landscapes. Similarly a comparative framework supports the analysis of change in the forest landscapes of the Himalaya. In the Middle Hills of Nepal, large-scale disturbances such as fire and landslides are necessary for the regeneration of various forest species, while other disturbances such as chronic heavy grazing are seen as thwarting the continued establishment of forest diversity.

8

Biogeography of Land-Use Impacts in the Near East

Mark A. Blumler

Introduction

The Near East, roughly the region that extends from Greece and Egypt at the eastern end of the Mediterranean Sea to Iran (map 8.1), is an ancient land. It was the cradle of western civilization. Probably no other region on Earth has undergone such a long, intensive history of agropastoralist impact. The cultivation of crops such as wheat, barley, and lentils, and herding of animals (goats, sheep, and cattle) originated perhaps 10000 years ago in the Fertile Crescent, the curving arc of well-watered foothills flanking the Syro-Mesopotamian deserts and steppes (map 8.1). Subsequently, the presence of the great Nile, Euphrates, and Tigris Rivers allowed agriculture to spread into the adjacent deserts with the development of sophisticated irrigation technology, and enabled the flourishing of great civilizations with human populations that were very large for their time. As a result, the Near East remained technologically advanced compared to other regions for many millennia; in fact, the baton finally passed to Europe only during and after the Renaissance (Blumler 1996b).

In recent centuries, however, the Near East for the most part has been backward and poverty-stricken (with the partial exception of the oil-producing lands). It is almost universally accepted that this is a conse-

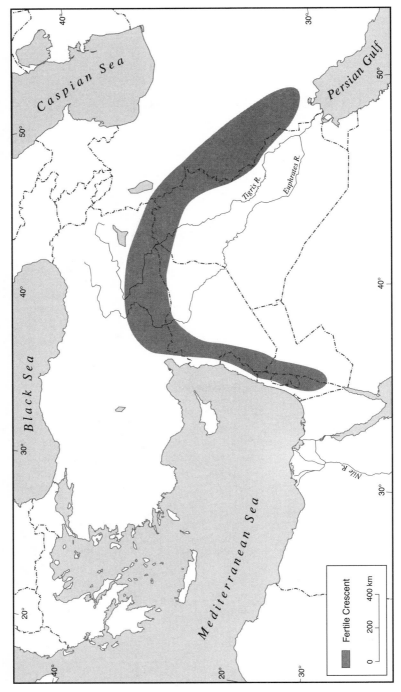

Map 8.1. The Near East, including the Neolithic heartland of the Fertile Crescent.

quence of the prolonged history of use and overuse that the Near East and Mediterranean region have suffered, which is assumed to have seriously degraded the land (Carter and Dale 1974; Le Houérou 1981; Lowdermilk 1943; Zohary 1962, 1973, 1983). In fact, the modern concern with land degradation can be traced back to the nineteenth-century research of George Perkins Marsh (1864), who painted a gloomy picture of degradation in Mediterranean countries. Certainly today much of the Near East has a used-up appearance, a biogeographical landscape of bare, rocky hills covered by scraggly, scrubby vegetation, alternating with agricultural settlement in the more fertile river valleys. And of course, "the struggle between the desert and the sown" was a concern even in Biblical times (Reifenberg 1955).

This perception of long-term degradation strongly influences policy in the area of conservation-with-development. For example, Jewish organizations that participated in the resettlement of Palestine, and the eventual creation of the state of Israel, believed the land to be seriously deforested (Orni 1963). Consequently, they (and later the Israeli government) have sponsored tree-planting (mainly of Aleppo pine, *Pinus halepensis*) on such a scale as to effect the radical transformation of large areas of wildland. Israelis support this practice with patriotic fervor, and it generates considerable positive publicity and monetary donations among conservation-minded Westerners. But Israeli plant ecologists understand that most plantations are on soils that never supported pine forests naturally (Rabinovich-Vin 1983; Zohary 1962). And sadly, as Israeli ecologists know but find it difficult to say publicly, tree planting has converted formerly species-rich landscapes into biological deserts, supporting few living things other than the trees themselves (M. Blumler, unpublished data; P. Kutiel, personal communication; A. Mann, unpublished data; Z. Naveh, personal communication; D. Zohary, personal communication).

To develop sound policy for conservation-with-development, then, it is important to reevaluate our assumptions about the effects of the past. A historical perspective on human impacts on nature should be illuminating. True, our impacts today are unlikely to be identical to those of humans in the past, who operated within different political and social structures and employed a more primitive technology. So any comparison with the past may be misleading. Moreover, if one's historical perspective is constructed carelessly, it can do more harm than good. But on the other hand, a historical study can give us a sense of long-term patterns that may not be apparent in the short-term.

Nowhere on Earth is an historical perspective so necessary, or most people's perceptions of the influence of the past more distorted, as in the Near East. The popular picture of unremitting decline and degradation of natural resources, although not entirely invalid, is overdrawn, and for the most part based on (mis)perception rather than evidence. In particular, most reports about conditions in the Near East derive from the traditional equilibrium view of nature, and associated linear models of vegetation change under human impact. When the evidence is reevaluated from the nonequilibrium perspective of the new ecology, it turns out that biogeographical landscapes have been remarkably resilient in the face of undeniably severe human impacts. I develop this argument in this chapter. After briefly describing the natural environment and landscapes, I discuss the underlying basis for the general perception of severe degradation. I then summarize some empirical evidence from Israel that suggests this perception is not entirely valid. Finally, I discuss some implications for conservation-with-development.

Regional Description

Climate in the Near East ranges from quite moist on some windward mountain slopes and seacoasts, such as in Pontic Turkey, to hyperarid in the south (Egypt, Persian Gulf lowlands). The seasonal regime conforms to the classic Mediterranean pattern of mild, rainy winters and hot, dry summers, an unusual climatic type in that most parts of the globe experience a precipitation maximum during the warm season. This mediterranean-type climate extends all the way from the Atlantic shores of Portugal and Morocco to the Tien Shan and Pamir Mountains of Central Asia, albeit with considerable variation in amount of rainfall, the degree to which it is restricted to the cool season, and temperature pattern (Blumler 1984). In much of the Near East, summers are hotter and drier than anywhere else on Earth; thus in a sense the Near East is the most mediterranean-type region. In contrast, the much studied ecosystems of southern France, often taken to typify the mediterranean environment (Braun-Blanquet et al. 1951), flourish under a regime that includes a significant percentage of summer rain and therefore is arguably not mediterranean at all. Consequently, the results of French ecological studies may be misleading if applied indiscriminately to the Near East or summer dry parts of the Mediterranean Basin (Blumler 1984, 1992b, 1993).

Geologically, calcareous rock such as limestone underlies much of the Near East and Mediterranean region. Limestone weathers almost exclusively through dissolution in acidic water, a process that is highly irregular because it depends on the circulation of water through the rock. Therefore, limestone can give rise to either fertile or shallow soils, depending on circumstances; the result is a highly heterogeneous microtopography, with deep and shallow soils, and rock outcrops, in close proximity. In turn, this provides an extreme heterogeneity of niches available to plants at a microscale (Blumler 1992a, 1992b, Blumler et al., n.d.) Soils are also much affected by aeolian transport of dust out of the great deserts to the south (Macleod 1980; Mizota et al. 1988; Yaalon and Ganor 1973, 1975). This calcareous dust is nutrient-rich and serves as a sort of natural fertilizer.

Surprisingly, perhaps, the diversity of plant species is extraordinarily high (Blumler 1992a, 1992b; Naveh and Whittaker 1979). At a local scale—typically scientific measurements are taken in rectangular one-tenth-hectare plots—diversity of herbaceous vegetation (grassland, open woodland) may approach that of the tropical rain forest, and appears to be higher than in any grass-dominated ecosystems elsewhere in the world. Local diversity in these grassland and woodland landscapes is much higher than in any plant community in the United States, for example (Blumler 1992b). Diversity also is very high when extrapolated to the entire region (Davis et al. 1971; Zohary and Feinbrun-Dothan 1966–1988).

Particularly abundant and diverse are annual plants (herbs that complete their life cycle from germination to maturity in less than a year). These plants are well-attuned to the climatic rhythm, germinating with the autumn rains, flowering often spectacularly in spring, and maturing as the summer drought intensifies. Thus they avoid the drought season as dormant seed. Many perennial plants also go dormant at this time, persisting as bulbs, corms, or tubers in the soil. Other perennials, on the other hand, have adaptations to resist drought. Some species drop most of their leaves and continue photosynthesis at a reduced rate through their stems; others have tough, waxy leaves ("sclerophylls") that do not transpire water as rapidly as softer leaves; still others tap moisture sources within water-retentive rocks (Blumler 1992b; Di Castri and Mooney 1973; Di Castri et al. 1981; Orshan and Zand 1962). Adaptations to tolerate drought inherently translate into reduced rates of photosynthesis, so perennial plants grow more slowly on a daily basis than do the annuals. This gives annuals a competitive advantage that they lack in other environments (Blumler 1992b, 1993). Vegetation generally has a scrubby appearance with low-growing shrubs and herbs in various combinations, designated "maquis," "batha," "steppe," and so on; only unusually favorable sites are clothed with tall forest (Blumler 1992b; Zohary 1962, 1973). The extensive areas of open woodland known as the "park-forest" landscape comprise perhaps the most interesting environment, since it is characterized by the highest diversity and is the natural habitat of the ancestors of the earliest domesticated plants.

Traditionally, pastoralism has been adopted in places that are too dry, too cold, or too mountainous for farming, whereas agriculture has been characteristic wherever moisture and soil fertility allow. However, at times the boundary between desert and sown land shifts. During periods of strong state control, as in Roman times, farming expands outward, whereas pastoral nomads pushed farmers back a great distance at other times (Noy-Meir and Seligman 1979; Smith 1931). Other factors also enter in, such as disease (malaria can depopulate fertile floodplains) and climate change. Despite the supposedly degraded nature of the landscapes, agriculture continues to thrive; for example, in Israel and Syria farming now extends into the desert to the limits established under the Romans (Evenari et al. 1982).

Human Impacts: Traditional Perceptions and Ecological Theory

Most scholars assert that the Near East suffers from severe deforestation, overgrazing, vegetation "regression" (replacement of taller, woodier plants by smaller, short-lived herbs, in a reversal of succession), massive soil erosion, and desertification. These claims are based upon perceptual models rather than empirical data, however (table 8.1). Surprisingly, good supporting evidence exists only for the reports of accelerated soil erosion. For instance, it is often claimed that pollen diagrams (see chapter 6) demonstrate deforestation in the Near East during recent millennia (Bottema 1982 [1985], 1991). But in fact, they consistently show low percentages of arboreal pollen during the last glacial period, with a subsequent increase, due presumably to regrowth of woods, that ends in the early to mid-Holocene; thereafter, most diagrams show either stasis, or fluctuations with no indication of trend, in arboreal pollen (Blumler 1995b). Of course, lack of evidence, or contrary evidence, does not necessarily mean that traditional viewpoints are incorrect, but it does indicate the need for reassessment. This is particularly so given that recent empirical data from other places, such as Sub-Saharan Africa, have cast doubt on the traditional picture of degradation (table 8.1), and have led to increasing appreciation of the validity of the nonequilibrium view of nature in arid and semiarid lands (Behnke and Scoones 1992; Mace 1991; Turner, this volume; Walker et al. 1981; Westoby et al. 1989). Finally, the extraordinarily high species diversity of Near Eastern vegetation would argue against such severe degradation as most scholars have depicted.

In the equilibrium view, the presumption is that vegetation change (succession/regression, deforestation, grazing impacts), soil erosion, and development are essentially linear processes, with progressively greater population pressure and technological development over the long-term producing progressively more negative effects on the environment. Beliefs about human impacts on vegetation are based explicitly on traditional succession theory (Clements 1916), an equilibrium model in which distur-

Table 8.1. Perceptions of degradation in the Near East, and the empirical evidence.

Perceived type of degradation	Supporting empirical evidence	Contrary evidence and commentary
Deforestation	Very little.	Most reports of deforestation (e.g., Mikesell 1969; Thirgood 1981) fail to consider the fact that forests can and often do regenerate after cutting; empirical studies provide no good evidence for significant long-term deforestation (Rackham 1982, 1990; Willcox 1991).
Overgrazing	None.	Empirical studies in other dry lands (e.g., E. Africa) demonstrate that it is very difficult to overgraze, because of inherent negative feedbacks (Behnke and Scoones 1992; Coughenor et al. 1985; De Leeuw and Tothill 1990; Mace 1991; Walker et al. 1981; Westoby et al. 1989).
Vegetation regression	Scanty, and only from infertile or rocky substrates (Litav 1967; Litav et al. 1963).	Regression is *assumed* based on evidence from southern France and other regions that are not climactically equivalent to the Near East, and which consequently exhibit a different successional pattern.
Soil erosion	Considerable—although whether the erosion has actually degraded the land or not is another question.	Soil generally moves elsewhere within the region, and only a small percentage is actually lost to the sea; aeolian input from desert regions partially counteracts erosion from agricultural areas; erosion may *favor* trees and other woody plants.
Desertification	Only in specific, small areas.	Empirical studies in other regions (e.g., Sahel) demonstrate that desertification has been greatly exaggerated, and perhaps is a nonissue (Helldén 1991; Thomas 1993; Thomas and Middleton 1994; Tucker et al. 1991).

Note: For detailed reviews on these topics, with evaluation of the available empirical evidence, see Blumler (1984, 1992b, 1993, 1994a, 1995, 1995b).

bance upsets the natural order of things but, given sufficient relief from human influences, vegetation progressively recovers its potential highest ("climax") state (figure 8.1). Succession diagrams based on this theory were developed for the Mediterranean region, and subsequently applied to the Near East. Such diagrams depict a linear sequence from bare ground to annual plants, perennial herbs, low shrubs, tall shrubs, and finally, climax forest; they also predict that under human impacts vegetation retrogresses in the opposite direction (Colloques Phytosociologiques 1977; Jackson 1985; Mikesell 1969; Tomaselli 1977). Even when not formally codified in a diagram, most authorities have a working assumption of the linearity of vegetation regression under human impact (Hughes 1983; Thirgood, 1981). Models that predict grazing impacts upon soil erosion

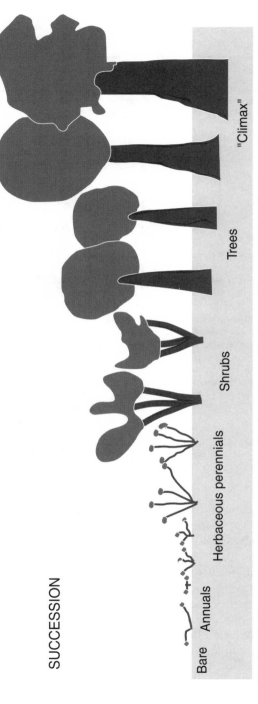

SUCCESSION

Bare   Annuals   Herbaceous perennials   Shrubs   Trees   "Climax"

Fig. 8.1. The Clementsian model of succession after disturbance in natural landscapes (Clements 1916). Similar, but more beautiful, diagrams can be found in all ecology texts. Such diagrams idealize and therefore encourage acceptance of the model, whereas the sketch here trivializes it, and therefore makes it easier to perceive the underlying assumptions of progress, linearity, equilibrium, and bioutopia (climax).

rates, such as those of Thornes (1985, 1990) also are predicated upon this assumption (Blumler 1993).

But this equilibrium view is not in accord with the new ecology (Blumler 1993; Connell and Slatyer 1977). Today it is increasingly recognized that natural disturbances are pervasive and that human impacts tend to be complicated rather than unilinear (Blumler 1994a, 1996a; Botkin 1990; Sprugel 1991). It also is recognized that succession theory has serious weaknesses. In particular, the theory assumes that competition for light is all that matters in the absence of disturbance: Taller plants eventually shade out smaller ones, and "win" the race to climax. This is not too out-of-line with reality in humid climates, but breaks down badly in dry regions, especially where there is seasonal drought: There, competition for water in the soil may be more important than competition for light (Blumler 1994b; compare to Litav et al. 1963). Therefore plants of relatively small stature may compete strongly with trees; theoretically, severe seasonal drought might even enable annual plants to be "climax," because they are unaffected by the drought yet grow faster than other plants during the rainy season (Blumler 1984, 1992b).

In this light, it is instructive to examine a few of the works that have been most influential in fostering the traditional picture of regional degradation. A number of treatises on Near Eastern soil erosion were published after the American Dust Bowl of the 1930s, of which the most widely-read was *Topsoil and Civilization* (Carter and Dale 1974). The authors of these works collected anecdotal (observational and historical) accounts, but made no attempt to gather quantitative empirical data (even their empirical studies of Dust Bowl erosion are now regarded as sloppy and inaccurate [Bowden 1977]). Nor was there any attempt to be judicious, since the purpose for painting a gloomy picture of the Near East was to warn Americans about downward trends in their own regions. For example, Carter and Dale claimed an ancient loss of agricultural productive capacity in Italy, Sicily, North Africa, and Syria, but more careful studies have demonstrated otherwise (Bréhaut in Cato [1933]; Braudel 1972; Murphey 1951; White 1970). Accelerated soil erosion certainly occurred in many places, but only in an equilibrium perspective does it necessarily follow that a degraded landscape was the result. For one thing, soil moisture is more likely to be the limiting factor than nutrient status in dry climates (Noy-Meir 1973; Noy-Meir and Seligman 1979), and shallow soils may have better moisture status than deep soils if underlain by water-retentive rocks. On the other hand, the Nabataeans *intentionally* channeled soil and water *off* slopes in order to facilitate agriculture, in an environment so dry that water could only become useful for plant growth if concentrated (Evenari et al. 1982).

The essential authority on Near Eastern vegetation is the late Michael Zohary, who wrote the two-volume *Geobotanical Foundations of the Middle East* (1973). Like other phytosociologists, Zohary accepted Clements' views on succession. He also accepted that plant species occur together repeatedly in characteristic groups (the literal meaning of "phytosociology"), another Clementsian hypothesis that was tested and rejected by English-speaking ecologists a long time ago (Gleason 1926; Whittaker 1956). Phytosociologists, on the other hand, do not test their fundamental beliefs; instead (and because the influence of humans is so pervasive in the Mediterranean and Near East), any deviation from Clementsian expectation is simply ascribed to human impacts. Supporting evidence is not felt to be necessary. For instance, Zohary argued that the presence of even a single tree proved the former existence of dense forests. He mapped a steppe forest (*Pistacia atlantica–Amygdalus* species) climax in Khuzistan, despite the fact that the individual trees are often *kilometers* apart and occur only around rocks in a landscape of generally deep soils (Wright et al. 1967). Similarly, Zohary argued that *Zizyphus lotus* could not naturally have grown at higher elevations than Mediterranean oaks—despite the fact that the species clearly flourishes at such elevations today—because it is from a genus with tropical affinities. Therefore, its present distribution must be due to human impacts on the vegetation, and represent an example of degradation. Examples of such reasoning—plausible in an equilibrium perspective but not in a nonequilibrium one—permeate Zohary's otherwise magnificent work.

A more widely-read book on deforestation is Thirgood's *Man and the Mediterranean Forest* (1981). Thirgood attempted to show, through historical study, that Mediterranean forests have been greatly reduced in extent and degraded. As is typical of works of this sort, Thirgood emphasized episodes of deforestation, while paying insufficient attention to the often excellent ability of forests to regenerate. Nonetheless, it is clear from his own presentation of the evidence that regeneration of cut-over forests was characteristic. In this regard, it is important to consider that until recently the technology did not exist to clear-cut huge swaths of forest at a single time. Ancient "deforestation" was usually piecemeal, gradual, and left trees standing that could serve as seed sources for new individuals. Moreover, it is now recognized (although not by Thirgood) that many if not most forests are naturally subject to catastrophic destruction from blowdowns, fire, pest outbreaks, and so on, so that even large clear-cuts should not eliminate forest unless regeneration is prevented by further development. Finally, Thirgood believed that goat grazing has had a disastrous effect in preventing tree establishment. He reported that the 141-square-mile Papho forest reserve was carrying 7500 goats, a supposed indication of the severe pressure to which the vegetation is subject in Cyprus. However, these numbers work out to the equivalent in animal units of 10–11 steers/square mile, or 5 A/AUM—a light stocking rate! Exaggeration of both the density of stocking and the depredations of goats seems to be characteristic of northern European reports on Third World conditions (compare Coughenor et al. 1985).

Human Impacts and Conservation: Some Empirical Results

My Israeli colleagues Z. Naveh and L. Olsvig-Whittaker and I undertook a series of studies of vegetation and diversity dynamics with and without human impacts. Most studies were in an open oak (*Quercus ithaburensis*) woodland at Neve Ya'ar in the Lower Galilee, with average annual rainfall of 578 millimeters, productive brown rendzina soil formed over calcrete (which also outcrops as scattered, loaf-shaped rocks), and about 150 species/one-tenth hectare. We fenced off pastures that were placed under differing levels of grazing intensity; we repeatedly sampled these along permanent transects and determined the total number of plant species present to determine quantitatively how vegetation changes over time when grazing is relaxed or intensified. In addition, we quantified the tendencies of species to occur in specific microsites, because we suspected that environmental heterogeneity plays a major role in maintaining diversity. Some results are highly preliminary, coming from a single site studied for a few years. Much more research is needed to verify their generality. The seven main results from these studies are briefly summarized below.

1. In the absence of grazing and other disturbance, large-seeded annual "wild cereal" grasses such as wild barley (*Hordeum spontaneum*) and wild oats (*Avena sterilis*) become overwhelmingly dominant in deep soil.

2. Most woody plants and perennial herbs occur preferentially around rock outcrops, and most if not all require disturbance, such as grazing, to establish in deep soil. This result is the complete opposite to predictions based on traditional succession theory. Annual plants should give way to perennials in the absence of disturbance, and retreat to marginal habitats such as rock outcrops. Numerous other studies have generated similar results (Litav 1965; Litav et al. 1963; Naveh and Whittaker 1979; Noy-Meir 1990; Noy-Meir and Seligman 1979; Rackham 1982, 1990; Zohary 1969), but scientists sometimes are reluctant to accept them because they are so contrary to equilibrium expectation. From the nonequilibrium perspective of the new ecology, however, the results are not so surprising. It is likely that

fast-growing annual plants are able to use up much of the soil water during their growing season, leaving little available for perennials to draw upon during the long, rainless summer. In addition, precisely because they must be able to deal with the summer drought, perennials do not grow as rapidly as annuals do. Therefore, dense stands of annuals can be strongly competitive during the rainy season. With grazing, the annual stands may be thinned out, allowing occasional establishment of a perennial in the openings. On the other hand, rocks can shelter plants against competition, and often maintain high moisture levels through the summer by mulching against evaporation.

This illustrates why French results should not be extrapolated to regions with rainless summers. In southern France, a significant quantity of rainfall may fall after the annuals die off at the beginning of the summer—enough to allow the establishment of perennial seedlings during the dormant period of the annuals. But summer establishment of seedlings is not a viable strategy in Israel and the rest of the Fertile Crescent.

3. Most species (annuals, too) were found primarily around rocks.
4. Suites of species are adapted to different microsites such as microscopic rock depressions (*Campanula, Centaurium*), shallow rock pockets (*Poa, Sedum*), deep rock pockets (*Theligonum, Mercurialis*), rock crevices (*Aristolochia, Dianthus*), shaded rocks (*Cyclamen, Plumbago*), and ant mounds (*Silybum*) (Blumler 1992b).

These findings were no surprise, since it is generally accepted that environmental heterogeneity increases species diversity, and the rocks have a high diversity of microhabitats.

5. Diversity declines when there is no grazing or other disturbance. It also appears to decline at very high levels of grazing and, presumably, other disturbances.
6. Suites of species are adapted to different types of disturbance.
7. Rock outcrops slow the decline in diversity that occurs under undisturbed conditions by serving as a temporary refuge for species that normally do better in open ground. Because the two contrasting environments interfinger so thoroughly at Neve Ya'ar, the resilience of the system is increased and diversity is enhanced considerably.

These results suggest that variations in disturbance regime favor diversity. Ungrazed conditions favor a few species, such as the wild cereal grasses; heavy grazing favors others. Theoretically, then, shifting back and forth between no grazing and heavy grazing should allow species favored by each to be maintained in the environment (especially if the rocks are enabling species to hold out under unfavorable conditions). This argument is similar to that which Connell

(1978) put forth to explain the tendency of species diversity to be maximized at some intermediate frequency and intensity of disturbance. Connell's intermediate disturbance hypothesis is popular, but suffers from the vagueness of the term "intermediate" and does not seem to explain patterns of diversity fully. My studies indicate that intermediate disturbance is better than either no disturbance or severe disturbance because, in a sense, it incorporates both of them temporally (Blumler 1992a). I suggest a modification of the hypothesis by arguing that "disturbance-heterogeneity" (variety of disturbance types and regimes), and not simply intensity of disturbance, is the key to species diversity (figure 8.2). As figure 8.2 shows, it is likely that each type of disturbance (fire, drought, plowing, etc.) favors a particular suite of species, and all species remain in the system as long as the different disturbance types cycle through with sufficient frequency. This elaboration of Connell's ideas takes his nonequilibrium explanation for diversity and makes it more so, by extending it to additional dimensions.

Discussion: Implications for Conservation-with-Development

Under the old ecology, disturbance was seen as detrimental to nature, and therefore almost inevitably all types of development were, too. Moreover, the sorts of plants that would take over in the absence of disturbance (or, conversely, in its presence) could be predicted from a simple linear succession diagram. Now, however, it seems that these principles do not hold. Disturbance, especially if of a varied nature, may actually favor biodiversity; it is not necessarily true that intensive development favors small annuals, or that relaxation of human impacts favors trees. One might even hypothesize, from the pronounced tendency of Near Eastern trees and other woody plants to occur around rocks, that they are favored by soil erosion. Improved understanding of successional relationships in drylands is perhaps the most crucial scientific need in improving conservation policy recommendations.

Biodiversity is a major concern today. Estimates of the rate of species loss proliferate, based primarily on species-area equations encoded in *The Theory of Island Biogeography* (MacArthur and Wilson 1967; for example, Wilson, 1992). It is well-established that as the size of an area increases, so does its diversity at a fairly regular rate (the number of species approximately doubles with each tenfold increase in area); therefore, with habitat destruction, the number of species should decrease. However, the estimates of current species extinction rates assume an instantaneous response: Reduce the rain forest to one-tenth of its former acreage, and

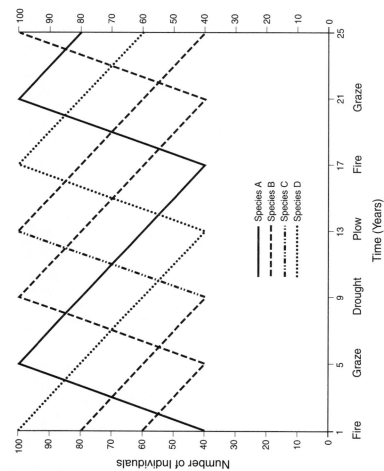

Fig. 8.2. The disturbance-heterogeneity hypothesis (Blumler 1992a). If a species is favored by one type of disturbance but not by others, it will increase in abundance after the former but decrease after the latter. As long as the type of disturbance that it "needs" cycles through frequently enough, it can remain in the landscape. If there are many species adapted to different sorts of disturbances, then overall species diversity can be greatly enhanced if disturbances are also diverse.

the number of species is reduced by 50%. This overlooks the well-known phenomenon of "relaxation"—that extinction is typically *not* instantaneous but protracted, so that species richness gradually declines to the predicted value. Relaxation can take thousands of years. This is good news to conservationists, since it suggests that a window of time is available to preserve many threatened taxa. It is likely that providing local refuges, such as the rock outcrops of our Neve Ya'ar studies, is the key to prolonged relaxation.

Large animal diversity in the Near East is low, the result of extinction events occurring over the long-term. Since such animals would not be able to hide in small rock outcrops, it may be that refuges were not available for them; but it also is likely that some species (such as lions) were intentionally extirpated. Among plants, on the other hand, the species subjected to the most sustained exploitation arguably has been the cedar of Lebanon (*Cedrus libani*). Not only the Phoenicians but also the rulers of desert civilizations to the west (Egypt) and east (Mesopotamia) deforested Mount Lebanon (Mikesell 1969). In addition, goat grazing has exerted intense pressure. Yet huge stands are still present in Turkey (Zohary 1973), and it is not even extinct in Lebanon. The key, again, may be the limestone rocks that provide crannies in which seedlings can hide from the goats. Despite the soil erosion that surely has taken place (or perhaps because of it), regeneration is rapid when the goats are removed.

This illustrates that although the human imprint on the Near Eastern landscape is major, effects are often conflicting and diffuse rather than unilinear and utterly degradational. For instance, people destroy woody plants when they gather them for fuel or construction material, but may *favor* their establishment and survival when domestic animals graze and weaken the competing herbs, or when soil erosion exposes rocks that then protect woody seedlings and provide them a favorable, summer-moist microclimate (Blumler 1993). Eig (1926) pointed out long ago that the traditional Near Eastern plow also does not have the negative effects on woody plants that most scholars have assumed, because it barely scratches the soil surface. Rather, its major effect is to destroy the seedlings of annual plants, which in turn enables some herbaceous perennial and woody weeds to flourish and spread (see also Zohary, 1950). The evolutionary changes induced in Near Eastern plants under agropastoralism also seem to be complicated, diffuse, and to some degree contradictory, rather than unilinear (Blumler 1994a). In part, the resilience of Near Eastern landscapes appears to be natural and perhaps unique, because of the characteristic interfingering of rock outcrops and deep soil. It also reflects in part the mixed and varying nature of the land use, which adds disturbance heterogeneity to the microsite heterogeneity.

Space precludes discussion of soil erosion in the detail that it deserves. However, it is worth emphasizing that assumptions similar to those about vegetation underlie the gloomy statements that pervade the popular literature on soils for the Near East and the Mediterranean region—for example, that deeper soil is better soil, older soil is better soil, natural erosion is slow and gradual, or that all eroded soil is lost. These assumptions reflect turn-of-the-century beliefs about soils, which were consciously aligned with Clements' views on vegetation and succession (Drury and Nisbet 1971; McIntosh 1985). But soil scientists now know that all these assumptions are false (Blumler 1995b). In dry regions, deep soils are not necessarily the ones with the best moisture availability. Old soils tend to become depleted of bases and available phosphorous; consequently, a certain amount of erosion is necessary to maintain fertility. Natural erosion seems to be episodic: gradual for much of the time, with pulses of catastrophic movement off slopes. Most eroded soil is moved to some other terrestrial location, and farmers tend to respond by shifting the location of agriculture and the crops grown. On slopes, for instance, erosion may cause soils to become unsuitable for cereals (annual grasses), but favor the cultivation of woody plants such as the vine and olive. Soil scientists shifted away from traditional views a long time ago (Jenny 1941; Jenny et al. 1969; Noy-Meir 1973), but that fact is not reflected in the literature on land degradation.

All this suggests that conservation-with-development is possible, particularly if care is taken to ensure that impacts are varying rather than uniform. I have noticed, for instance, that traditional Arab grazing tends to involve moving herds from place to place in a pattern that is intelligent, in that previously hard-hit pastures are unlikely to be grazed if a less utilized area is available nearby, but is anything but systematic. Some of these areas are farmed from time to time, and the rockier places that support good shrub cover may be burned occasionally to open them up. Again, such practices seem anything but systematic. In contrast, and in accordance with the equilibrium view of nature, scientific and government management prescriptions typically emphasize finding a specific "best" management strategy for a given biogeographical landscape, usually incorporating the elimination of disturbance as much as possible. The point is that perhaps there is no single best strategy for any given biogeographical landscape after all, except in the immediate short-term; rather, it may be best policy to encourage flexible, varying land use. Of course, this is particularly unlikely to be the sort of prescription adopted by government agencies, whose bureaucracies are accustomed to detailed rules of procedures (figure 8.3).

To end the confusion once and for all, the Park Service has decided to paint all stalactites RED and stalagmites BLUE.

Fig. 8.3. Bureaucracies prefer uniformity and strict rules rather than flexibility of response. Copyright © Tribune Media Services. All rights reserved. Reprinted with permission.

Concluding Comment

The West has long been oriented towards the finding of solutions to problems. As a purely pragmatic approach, this is fine. Often, however, problem-solving incorporates the notion of progress—the assumption that by solving a series of problems, one can improve conditions in more or less linear fashion. This is particularly likely to be so when Western-trained conservation managers, scientists, and others idealistically tackle Third World development. I would argue, however, that prediction of development impacts on the environment requires recognition that nonlinearity is characteristic. From the perspective of long-term history, problems are never really solved, at least in the sense that their solution generally creates new, and often unforeseen, problems. *Change* is characteristic, but permanent improvement (or degradation) may not be. This is a view that is closer to that of the *I Ching* (Ritsema and Karcher 1995) and other eastern philosophies, and one that I think is needed at this particular point in our history. The *I Ching* ("Book of Changes") was created about 3000 years ago to help guide individuals in choosing the best course of action in circumstances that were not at all clear. In the *I Ching* good does not necessarily lead to more good, or bad to bad; reality is complex, and the effects of one's actions have ramifications that may be rather surprising. This encourages flexibility of response. And I would argue that we also need to encourage flexibility of response rather than a search for solutions as we seek to develop meaningful conservation-with-development strategies.

References

Behnke, R. H., and I. Scoones. 1992. *Rethinking Range Ecology: Implications for Rangeland Management in Africa.* Overseas Development Institute Dryland Networks Programme Issues Paper. London: International Institute for Environment and Development

Blumler, M. A. 1984. *Climate and the Annual Habit.* M.A. thesis, University of California–Berkeley.

Blumler, M. A. 1992a. *Microsite/Disturbance Heterogeneity and Species Diversity in a Mediterranean Oakpark.* Paper presented at the Association of American Geographers Annual Meeting, San Diego, April 19, 1992.

Blumler, M. A. 1992b. *Seed Weight and Environment in Mediterranean-type Grasslands in California and Israel.* Ph. D. thesis, University of California–Berkeley.

Blumler, M. A. 1993. Successional Pattern and Landscape Sensitivity in the Mediterranean and Near East. In *Landscape Sensitivity,* eds. D. S. G. Thomas and R. J. Allison, pp. 287–305. Chichester: John Wiley and Sons.

Blumler, M. A. 1994a. Evolutionary Trends in the Wheat Group in Relation to Environment, Quaternary Climate Change and Human Impacts. In *Environmental Change in Drylands,* eds. A. C. Millington and K. Pye, pp. 253–269. Chichester: John Wiley and Sons.

Blumler, M. A. 1994b. *Succession Theory and Vegetation Management.* Paper presented at the Applied Geography Conference, Akron, October 13, 1994.

Blumler, M. A. 1995a. *Nonlinear Dynamics of Human-Environment Interactions in the Near East.* Paper presented at the Association of American Geographers Annual Meeting, Chicago, March 16, 1995.

Blumler, M. A. 1995b. *Palynology, Ecological Theory and Forest History in the Mediterranean and Near East.* Paper presented at the Institute of British Geographers Annual Conference, Newcastle, January 4, 1995.

Blumler, M. A. 1996a. Ecology, Evolutionary Theory, and Agricultural Origins. In *The Origins and Spread of Agriculture and Pastoralism in Eurasia,* ed. D. R. Harris, pp. 25–50. London: UCL Press.

Blumler, M. A. 1996b. *Wheat, Bread, the Waterwheel, and the Industrial Revolution.* Paper presented at the Society for Economic Botany Annual Meeting, London, July, 1996.

Blumler, M. A., L. Olsvig-Whittaker, and n.d. Naveh. Z. *Between a Rock and a Hard Place: Succession, Vegetation Resilience, and Microsite Heterogeneity on Calcareous Substrate in Israel* In preparation.

Bottema, S. 1982 (1985). Palynological Investigations in Greece with Special Reference to Pollen as an Indicator of Human Activity. *Palaeohistoria* 24: 257–289.

Bottema, S. 1991. Développement de la végétation et du climat dans le bassin Méditerranéen Oriental à la fin du Pléistocène et pendant l'Holocène. *L'Anthropologie* 95: 695–728.

Bowden, M. J. 1977. Desertification of the Great Plains: Will It Happen? *Economic Geography* 53: 397–406.

Braudel, F. 1972. *The Mediterranean and the Mediterranean World in the Age of Philip II.* New York: Harper and Row.

Braun-Blanquet, J., N. Roussine, and R. Nègre. 1951. *Les groupements Végétaux de la France Méditerranéenne.* Montpellier: Centre National de la Récherche Scientifique.

Carter, V. G., and T. Dale. 1974. *Topsoil and Civilization.* Norman: University Oklahoma Press.

Cato. 1933. *On Agriculture.* Trans. E. Bréhaut. New York: Columbia University Press.

Clements, F. E. 1916. *Plant Succession: An Analysis of the Development of Vegetation.* Washington, D.C.: *Carnegie Institute of Washington.*

Colloques Phytosociologiques. 1977. *La Vegetation des Pelouses Seches à Therophytes.* Volume 6.

Connell, J. H. 1978. Diversity in Tropical Rain Forests and Coral Reefs. *Science* 199: 1302–1310.

Connell, J. H. and R. O. Slatyer. 1977. Mechanisms of Succession in Natural Communities and Their Role in Community Stability and Organization. *The American Naturalist* 111: 1119–1144.

Coughenor, M. B., J. E. Ellis, D. M. Swift, D. L. Coppock, K. Galvin, J. T. McCabe, and T. C. Hart. 1985. Energy Extraction and Use in a Nomadic Pastoral Ecosystem. *Science* 230: 619–625.

Davis, P. H., P. C. Harper, and I. C. Hedge, eds. 1971. *Plant Life of South West Asia.* Edinburgh: Royal Botanical Garden.

De Leeuw, P. N., and J. C. Tothill. 1990. The Concept of Rangeland Carrying Capacity in Sub-Saharan Africa—Myth or Reality. *Pastoral Development Network.* Paper 29b.

Di Castri, F., and H. A. Mooney, eds. 1973. *Mediterranean-Type Ecosystems: Origin and Structure.* Berlin: Springer-Verlag.

Di Castri, F., D. W. Goodall, and R. L. Specht, eds. 1981. *Mediterranean-Type Shrublands.* New York: Elsevier.

Drury, W. H., and I. C. T. Nisbet 1971. Inter-relations between Developmental Models in Geomorphology, Plant Ecology, and Animal Ecology. *General Systems* 16: 57–68.

Eig, A. 1926. *A Contribution to the Knowledge of the Flora of Palestine. Agricultural Experiment Station Bulletin 4.* Jerusalem: Zionist Organization and Hebrew University Institute of Agriculture and National History.

Evenari, M., L. Shanan, and N. Tadmor. 1982. *The Negev: The Challenge of a Desert.* 2nd edition. Cambridge: Harvard University Press.

Gleason, H. A. 1926. The Individualistic Concept of the Plant Association. *Bulletin of the Torrey Botanical Club* 53: 7–26.

Helldén, U. 1991. Desertification—Time for an Assessment? *Ambio* 20: 372–383.

Hughes, J. D. 1983. How the Ancients Viewed Deforestation. *Journal of Field Archaeology* 10: 437–445.

Jackson, L. E. 1985. Ecological Origins of California's Mediterranean Grasses. *Journal of Biogeography* 12: 349–361.

Jenny, H. 1941. *Factors of Soil Formation.* New York: McGraw-Hill.

Jenny, H., R. Arkely, and A. M. Schultz. 1969. The Pygmy Forest Podsol Ecosystem and Its Dune Associates of the Mendocino Coast. *Madroño* 20: 60–74.

Le Houérou, H. N. 1981. Long-Term Dynamics in Arid-Land Vegetation and Ecosystems of North Africa. In *Arid-Land Ecosystems: Structure, Functioning, and Management,* eds. D. W. Goodall, R. A. Perry, and K. M. N. Howes, pp. 357–384. New York: Cambridge University.

Litav, M. 1965. Effects of Soil Type and Competition on the Occurrence of *Avena Sterilis* L. in the Judean Hills (Israel). *Israel Journal of Botany* 14: 74–89.

Litav, M. 1967. Micro-environmental Factors and Species Interrelationships in Three Batha Associations in the Foothill Region of the Judean Hills. *Israel Journal of Botany* 16: 79–99.

Litav, M., G. Kupernik, and G. Orshan. 1963. The Role of Competition as a Factor Determining the Distribution of Dwarf Shrub Communities in the Mediterranean Territory of Israel. *Journal of Ecology* 51: 467–480.

Lowdermilk, W. C. 1943. Lessons from the Old World to the Americas in Land Use. In *Annual Report of the Smithsonian Institution,* pp. 413–427. Washington D.C.: Government Printing Office.

MacArthur, R., and E. O. Wilson. 1967. *The Theory of Island Biogeography.* Princeton: Princeton University Press.

Mace, R. 1991. Overgrazing Overstated. *Nature* 349: 280–281.

Macleod, D. A. 1980. The Origin of the Red Mediterranean Soils in Epirus, Greece. *Journal of Soil Science* 31: 125–136.

Marsh, G. P. 1864. *Man and Nature, or Physical Geography as Modified by Human Action.* New York: Charles Scribner and Company.

McIntosh, R. P. 1985. *The Background of Ecology.* Cambridge: Cambridge University Press.

Mikesell, M. 1969. The Deforestation of Mount Lebanon. *Geographical Review* 59: 1–28.

Mizota, C., M. Kusakabe, and M. Noto. 1988. Eolian Contribution to Soil Development on Cretaceous Limestones in Greece as Evidenced by Oxygen Isotope Composition of Quartz. *Geochemical Journal* 22: 41–46.

Murphey, R. 1951. The Decline of North Africa since the Roman Occupation: Climatic or Human? *Annals of the Association of American Geographers* 41: 116–132.

Naveh, Z., and R. H. Whittaker. 1979. Structural and Floristic Diversity of Shrublands and Woodlands in Northern Israel and other Mediterranean Countries. *Vegetatio* 41: 171–190.

Noy-Meir, I. 1973. Desert Ecosystems: Environment and Producers. *Annual Review of Ecology and Systematics* 4: 25–52.

Noy-Meir, I. 1990. The Effect of Grazing on the Abundance of Wild Wheat, Barley and Oat. *Biological Conservation* 51: 299–310.

Noy-Meir, I., and N. Seligman. 1979. Management of Semi-arid Ecosystems in Israel. In *Management of Semi-Arid Ecosystems,* ed. B. H. Walker, pp. 113–160. Amsterdam: Elsevier.

Orni, E. 1963. *Reclamation of the Soil.* Jerusalem: Israel Digest.

Orshan, G., and G. Zand. 1962. Seasonal Body Reduction of Certain Desert Half-Shrubs. *Bulletin of the Research Council of Israel* 11D: 35–42.

Rabinovitch-Vin, A. 1983. Influence of Nutrients on the Composition and Distribution of Plant Communities in Mediterranean-type Ecosystems of Israel. In *Mediterranean-Type Ecosystems: the Role of Nutrients,* eds. F. J. Kruger, D. T. Mitchell, and J. U. M. Jarvis, pp. 74–85. Berlin: Springer-Verlag.

Rackham, O. 1982. Land-Use and the Native Vegetation of Greece. In *Archaeological Aspects of Woodland Ecology,* eds. M. Pull and S. Limbrey, pp. 177–198. Oxford: British Archeological Report.

Rackham, O. 1990. The Greening of Myrtos. In *Man's Role in the Shaping of the Eastern Mediterranean Landscape,* eds. S. Bottema, G. Entjes-Nieborg, and W. van Zeist, pp. 341–348. Rotterdam: Balkema.

Reifenberg, A. 1955. *The Struggle between the Desert and the Sown.* Jerusalem: Jewish Agency.

Ritsema, R., and S. Karcher, trans. 1995. *I Ching.* New York: Barnes and Noble.

Smith, G. A. 1931. *The Historical Geography of the Holy Land.* Jerusalem: Ariel Press.

Thirgood, J. V. 1981. *Man and the Mediterranean Forest.* London: Academic Press.

Thomas, D. S. G. 1993. Sandstorm in a Teacup? Understanding Desertification. *Geographical Journal* 159: 318–331.

Thomas, D. S. G., and N. J. Middleton. 1994. *Desertification: Exploding the Myth.* Chichester: John Wiley and Sons.

Thornes, J. B. 1985. The Ecology of Erosion. *Geography* 70: 222–235.

Thornes, J. B. 1990. The Interaction of Erosional and Vegetational Dynamics in Land Degradation: Spatial Outcomes. In *Vegetation and Erosion,* ed. J. B. Thornes, pp. 41–53. Chichester: John Wiley and Sons.

Tomaselli, R. 1977. Degradation of the Mediterranean Maquis, UNESCO. *Programme on Man and the Biosphere Technical Notes* 2: 33–72.

Tucker, C. J., H. E. Dregne, and W. W. Newcomb. 1991. Expansion and Contraction of the Sahara Desert from 1980 to 1990. *Science* 253: 299–301.

Walker, B. H., D. Ludwig, C. S. Holling, and R. M. Peterman. 1981. Stability of Semi-arid Savanna Grazing Systems. *Journal of Ecology* 69: 473–498.

Westoby, M., B. H. Walker, and I. Noy-Meir. 1989. Opportunistic Management for Rangelands Not At Equilibrium. *Journal of Range Management* 42: 266–273.

White, K. D. 1970. *Roman Farming.* Ithaca: Cornell University Press.

Whittaker, R. H. 1956. Vegetation of the Great Smoky Mountains. *Ecological Monographs* 26: 1–80.

Willcox, G. 1991. Exploitation des espèces ligneuses au Proche-Orient: données anthracologiques. *Paléorient* 17(2): 117–124.

Wilson, E. O. 1992. *The Diversity of Life.* Cambridge: Harvard University Press.

Wright, H. E., J. H. McAndrews, and W. van Zeist. 1967. Modern Pollen Rain in Western Iran and Its Relation to Plant Geography and Quaternary Vegetational History. *Journal of Ecology* 55: 415–443.

Yaalon, D. H., and E. Ganor. 1973. The Influence of Dust on Soils During the Quaternary. *Soil Science* 116: 146–155.

Yaalon, D. H., and E. Ganor. 1975. Rates of Aeolian Dust Accretion in the Mediterranean and Desert Fringe Environments of Israel. *Congress of Sedimentology* 2: 169–174.

Zohary, D. 1969. The Progenitors of Wheat and Barley in Relation to Domestication and Agricultural Dispersal in the Old World. In *The Domestication and Exploitation of Plants and Animals,* eds. P. J. Ucko and G. W. Dimbleby, pp. 47–66. London: Duckworth.

Zohary, M. 1950. The Segetal Plant Communities of Palestine. *Vegetatio* 2: 387–411.

Zohary, M. 1962. *Plant Life of Palestine.* New York: Ronald Press.

Zohary, M. 1973. *Geobotanical Foundations of the Middle East.* Stuttgart: Gustav Fischer.

Zohary, M. 1983. Man and Vegetation in the Middle East. In *Man's Impact on Vegetation,* eds. W. Holzner, M. J. A. Werger, and I. Kusima, pp. 287–296. The Hague: Dr W. Junk.

Zohary, M., and N. Feinbrun-Dothan. 1966–1988. *Flora Palaestina.* Jerusalem: Israel Academy of Sciences and Humanities.

9

The Interaction of Grazing History with Rainfall and Its Influence on Annual Rangeland Dynamics in the Sahel

Matthew D. Turner

Introduction: Equilibrium Models and Pastoral Development

Development programs directed at the dryland livestock sector have always had a strong range management emphasis (Anderson 1984; Goldshmidt 1981; Wyckoff 1985), reflecting the long-standing concern about the role of livestock in the ecological degradation of dryland landscapes in Africa (Brown 1971; Chevalier 1934; Cloudsley-Thompson 1974; Lamprey 1983; Sinclair and Frywell 1985; Stebbing 1935; Warren and Maizels 1977). Attempts to introduce western-style range management practices into agropastoral production systems have been met with resistance by livestock-rearing peoples (Baker 1975; Ranger 1985; Tignor 1971). State-sanctioned development projects have often had to depend on market and/ or military means to elicit local "cooperation" in range management programs of destocking, ranching, sedentarization, and regional stratification of production (Ndalaga 1982; Neumann 1995; Oxby 1982; Peluso 1993). Such programs have failed on both ecological and economic grounds (de Haan 1994; Homewood and Rogers 1987; Horowitz 1979; Sandford 1983; Thébaud 1988). As a result of these and other failures, development attention has shifted away from the dryland livestock sector since the mid-

1980s to other areas considered to be of higher investment potential (Derman 1984; International Livestock Centre for Africa 1987).

Despite the important role played by range management concepts in shaping pastoral development programs, their failure did not provoke a reevaluation of the relevance of western range management to development in the dryland landscapes of Africa. Widespread failure was generally attributed instead to incomplete adoption of management prescriptions by livestock producers due to their behavioral attributes (Brown 1971), common property externalities (Picardi and Siefert 1976; Simpson and Sullivan 1984), or poor project design/management (de Haan 1994; Shanmugaratnam et al. 1992). Early critiques of range management's influence on pastoral development came not from physical scientists but from social scientists (Hjort 1982; Homewood and Rodgers 1984; Horowitz and Little 1987; Sandford 1982). These critics were the first to publicly question mainstream assertions of widespread overgrazing in African drylands, emphasizing the dominant influence of climate variability on rangeland productivity.

Such arguments in fact ran parallel to increased questioning within the fields of ecology and biogeography of long-dominant, equilibrium-based successional models (Blumler 1993; Botkin 1990; DeAngelis and Waterhouse 1987; Wiens 1977). While most often associated in North America with Clements (1916), other ecological traditions that have influenced range management in Africa, such as the phytosociological and ecosystems schools, also view vegetative structures evolving toward internally-regulated equilibriums or "climax" types (Braun-Blanquet 1932; Odum 1971; Worster 1990). Such views have significantly influenced range management as practiced in dryland Africa (Boudet 1975; Breman and de Ridder 1991; Pratt and Gwynne 1977; Stoddart et al. 1975). As described by Westoby et al. (1989), the dominant successional paradigm in range management views both grazing pressure and poor rainfall as producing changes (for example, increasing forb fraction of biomass) that move the rangeland away from the more productive climax. Rainfall and grazing pressure are viewed differently because the former is uncontrollably variable, and the latter is controllable through management. In response to drought, for example, management should reduce grazing pressure to counteract drought-induced pressure toward poorer range condition.

Nonequilibrium Range Ecology

The rationale for the more socially coercive aspects of pastoral development such as destocking, ranching, and forced sedentarization can be traced to the successional view that tighter control over stocking rates will result in more productive rangelands. Building upon the early critiques by social scientists and ecological research that demonstrates the strong influence of rainfall fluctuations on the interannual variability of rangeland productivity and species composition (Cissé 1986; Grouzis 1988; Hiernaux et al. 1988; Penning de Vries and Djitèye 1982; Walker et al. 1981), this view began to be more widely criticized within the dryland range management literature in the late 1980s (Ellis and Swift 1988; Westoby et al. 1989). Particularly in East and Southern Africa, the nonequilibrium dynamics of pastoral grazing systems is now receiving significant attention (Behnke et al. 1993; Scoones 1994). The major argument, substantiated by comparing historical series of livestock densities to ecological indicators or to carrying capacities, is that domestic livestock populations in arid and some semiarid environments rarely reach levels at which grazing reduces forage quantity/quality (through direct consumption or reductions in productivity) to levels that would limit the growth of such populations (Abel and Blaikie 1989; Ellis and Swift 1988; McCabe 1987; Scoones 1992; Tapson 1991). In these environments frequent drought, not density dependence, limits livestock populations because of recurrent episodes of emigration, forced sales, reduced fecundity, and higher mortality. Interannual variations in rangeland production and domestic livestock populations are therefore driven not by their interaction but by interannual variation in rainfall, an external factor to the grazing system.

By highlighting the limitations of equilibrium theory and by adding greater historical breadth to environmental analyses, nonequilibrium range ecology has had an important positive influence on ecological perspectives in arid and semiarid landscapes of Africa. As a result, stocking control prescriptions cannot be, as in the past, automatic responses to poor range condition. However, continued controversy and confusion about its broader implications for range management (Behnke et al. 1993; de Queiroz 1993; Overseas Development Institute 1990; Overseas Development Institute 1993; Scoones 1994) are indicative of the oversimplified models of grazing impact used by its proponents and critics alike, a legacy in fact of successional paradigm conventions. There are five major sources of confusion evident in such debates:

1. Different vegetative parameters of concern evaluated with respect to divergent management goals. More often than not, references to the degradation of rangeland are based on a particular statistic (e.g., average, variability, range) of one or several possible vegetative parameters such as species composition, species diversity, aboveground biomass, vegetative cover, and forage quality. The choice of these parameters, as well as the interpretation of their change, is

affected by the management goals assigned to the range site by the analyst (Scoones 1989a).

2. A continued reliance on coarse indicators of grazing-induced ecological stress. Changes in livestock population and comparison of stocking rates to carrying capacities are examples of such indicators. Considerable grazing impact on rangeland could very well occur at stocking densities well below those in which density-dependent mortality occurs. Because of the dearth of controlled grazing experiments in Africa, carrying capacities are nothing but simple rules of thumb—usually estimated by calculating the stocking density that will consume 50% of mean aboveground herbaceous production (Bartels et al. 1990; de Leeuw and Tothill 1990). Clarification of the "overgrazing question" can only begin once analysts shift their focus from such indicators to a more explicit consideration of grazing-initiated processes that are important at different spatial and temporal scales (see points below).

3. Consistent with the carrying capacity approach, there is a tendency to treat grazing pressure as a singular, aggregated stress on particular vegetative parameters. Grazing in fact influences any parameter through multiple pathways (Vale 1982). For example, the influence of grazing on herbaceous species composition can occur through grazing selectivity, differential abilities of different plant species to escape nonselective grazing, and different competitive abilities under livestock-induced changes in soil moisture and nutrient status. As will be further explored in this chapter, it is important to understand the relative importance of these pathways in order to gain a better understanding of the dynamic and spatially-differentiated nature of grazing effects on rangeland vegetation.

4. A tendency to overlook interannual effects of grazing. As argued by proponents of the nonequilibrium school, interannual variation in rainfall is indeed what drives observed interannual patterns in production and species composition of arid rangeland ($<$ 500 millimeters/year). This does not imply that grazing has little or no effect on arid rangelands. Interannual effects of grazing may have slow progressive effects on vegetative parameters such as nutrient depletion (van Keulen and Breman 1990), or display significant threshold or synergistic (grazing with external factor) effects such as soil structural changes and changes in tree/shrub cover (Friedel 1991; Laycock 1991; Savage 1991; Veblen and Lorenz 1988; Westoby et al. 1989). As shown graphically by Smith and Pickup (1993), such impacts would result in gradual or stepped changes in the vegetative parameter,

changes which would not be discernable from simply monitoring interannual patterns alone.

5. Inadequate consideration of the importance of spatial scale in evaluating grazing impact (Forman and Godron 1986; Pickett and White 1985; Vale 1982). Different causal pathways, working at different biological organization levels, affect vegetative parameters in a spatially-differentiated fashion (Brown and Allen 1989; Coughenour 1991; Levin 1993). For range management, a more spatially-differentiated, multiple-scaled perspective is critical, not only because of the spatially-differentiated nature of grazing effects, but also because of the sensitivity of livestock production (Scoones 1989b) and rangeland resilience (Hiernaux 1995) to changes in the spatial heterogeneity of rangeland vegetation.

These points illuminate reasons why arguments made by the "nonequilibrium range ecology" school for density-independent regulation of livestock populations and for the important role of rainfall in influencing year-to-year vegetative variation, lend insufficient support for concluding a benign influence of livestock on African rangelands. The relaxation of equilibria assumptions does not promote a "laissez-faire" approach towards resource management, but instead exposes many factors and interactions, working at different spatiotemporal scales, to inspection (Zimmerer 1994). Rather than treat any spatiotemporal variation as a sign of livestock disturbance, dryland research must analyze management impacts within a web of abiotic and biotic processes affecting the ecological parameters of concern.

This chapter will briefly describe results from an ecological study of livestock grazing on the herbaceous layer of Sahelian annual rangeland. In so doing, the following biophysical features of Sahelian rangelands will be described and their management implications discussed:

1. The high spatial and temporal heterogeneity of rangeland production

2. The complex nature by which livestock activities affect rangeland production with mediating processes active at different spatial and temporal scales

3. The strong interaction (contingency) of all of these livestock-induced processes with rainfall, which can be viewed as the major "forcing variable" affecting rangeland dynamics

Sahelian Annual Rangelands

The Sahel of West Africa is the area lying south of the Sahara that receives a long-term annual average of between 100 and 600 millimeters of rain falling in a unimodal fashion during the months of June through September. Species composition of the herbaceous strata, which is dominated by annual grasses and forbs, is interannually variable, determined in large measure by the magnitude and pattern of rainfall (Cissé 1986). Soils are very infertile. Previous research has demonstrated that for at least the southern Sahel (400–600 millimeters annual mean rainfall) rangeland production is limited as often by nutrient availability (nitrogen and phosphorus) as by soil moisture (Penning de Vries and Djitèye 1982).

As one moves north within the Sahel, rainfall declines and forage quality increases, because of the growing importance of soil moisture in limiting growth. Since forage quality rather than forage quantity (energy) most often limits livestock production in this region, livestock managers to the south have historically moved their herds hundreds of kilometers to the northern Sahel (transhumance) during the rainy season to take advantage of the less dense but higher quality forage found on unobstructed pastures above the cultivation limit. Livestock actions can affect vegetative parameters of rangeland by impacting proximate factors such as soil moisture, seed stock, leaf mass, and nutrient availability. Livestock actions are linked to these proximate factors through a myriad of causal chains that work at different spatial and temporal scales. The importance of these chains vary across edaphic zones.

Figure 9.1 presents a highly-stylized schematic of important pathways that mediate between livestock actions and the proximate factors that affect herbaceous productivity on coarse-sand substrate. Within this highly variable environment, the relative importance of proximate factors in affecting rangeland production will vary. For example, during higher rainfall years (> 350 millimeters) on surfaces with lower runoff (≤ 10% net runoff), one would expect that causal chains affecting nutrient availability have a much more observable impact on production than those affecting moisture availability. In fact, given the high interannual variability of rainfall in the Sahel, grazing influences on vegetation interact strongly with same-season rainfall and recent rainfall history. Grazing effects as observed at any one site are likely to be highly variable from year to year. To gain a fuller, more dynamic understanding of grazing impact, one must study the important pathways that link livestock actions to proximate factors, and the conditions under which the different proximate factors are most limiting to vegetative growth.

This approach was taken to understand both the short- and long-term

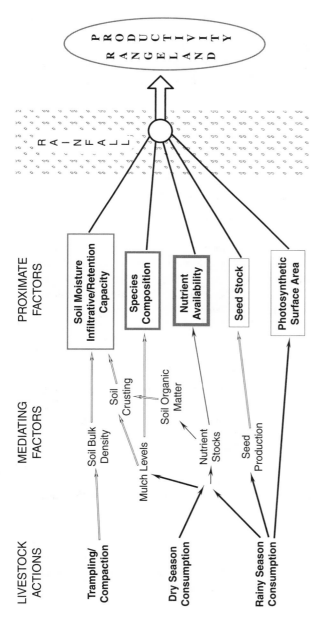

Fig. 9.1. Major causal pathways through which livestock actions are thought to affect proximate factors that may influence the productivity of Sahelian annual rangelands, depending on rainfall. These pathways work at different spatial and temporal scales. Those causal linkages with generative times that are same-season, multiple-year, and long (> 20 years) are denoted by broad dark arrows, broad light arrows, and narrow dark arrows, respectively. The thickness of the borders encircling each proximate factor designates the maximum spatial scale at which livestock actions influence the factor, varying from the grazed patch (~ 10 m²) in the case of photosynthetic surface area, to the area enclosed by a grazing-orbit circumference (~ 60 km²) in the case of nutrient availability.

effects of grazing on herbaceous production within an important transhumance corridor of Mali. Three proximate factors, operating at contrasted temporal and spatial scales, will be the focus of this study: (1) defoliation by grazing typified by a quick generative period (same-season) affecting production at the grazing patch level with very weak persistence; (2) shifts in species composition typified by mixed generative periods affecting production at the community level with multiple-year persistence; and (3) changes in nutrient availability through livestock-mediated redistribution typified by very long generative periods affecting production at the landscape level with long persistence (decades).

The Study Region: The Inland Niger Delta of Mali

Ecological research was conducted along the dry western border of the Inland Niger Delta of Mali (map 9.1). The delta is a 20000-square-kilometer floodplain south of Lake Debo fed by the Niger and Bani Rivers. Mean annual rainfall varies from 400 millimeters in the north to 650 millimeters in the south, received from June through October (peaking in August). Coefficients of variation of annual rainfall within the region vary from 25% to 40% (Sivakumar et al. 1984). It is one of the most important dry-season pasture reserves in West Africa, with approximately 1.5 million cattle and 2.5 million small ruminants converging on the floodplain at the beginning of the dry season (October–November), and remaining there until the first rains in June (International Livestock Centre for Africa 1981; Resource Inventory and Management Limited 1987). An examination of the movements of livestock to and from the floodplain suggests that the dry, rain-fed border of the floodplain is the most highly stressed zone along the yearly transhumance cycle, experiencing high grazing pressure during the beginning and end of the rainy season. Numerous observers have noted that this has led to environmental degradation (Boudet 1972; Gallais 1967; International Livestock Centre for Africa 1981), especially along the dry western border where a comparison of 1952 and 1975 aerial photography found a significant increase in bare, highly-reflective soil surfaces (Haywood 1980).

The 72-square-kilometer study area (map 9.1) is within the western border region. Typical of Sahelian "thorny scrub," the herbaceous layer is composed of annual forbs and grasses overlain by a sparse lignaceous layer of trees and shrubs, dominated by Mimosaceae. Species composition is heavily influenced by edaphic condition. The study area contains two major soil types: poorly-developed, well-drained sandy-loam soils and poorly-drained, silty-clay soils (map 9.2). The sandy-loam soils are found

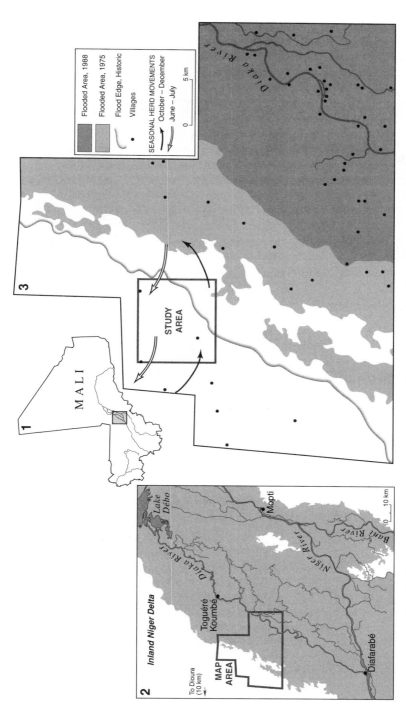

Map 9.1. Location of study area with respect to the historic extents of the floodplain of the Inland Niger Delta of Mali.

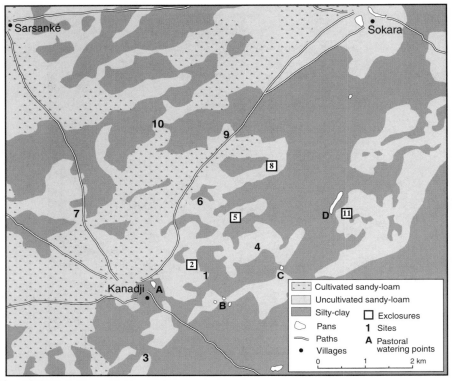

Map 9.2.   Locations of pans frequented by livestock (designated by letters), and sites where plant and soil samples were collected in relation to villages, paths, cultivated areas, and major edaphic zones.

primarily on eroded dunes and knolls lying two to five meters above nearby silty-clay depressions.

Overlying this edaphic mosaic are long-term grazing gradients established around four ephemeral pans distributed linearly along a northeast-southwest axis. The three ephemeral pans to the northeast (B, C, and D on map 9.2) are used to water livestock during the rainy season. Herders camp within 300 meters of watering pans, with their herds dispersing from camps to graze. Pan A, on the other hand, is more persistent and is used primarily by transhumance herds during the beginning of the dry season. Therefore, centers of livestock presence are pans B, C, and D during the rainy season and pan A during the early dry season. This general pattern, as described in historical terms by local villagers and Fulani clan members, is confirmed by the spatial distribution of manure deposition during the three years of the study. Manure deposition declined exponentially with distance from the nearest ephemeral pan (B, C, and D) and from pan A during the rainy and dry seasons, respectively (Turner n.d.a).

Vegetation, soil, and rainfall parameters were monitored over a two-year period along four permanent 50-meter transects at each of eleven sandy-substrate sites (labeled 1–11 in map 9.2). Although their soil texture, aspect, and topographic position are similar, these sites varied with respect to two distances: D1, the distance (1 to 4.3 kilometers) from pan A which is negatively correlated with long-term livestock pressure during the early dry season; and D2, the distance (0.6 to 3.7 kilometers) from the nearest ephemeral pan (B, C, and D) which is negatively correlated with long-term livestock pressure during the rainy season. This study was conducted during a period of historically low stocking rates as a result of a low regional livestock population and a shift of the floodplain border 12 kilometers to the east, both of which are due to the low rainfall received in the region during the 1970s and 1980s. Therefore vegetative patterns revealed in the transect work will reflect only the long-term, persistent effects of grazing. A full description of methods and results of the transect work is presented elsewhere (Turner n.d.a; Turner n.d.b). Short-term effects of heavy grazing on vegetative production were studied through clipping studies within exclosures established at four of the transect sites (see map 9.2). Although the results of this work will not be described in any detail, their implications for range management will be referred to later in this chapter. A fuller presentation of methods and results of clipping work is presented elsewhere (Hiernaux and Turner 1996; Turner 1992).

Spatiotemporal Variability of Aboveground Herbaceous Production

The occurrence of significant vegetation change across the landscape was demonstrated in the monitoring of range sites over two years of above average rainfall in 1988 and 1989 (567 millimeters and 512 millimeters, respectively). These two years followed a significant drought year (215 millimeters), when rainfall received at local precipitation stations was in the lowest twentieth percentile of the preceding decade (1978–1987), the poorest rainfall decade recorded in the region since 1920. Herbaceous vegetation changed dramatically with the increase in rainfall in 1988, with large increases in peak herbaceous mass and shifts in species composition, the most notable being a shift from forbs with short life cycles to grasses (table 9.1). Despite similar amounts of rainfall received and biomass produced, species composition of herbaceous vegetation was dramatically different in 1989 compared to 1988 (table 9.2). This was because of a com-

Table 9.1. Mean and intersite standard deviations (s.d.) of peak standing herbaceous mass (PHM) and forb percentage of peak herbaceous mass within exclosures at four of eleven transect sites (1, 4, 8, 11) during 1987–1989.

Year	Rainfall (mm)	PHM (g/m²)		Forb % of PHM	
		Mean	s.d.	Mean	s.d.
1987	215	29	11	69	10
1988	565	214	49	29	8
1989	513	196	48	10	5

bination of differences in seed stock (1988 following a drought year) and rainfall distribution, with the 1989 rainfall pattern more uniform, benefitting the establishment of species with homogeneous germination patterns such as *Schoenefeldia gracilis* and *Dactyloctenium aegyptium* (Breman and Cissé 1977). Species diversity declined over the two-year period, contrary to the recovery trends seen at other sites (Hiernaux et al. 1988).

Typical of Sahelian annual grasslands, standing herbaceous mass is spatially heterogeneous with a large fraction of vegetation patchiness at sub-50-meter scales (Turner n.d.b). Not only is spatial variation in standing herbaceous mass high, but spatial patterns of production are not persistent across years. 1988 and 1989 herbaceous mass densities were found to be either weakly correlated (3-plot running averages, transect mean comparisons) or not significantly correlated (site means), depending on the spatial scale of comparison (Turner n.d.b). Therefore, spatial patterns of production can shift from year to year, even across years with quite similar rainfall. The low correlation between the spatial distributions of herbaceous mass in 1988 and 1989 suggests that: (1) interannual grazing effects on productivity are very small compared to other factors; and/or (2) there are strong interactions between grazing history, rainfall history, and same-season rainfall. A clarification of this question is the focus of the next section.

Interannual Effects of Grazing on Herbaceous Production

As described above, proximate factors influenced by pathways with generative times of several years or longer include soil moisture, species composition, and nutrient availability (figure 9.1). On sandy-loam soils such as those at the eleven study sites, livestock grazing is thought to affect soil moisture—not by soil compaction as for soils with finer particle sizes, but by increasing the exposure of soils to raindrop impact and by reducing soil carbon, both of which increase the likelihood of the formation of thin surface crusts (0.1 millimeters) which reduce the soil's infiltration capacity

Table 9.2. Means and coefficients of variation (CV %) of the percentage contribution to peak standing herbaceous mass of major species at transect sites during 1988 and 1989.

Species	1988		1989	
	Mean	CV %	Mean	CV %
Forbs				
Tribulus terrestris	10.9	182	0.0	0
Other nonlegumes[1]	12.0	68	0.1	663
Zornia glochidiata	6.7	109	12.4	82
Other legumes[2]	1.8	212	0.0	0
Grasses				
Dactyloctenium aegyptium	11.3	56	18.1	63
Schoenefeldia gracilis	10.5	111	43.0	43
Chloris prieurii	9.8	98	4.8	124
Cenchrus biflorus	17.4	104	11.6	86
Eragrostis pilosa	9.6	93	2.9	145
Eragrostis tremula	5.0	104	3.1	169
Panicum laetum	3.8	181	2.4	244
Other grasses[3]	1.0	314	1.6	162

Notes:

1. Other nonlegume forb species include *Amaranthus graecizans, Blepharis linariifolia, Borreria stachydea, Conchorus tridens, Citrullus* sp., *Gisekia pharnacioides, Ipomea coscinosperma, Ipomea vagans, Leptadenia hastata, Merremia coptica, Mollugo nudicaulis,* and *Polycarpaea linearifolia.*
2. Legumes present other than *Zornia* are primarily *Indigofera senegalensis* with some *Indigofera pilosa.*
3. Listed in decreasing importance, other grasses include *Aristida adscensionis, Brachiaria xantholeuca, Tragus berteronianus, Brachiaria ramosa, Digitaria ciliaris,* and *Sporobolus* sp.

(Hoogmoed and Stroosnijder 1984; Penning de Vries and Djitèye 1982). High livestock presence is thought to affect species composition by stimulating short-cycle forbs over longer-cycle grasses due to the fact that the former are more able to escape defoliation pressure or increased moisture limitation (Penning de Vries and Djitèye 1982; Valenza 1981). The most important means by which livestock grazing could affect nutrient availability is through the ingestion of vegetation with subsequent excretion of a significant fraction of nitrogen and phosphorus as urine and faeces. This leads to a spatial redistribution of nutrient availability (from grazed zones toward resting points) along with partial loss of nitrogen (urine) from the system through volatilization (Powell et al. 1995; Schimel et al. 1986; Woodmansee and Duncan 1980).

If these processes were active, one would expect spatial gradients of historic livestock pressure to lead to corresponding gradients in bare crusted patches, species composition, and nutrient availability. Regression modelling work confirms these predictions, but at the same time demonstrates the importance of rainfall and rainfall history in influencing the nature of these relationships and their effects on spatial patterns of herbaceous production (Turner n.d.b). The surface fraction of bare crusted patches was found to be positively associated with historic rainy-season livestock presence during both years (declining 5% for every kilometer of D2). The overall importance of soil crusting was, however, significantly lower in 1989 (10% compared to 16%), reflecting differences in rainfall history.

Logistic regression analysis of 1988–1989 species composition data found that fractional contributions to peak herbaceous mass of different species vary significantly along the axes running parallel to historic dry-season and rainy-season grazing pressure, although the distribution of some species fractions changed dramatically between the two years of the study (Turner n.d.b). Forbs as a class exhibit a complex distribution in relation to D1 and D2 as a result of differing spatial patterns found for nonlegumes and legumes (Turner n.d.b). Peak herbaceous mass of vegetated surface area was in turn found to be negatively correlated to forb fraction of vegetation mass in 1988. This correlation, however, declined significantly in 1989 because of the reduced importance of forbs within the study area (table 9.2).

Phosphorus concentrations in plant tissue samples were found to decline with D2. No corresponding relationship between livestock presence and nitrogen concentrations was found. These observations are consistent with grazing-induced redistribution of nutrients, resulting in the establishment of a gradient in phosphorus availability over many decades. The lack of an observable nitrogen availability gradient is not surprising given its open cycle (Penning de Vries and Djitèye 1982): A significant fraction of ingested nitrogen is lost from the system through volatilization and is therefore not deposited toward rainy season encampment points. Once forb mass fraction and soil texture (coarse sand fraction) were controlled for, herbaceous mass on vegetated surfaces in 1988 was found to be negatively associated with D2. In other words, the production residual was positively associated with the observed phosphorus gradient. However, no such relationship was found in 1989.

The etiology of observed spatiotemporal variability of Sahelian range production is complex. In this harsh environment, vegetative productivity is limited by a number of different factors fluctuating in space and time. The effect of each on production is contingent on the level of others. Due to its relatively high interannual variability within a range relevant to production, rainfall is the major factor driving the dynamics of rangeland production. The productivity consequences of historic livestock grazing, through its effects on nutrient availability, infiltration capacity, and species composition, are shaped by rainfall and rainfall history. In general, it is during higher rainfall years and, most particularly, transition years from low to high rainfall that grazing history has its strongest impact on production. Clipping studies conducted within the same study area demonstrate that the magnitude and even directionality of short-term effects of grazing are also influenced by rainfall (Hiernaux and Turner 1996). Over both the short- and long-term, herbaceous vegetation is found to be most sensitive to grazing during the growing season (Hiernaux and Turner 1996; Turner n.d.b).

Implications for Agropastoral Development and Resource Management

These findings support nonequilibrium arguments that the dynamics of annual rangeland production and species composition are influenced most strongly by interannual fluctuations in rainfall, and that Sahelian annual rangelands are generally less fragile and more resilient to grazing than commonly assumed (Dodd 1994; Hiernaux 1993). Even in 1988, a year in which conditions were such that longer-term grazing effects were most visible across the landscape, grazing, through its effects on proximate factors, was associated with at most 17% of intersite variation in peak herbaceous production (Turner n.d.b). Common views of widespread anthropogenic degradation of Sahelian landscapes derive not only from an overreliance on equilibrium-based concepts, but also an ignorance of the inherent spatiotemporal heterogeneity and resilience of these dryland landscapes. The high spatial heterogeneity of vegetative cover has often been taken as a sign of persistent degradation. Past and contemporary revegetation, reseeding, and rangeland hydrology programs on the subcontinent can be seen as often ill-fated attempts to homogenize the cover of natural vegetation. Revegetation failures have reinforced notions of persistence of grazing-induced degradation when in fact vegetation patchiness is temporally dynamic and often can be explained by edaphic and topographic characteristics (Thiéry et al. 1995).

In such resource-poor environments, rural subsistence strategies often depend on the same heterogeneity and patchiness in resources that are interpreted by scientists as signs of degradation. If all water and nutrients were to be distributed uniformly across the landscape, agricultural production in many such areas would decline. Within settled zones, rural communities not only take advantage of heterogeneity resulting from natural processes but actually actively work to concentrate water and nutrients into patches so as to reach economic and biophysical thresholds in agricultural production (for example, water harvesting, livestock manuring, and green manuring). Resource management and conservation projects should therefore recognize that local human and wildlife populations may depend heavily on certain resource-rich patches within the area that may be functionally dependent on resource-poor ("degraded") areas.

By demonstrating the sensitivity of herbaceous production to heavy persistent livestock grazing during the rainy season, the preceding case points to the need for more active political and economic support of the pastoral sector. The practice of transhumance has declined over the past two decades of drought and government neglect, a result of agricultural encroachment onto key pastoral areas (transhumance corridors, dry-season pastures, water points) and shifts in livestock ownership away from traditional livestock managers. As a result of these changes, a higher fraction of the regional livestock population is found within more densely-populated agricultural zones during the rainy season. In the present situation the practice of transhumance cannot be reinvigorated without strong support from national governments, nongovernmental organizations and local communities. North-south transhumance corridors need to be protected and, where destroyed, recreated. The security of contracts (entrustment, wage) between these groups and new livestock owners often need to be improved through measures that will render these contracts more enforceable and livestock transactions more transparent.

Implications for Rangeland Assessment

This case also calls into question standard approaches of range inventory and ecological evaluation of agropastoral systems. Rather than a featureless scrubland driven by regional stresses of drought and grazing, the Sahelian landscape is dynamically heterogeneous, a continually shifting mosaic of patches differing in production and species composition. Assessments of grazing impact should not focus solely on standing plant mass and species composition, but also on how grazing affects the availability and spatial heterogeneity of plant growth factors that influence these vegetative parameters as mediated by rainfall and rainfall history. Since the importance of causal pathways linking livestock actions to rangeland productivity vary across edaphic zone, such assessments need to be edaphically stratified as well.

Range management has generally been approached by assigning carrying capacities to rangeland mapping units in order to estimate regional stocking rates. The concept of carrying capacity is problematic, not only because of a fluctuating resource base (Bartels et al. 1990; de Leeuw and Tothill 1990; Sandford 1982), but also because of large interannual and seasonal differences in the impacts of livestock-driven pathways on productivity. It is difficult to reduce the web of contingent causal connections into one measure without significant spatial and temporal qualification. Attempts to make such measures ecologically relevant lead to increasingly finer spatial and temporal divisions. On a practical level, this trend clashes with the spatial scale of available livestock population data, which is necessarily regional due to the mobility of livestock populations. Therefore, carrying capacities are usually left at a spatially- and annually-aggregated state so that they can be compared to regional livestock populations. Such comparisons represent very coarse measures of ecological stress, not only because of the necessary ecological abstraction of expressing carrying capacity at that spatiotemporal scale, but because of the highly aggregated distribution of key pastoral resources (e.g., water) and livestock populations within any region.

The spatiotemporal distribution of livestock within a region is in fact more important ecologically than the region's average stocking rate. A shift in range management emphasis is needed from the estimation and control of stocking rates to strategies that promote better grazing management (affecting spatiotemporal distribution of grazing pressure) and ecological monitoring of key grazing resources. Such a monitoring approach would focus stocking rate estimation and ecological monitoring on those key grazing resources identified to be the most prone to receive higher rates of grazing pressure during ecologically-sensitive periods (Scoones 1992). Such a monitoring system would identify ecological stresses earlier and with greater etiological precision than the regional carrying capacity approach. This alternative approach requires an understanding of specific production systems and therefore would most likely depend on local participation. By explicitly considering management as a means to reducing ecological stress, an advantage of such an approach is that it could identify a range of responses to overgrazing problems. This is in sharp contrast to overally-aggregated approaches that inherently lead to a destocking prescription.

Conclusions

Much of the debate about ecological change and the role of conservation in arid rangelands has focused on the relative importance of climate versus grazing as underlying factors forcing change. This debate is recurrent and has proven unproductive. One of the most notorious cases of confusion resulting from such dichotomized causal reasoning is the confusion surrounding the term "desertification" (Hellden 1991; Mainguet 1994; Thomas 1993). Are we referring to just anthropogenic change, climate change, or both? The argument made in this chapter differs from the overly-facile, even-handed characterization of "both" by stressing the interaction of climate and anthropogenic change over their independent effects. Due to the importance of interannual rainfall fluctuations in determining the relative importance of proximate factors in limiting growth each year, long- or short-term grazing effects are always contingent on rainfall in any given year. In this way, their interaction is so strong that it is difficult to talk about grazing effects independent of rainfall effects. The same livestock action will have remarkably different effects on the vegetation depending on rainfall and recent rainfall history.

The interactionist argument made in this chapter is consistent with previous work in biogeography and ecology that demonstrates the strong effect of climate on the nature of human and natural disturbances affecting perennial lignaceous vegetation (Minnich 1983; Savage 1991; Swetnam 1993; Vale 1982; Veblen and Lorenz 1988). Although the annual vegetation in the Sahel seems to start anew each rainy season, its productivity and composition is affected by the system's memory of rainfall and land-use history as stored in seed stock, soil fertility, soil structure, and lignaceous cover. Therefore, even for these seemingly ephemeral populations, history does matter (Gould 1986).

The way in which the characterization of Sahelian annual rangeland dynamics made here diverges from these previous studies is that rather than stressing the influence of climate on the nature of disturbances, the influence of climate on vegetative response to disturbance/change is stressed. This difference in emphasis reflects differences in time scales taken by the studies and in the relative interannual variability of climate and anthropogenic disturbance of the systems studied. Compared to these other biogeographical landscapes, the biologically-relevant variability in Sahelian rainfall (climate variable) is higher relative to the human-managed disturbance regime (grazing) since the regime's establishment in the study area over 100 years ago. Rainfall, by affecting the degree to which different growth factors limit plant growth, mediates short- and long-term grazing impact. Therefore, the climate-disturbance interaction in the Sahel is best conceptualized in terms of rainfall fluctuations that strongly influence the vegetative response to anthropogenic disturbance.

Specific implications of this perspective for conservation and sustainable development programs in arid lands of Africa have been outlined in this chapter. In juxtaposing these implications to previous forms of pastoral development, I hope to make two broader points. First, models and conceptualizations of ecological change do have a strong influence on conservation-motivated development. This contrasts with the view that conservation programs are simply a vehicle for outside usurpation of local resources. Political/economic motivations have underlain the involvement of many public and private entities in conservation programs. Still, the creation and implementation of each program requires the involvement of many actors, including conservation professionals. Although development programs involving settlement and destocking of pastoralists were consistent with the political aims of African governments, equilibrium range ecology supplied a critical, legitimating environmental rationale necessary for attracting broader financial and political support.

The second point is that nonequilibrium perspectives on vegetation dynamics should not lead to a relativist view of ecological disturbance, or a defeatist view with respect to understanding ecological change. History does matter, but the system's memory is shown here not to be stored in a black box but in measurable stocks and flows. Vegetation dynamics are complex and highly contingent. Still, process-oriented research can lead to generalized findings. Such findings may not produce target stocking or extraction rates, but do provide an understanding of the particular pathways to which vegetation is most sensitive under particular conditions. This chapter shows how one such finding, the higher sensitivity of herbaceous vegetation to heavy grazing during the rainy season, leads to specific management recommendations. In situations such as semiarid Africa, where pastures are not enclosed and forage production is highly variable, recommendations that concern grazing management may be more relevant and implementable than prescribed stocking rates. Unfortunately, previous range management research in the region provides very little of this process-oriented understanding. Instead, range research documentation is dominated by range inventories and carrying capacity estimates, an archive of limited usefulness today—a final unfortunate legacy of equilibrium-based approaches in range management.

References

Abel, N. O. J., and P. M. Blaikie. 1989. Land Degradation, Stocking Rates and Conservation Policies in the Communal Rangelands of Botswana and Zimbabwae. *Land Degradation and Rehabilitation* 1: 1–23.

Anderson, D. 1984. Depression, Dust Bowl, Demography and Drought: The Colonial State and Soil Conservation in East Africa During the 1930s. *African Affairs* 83: 321–344.

Baker, R. 1975. Development and the Pastoral Peoples of Karamoja, Northeastern Uganda: An Example of the Treatment of Symptoms. In *Pastoralism in Tropical Africa,* ed. T. Monod, pp. 187–205. London: Oxford University Press.

Bartels, G. B., B. E. Norton, and G. K. Perrier. 1990. *The Applicability of the Carrying Capacity Concept in Africa.* Pastoral Development Network Paper 30e. London: Overseas Development Institute.

Behnke, R. H., I. Scoones, and C. Kerven, eds. 1993. *Range Ecology at Disequilibrium*. London: Overseas Development Institute.

Blumler, M. A. 1993. Successional Pattern and Landscape Sensitivity in the Mediterranean and Near East. In *Landscape Sensitivity*, eds. D. S. G. Thomas and R. J. Allison, pp. 287–305. New York: Wiley and Sons.

Botkin, D. B. 1990. *Discordant Haromonies: A New Ecology of the Twenty-First Century*. New York: Oxford University Press.

Boudet, G. 1972. *Projet de developpement de l'élevage dans la région de Mopti*. Étude Agrostologique No. 37. Maisons-Alfort: Institut d'Élevage et de Médecine Vétérinaire des Pays Tropicaux.

Boudet, G. 1975. *Manuel sur les paturages tropicaux et les cultures fourragères*. Paris: Institut d'Élevage et de Médecine Vétérinaire des Pays Tropicaux (IEMVT).

Braun-Blanquet, J. 1932. *Plant Sociology*. New York: McGraw-Hill.

Breman, H., and A. M. Cissé. 1977. Dynamics of Sahelian Pastures in Relation to Drought and Grazing. *Oecologia* 28: 301–315.

Breman, H., and N. de Ridder. 1991. *Manuel sur les paturages des pays sahéliens*. Paris and Wageningan: Editions Karthala, and Technical Centre for Agricultural and Rural Cooperation.

Brown, B. J., and T. F. H. Allen. 1989. The Importance of Scale in Evaluating Herbivory Impacts. *Oikos* 54: 189–194.

Brown, L. H. 1971. The Biology of Pastoral Man as a Factor in Conservation. *Biological Conservation* 3(2): 93–100.

Chevalier, A. 1934. Les places dépourvues de végétation dans le Sahara et leur cause sous le rapport de l'écologie végétale. *Comptes rendus de l'Academie des Sciences (Paris)* 194 (5): 480–482.

Cissé, A. M. 1986. *Dynamique de la strate herbacée des pâturages de la zone sud-sahélienne*. Wageningen, the Netherlands: Projet Primaire du Sahel.

Clements, F. E. 1916. *Plant Succession: An Analysis of the Development of Vegetation*. Carnegie Institute of Washington Publications 242. Washington, D.C.: Carnegie Institute.

Cloudsley-Thompson, J. L. 1974. The Expanding Sahara. *Environmental Conservation* 1(1): 5–13.

Coughenour, M. B. 1991. Spatial Components of Plant-Herbivore Interactions in Pastoral, Ranching, and Native Ungulate Ecosystems. *Journal of Range Management* 44(6): 530–542.

de Haan, C. 1994. *An Overview of the World Bank's Involvement in Pastoral Development*. Pastoral Development Network Paper 36b. London: Overseas Development Institute.

de Leeuw, P. N., and J. C. Tothill. 1990. *The Concept of Rangeland Carrying Capacity in Sub-Saharan Africa–Myth or Reality?* Pastoral Development Network Paper 29b. London: Overseas Development Institute.

de Queiroz, J. S. 1993. *Range Degradation in Botswana: Myth or Reality?* Pastoral Development Network Paper 35b. London: Overseas Development Administration.

DeAngelis, D. L., and J. C. Waterhouse. 1987. Equilibrium and Nonequilibrium Concepts in Ecological Models. *Ecological Monographs* 57(1): 1–21.

Derman, W. 1984. USAID in the Sahel: Development and Poverty. In *The Politics of Agriculture in Tropical Africa,* ed. J. Barker, pp. 77–97. Beverly Hills: Sage Publications.

Dodd, J. L. 1994. Desertification and Degradation in Sub-Saharan Africa: The Role of Livestock. *BioScience* 44(1): 28–34.

Ellis, J. E., and D. M. Swift. 1988. Stability of African Pastoral Ecosystems: Alternate Paradigms and Implications for Development. *Journal of Range Management* 41: 450–459.

Forman, R. T., and M. Godron. 1986. *Landscape Ecology.* New York: Wiley and Sons.

Friedel, M. H. 1991. Range Condition Assessments and the Concept of Thresholds: A Viewpoint. *Journal of Range Management* 44(5): 422–426.

Gallais, J. 1967. *Le Delta intérieur du Niger.* (Mémoires de l'Institut Fondamental d'Afrique Noire #79). Dakar, Senegal: Institut Fondamental d'Afrique Noire.

Goldshmidt, W. 1981. The Failure of Pastoral Economic Development Programs in Africa. In *The Future of Pastoral Peoples,* eds. J. G. Galaty, D. Aronson, P. C. Salzman, and A. Chouinard, pp. 101–126. Ottawa: International Development Research Centre.

Gould, S. J. 1986. Evolution and the Triumph of Homology: Or Why History Matters. *American Scientist* 74: 60–69.

Grouzis, M. 1988. *Structure, productivité, et dynamique des systèmes écologiques Sahéliens (Mare d'Oursi, Burkina Faso), Collection Etudes et Thèses.* Paris: Editions de l'Office de la Recherche Scientifique et Technique Outre-Mer.

Haywood, M. 1980. *Changes in Land Use and Vegetation in the ILCA/Mali Sudano-Sahelian Project Zone.* ILCA Working Document 3. Addis Ababa, Ethiopia: International Livestock Centre for Africa.

Hellden, U. 1991. Desertification—Time for an Assessment. *Ambio* 20: 372–383.

Hiernaux, P. 1993. The Crisis of Sahelian Pastoralism: Ecological or Economic? Addis Ababa, Ethiopia: International Livestock Centre for Africa.

Hiernaux, P. 1995. *Spatial Heteroeneity in Sahelian Rangelands and Resilience to Drought and Grazing.* Paper read at Fifth International Rangeland Congress, July 23–28, 1995, Salt Lake City, Utah.

Hiernaux, P., L. Diarra, and A. Maiga. 1988. *Evolution de la végétation sahélienne après la sechéresse bilan du suivi des sites du Gourma en 1987.* Bamako, Mali: Centre International pour l'Elevage en Afrique.

Hiernaux, P., and M. D. Turner. 1996. The Effect of the Timing and Frequency of Clipping on Nutrient Uptake and Production of Sahelian Annual Rangelands. *Journal of Applied Ecology* 33: 387–399.

Hjort, A. 1982. A Critique of "Ecological" Models of Land Use. *Nomadic Peoples* 10: 11–27.

Homewood, K., and W. A. Rogers. 1987. Pastoralism, Conservation and the Overgrazing Controversy. In *Conservation in Africa,* eds. D. Anderson and R. Grove, pp. 111–128. Cambridge: Cambridge University Press.

Homewood, K. M., and W. A. Rodgers. 1984. Pastoralism and Conservation. *Human Ecology* 12(4): 431–441.

Hoogmoed, W. B., and L. Stroosnijder. 1984. Crust Formation on Sandy Soils in the Sahel: Rainfall and Infiltration. *Soil and Tillage Research* 4: 5–23.

Horowitz, M. M. 1979. *The Sociology of Pastoralism and African Livestock Projects* (AID Program Evaluation Discussion Paper 6). Washington, D.C.: United States Agency for International Development.

Horowitz, M. M., and P. D. Little. 1987. African Pastoralism and Poverty: Some Implications for Drought and Famine. In *Drought and Hunger in Africa,* ed. M. H. Glantz, pp. 59–84. Cambridge: Cambridge University Press.

International Livestock Centre for Africa. 1981. *Systems Research in the Arid Zones of Mali.* ILCA Systems Study 5. Addis Ababa, Ethiopia: International Livestock Centre for Africa.

International Livestock Centre for Africa. 1987. *ILCA's Strategy and Long-Term Plan: A Summary.* Addis Ababa, Ethiopia: International Livestock Centre for Africa.

Lamprey, H. F. 1983. Pastoralism Yesterday and Today: The Overgrazing Problem. In *Tropical Savannas: Ecosystems of the World,* ed. F. Bourliere, pp. 643–666. Amsterdam: Elsevier.

Laycock, W. A. 1991. Stable States and Thresholds of Range Condition on North American Rangelands: A Viewpoint. *Journal of Range Management* 44: 427–433.

Levin, S. A. 1993. Concepts of Scale at the Local Level. In *Scaling Physiological Processes: Leaf to Globe,* eds. J. R. Ehleringer and C. B. Field pp. 7–20. San Diego: Academic Press.

Mainguet, M. 1994. *Desertification: Natural Background and Human Mismanagement.* Berlin: Springer-Verlag.

McCabe, J. T. 1987. Drought and Recovery: Livestock Dynamics among the Ngisonyoka Turkana of Kenya. *Human Ecology* 15: 371–389.

Minnich, R. A. 1983. Fire Mosaics in Southern California and Northern Baja California. *Science* 219: 1287–1294.

Ndalaga, D. K. 1982. "Operation Imparnati": The Sedentarization of the Pastoral Maasai in Tanzania. *Nomadic Peoples* 10: 28–39.

Neumann, R. A. 1995. Local Challenges to Global Agendas: Conservation, Economic Liberalization and the Pastoralists' Rights Movement in Tanzania. *Antipode* 27(4): 363–406.

Odum, E. P. 1971. *Fundamentals of Ecology.* Philadelphia: W. B. Saunders.

Overseas Development Institute. 1990. *Comments on PDN papers 29A (Abel and Blaikie 1990) and 28B (Scoones 1989).* Pastoral Development Network Paper 29d. London: Overseas Development Institute.

Overseas Development Institute. 1993. *Responses to "Range Degradation in Botswana: Myth or Reality? (Paper 35b).* Pastoral Development Network Paper 35c. London: Overseas Development Institute.

Oxby, C. 1982. *Group Ranches in Africa.* Pastoral Development Network Paper 13d. London: Overseas Development Institute.

Peluso, N. L. 1993. Coercing Conservation? The Politics of State Resource Control. *Global Environmental Change* 3(2): 199–217.

Penning de Vries, F. W. T., and M. A. Djitèye. 1982. *La productivité des pâturages sahéliens.* Wageningen, the Netherlands: Centre for Agricultural Publishing and Documentation.

Picardi, A. C., and W. W. Siefert. 1976. A Tragedy of the Commons in the Sahel. *Technology Review* May: 42–51.

Pickett, S. T. A., and P. S. White, eds. 1985. *The Ecology of Natural Disturbance and Patch Dynamics.* New York: Academic Press.

Powell, J. M., S. Fernandez-Rivera, P. Hiernaux, and M. D. Turner. 1995. *Nutrient Cycling in Integrated Rangeland/Cropland Systems of the Sahel.* Paper read at Fifth International Rangeland Congress, July 23–28, 1995, Salt Lake City, Utah.

Pratt, D. J., and M. D. Gwynne, eds. 1977. *Rangeland Management and Ecology in East Africa.* Huntington, New York: R. E. Krieger Publishing Company.

Ranger, T. O. 1985. *Peasant Consciousness and Guerilla War in Zimbabwe: A Comparative Study.* Berkeley: University of California Press.

Resource Inventory and Management Limited. 1987. *Refuge in the Sahel.* St. Helier, Great Britain: Resource Inventory and Management Limited.

Sandford, S. 1982. Pastoral Strategies and Desertification: Opportunism and Conservatism in Dry Lands. In *Desertification and Development: Dryland Ecology in Social Perspective,* eds. B. Spooner and H. Mann, pp. 61–80. London: Academic Press.

Sandford, S. 1983. *Management of Pastoral Development in the Third World.* New York: John Wiley and Sons.

Savage, M. 1991. Structural Dynamics of a Southwestern Pine Forest under Chronic Human Influence. *Annals of the Association of American Geographers* 81(2): 271–289.

Schimel, D. S., W. J. Parton, F. J. Adamsen, R. G. Woodmansee, R. L. Senft, and M. A. Stillwell. 1986. The Role of Cattle in the Volatile Loss of Nitrogen from a Shortgrass Steppe. *Biogeochemistry* 2: 39–52.

Scoones, I. 1989a. *Economic and Ecological Carrying Capacity: Implications for Livestock Development in Zimbabwe's Communal Areas.* Pastoral Development Network Paper 27b. London: Overseas Development Institute.

Scoones, I. 1989b. *Patch Use by Cattle in Dryland Zimbabwe: Farmer Knowledge and Ecological Theory.* Pastoral Development Network Paper 28b. London: Overseas Development Institute.

Scoones, I. 1992. Land Degradation and Livestock Production in Zimbabwe's Communal Areas. *Land Degradation and Rehabilitation* 3: 99–113.

Scoones, I. ed. 1994. *Living with Uncertainty: New Directions in Pastoral Development in Africa.* London: Intermediate Technology Publications.

Shanmugaratnam, N., T. Vedeld, A. Mossige, and M. Bovin. 1992. *Resource Management and Pastoral Institution Building in West African Sahel.* World Bank Discussion Paper 175. Washington, D.C.: The World Bank.

Simpson, J. R., and G. M. Sullivan. 1984. Planning for Institutional Change in Utilization of Sub-Saharan Africa's Common Property Range Resources. *African Studies Review* 27: 61–78.

Sinclair, A. R. E., and J. M. Frywell. 1985. The Sahel of Africa: Ecology of a Disaster. *Canadian Journal of Zoology* 63: 987–994.

Sivakumar, M. V. K., M. Konate, and S. M. Virmani. 1984. *Agroclimatology of West Africa: Mali,* Pantancheru, India: International Crops Research Institute for the Semi-arid Tropics.

Smith, M. S., and G. Pickup. 1993. Out of Africa, Looking In: Understanding Vegetation Change. In *Range Ecology at Disequilibrium,* eds. R. H. Behnke, I. Scoones, and C. Kerven, pp. 196–226. London: Overseas Development Institute.

Stebbing, E. P. 1935. The Encroaching Sahara: The Threat to the West Africa Colonies. *Geographical Journal* 85: 506–524.

Stoddart, L. A., A. D. Smith, and T. W. Box. 1975. *Range Management.* 3d edition. New York: McGraw-Hill.

Swetnam, T. W. 1993. Fire History and Climate Change in Giant Sequoia Groves. *Science* 262(5135): 885–889.

Tapson, D. R. 1991. *The Overstocking and Offtake Controversy Reexamined for the Case of Kwazulu.* Pastoral Development Network Paper 31a. London: Overseas Development Institute.

Thébaud, B. 1988. *Elevage et développment au Niger.* Genève: Bureau International du Travail.

Thiéry, J. M., J. -M. D'Herbes, and C. Valentin. 1995. A Model Simulating the Genesis of Banded Vegetation Patterns in Niger. *Journal of Ecology* 83: 497–507.

Thomas, D. S. G. 1993. Sandstorm in a Teacup? Understanding Desertification. *Geographical Journal* 159: 318–331.

Tignor, R. L. 1971. Kamba Political Protest: The Destocking Controversy of 1938. *International Journal of African Historical Studies* 4(2): 237–251.

Turner, M. 1992. *Living on the Edge: FulBe Herding Practices and the Relationship Between Economy and Ecology in the Inland Niger Delta of Mali.* Ph. D. dissertation, University of California—Berkeley.

Turner, M. D. N.d.a. Long-term Effects of Daily Grazing Orbits on Nutrient Availability in Sahelian West Africa: 1. Gradients in the Chemical Composition of Rangeland Soils and Vegetation. *Journal of Biogeography* (in press).

Turner, M. D. N.d.b. Long-Term Effects of Daily Grazing Orbits on Nutrient Availability in Sahelian West Africa: 2. Effects of a Phosphorus Gradient on Spatial Patterns of Annual Grassland Production. *Journal of Biogeography* (in press).

Vale, T. R. 1982. *Plants and People: Vegetation Change in North America.* Washington, D. C.: Association of American Geographers.

Valenza, J. 1981. Surveillance continue des pâturages naturels sahéliens. *Revue de l'Élevage et Médecine Vétérinaire des Pays Tropicaux* 34(1): 83–100.

van Keulen, H., and H. Breman. 1990. Agricultural Development in the West African Sahelian Region: A Cure Against Land Hunger? *Agriculture, Ecosystems and Environment* 32: 177–197.

Veblen, T. T., and D. C. Lorenz. 1988. Recent Vegetation Changes along the Forest/Steppe Ecotone of Northern Patagonia. *Annals of Association of American Geographers* 78(1): 93–111.

Walker, B. H., D. Ludwig, C. S. Holling, and R. M. Peterman. 1981. Stability of Semi-arid Savannah Grazing Systems. *Journal of Ecology* 69: 473–498.

Warren, A., and J. Maizels. 1977. *Ecological Change and Desertification.* Paper No. A/CONF 74/7. New York: United Nations Conference on Desertification.

Westoby, M., B. Walker, and I. Noy-Meir. 1989. Opportunistic Management for Rangelands Not at Equilibrium. *Journal of Range Management* 42(4): 266–273.

Wiens, J. A. 1977. On Competition and Variable Environments. *American Scientist* 65: 590–597.

Woodmansee, R. G., and D. A. Duncan. 1980. Nitrogen and Phosphorus Dynamics and Budgets in Annual Grasslands. *Ecology* 61: 893–904.

Worster, D. 1990. The Ecology of Order and Chaos. *Environmental History Review* 14: 1–18.

Wyckoff, J. B. 1985. Planning Arid Land Development Projects. *Nomadic Peoples* 19: 59–70.

Zimmerer, K. S. 1994. Human Geography and the "New Ecology": The Prospect and Promise of Integration. *Annals of the Association of American Geographers* 84(1): 108–125.

10

Disturbances and Diverse Crops in the Farm Landscapes of Highland South America

Karl S. Zimmerer

Introduction: Diverse Crops and Agroecosystem Change

The biological diversity of food plants offers a multifaceted resource for the small farmers of developing countries. Millions of cultivators in Latin America, Africa, and Asia continue to rely on the diverse varieties of major food plants such as wheat, rice, maize, potatoes, sweet potatoes, manioc, and barley. Diverse minor crops include various beans, lentils, squash, tuber-bearers (taro, arrowroot, jicama, ulluco), and grains (millet, sorghum, quinoa, amaranth). The diverse types of food plants play a host of helpful roles for the small farmers, many of whom are peasant and indigenous people. Most notably their diverse crops serve as agronomic stocks that are well-suited to field sites located in marginal or suboptimal environments. Their diverse crops also supply valuable foodstuffs and are evoked locally in cultural expressions and identity (Brookfield and Padoch 1994; Cleveland et al. 1994; Oldfield and Alcorn 1987; Zimmerer 1991a, 1996, 1998).

The evolutionary future of the diverse crops is uncertain, however, and whether or not they continue to be cultivated is a widespread concern. Still the dilemma of the crops is complex and multidimensional. On the one hand, some of the small farmers of poorer countries find themselves no longer able or willing to sow diverse crops (Brush 1986, 1989; Zimmerer 1991b, 1996). Their curtailment is driving the loss of agricultural biodiversity known as genetic erosion. On the other hand, the current growers are supplying the plant-breeding and biotechnology industries worldwide with genetic raw materials that yield products worth billions of dollars each year. Heated controversies revolve around how the crops should be both conserved and utilized, and how the costs and benefits of use should be apportioned (Alcorn 1994; Brush and Stabinsky 1996; Fowler and Mooney 1990). Meanwhile, and at the center of mounting interest, fundamental questions persist about the ecogeography and human ecology of the diverse crops and the agroecosystems where they are grown.

This chapter presents new evidence and a new interpretation about one highly diverse complex of crops, the Andean potato species of the genus *Solanum*. It focuses on the ecological relations of potato biodiversity (or agrodiversity) with respect to the dynamics of farm landscapes in the Andes Mountains of South America. Findings demonstrate that the diverse Andean potatoes display a moderate to high degree of versatility in terms of ecological habitats and biogeographical distributions. Such ecological versatility is found to be shaped by the techniques of Andean farmers and thus bears the distinct mark of human ecological influences. Of equal importance, the character of natural disturbances in the Andean farm landscape is seen as a major factor in the evolution and maintenance of versatility of the potato crop. In particular, the broad spatial range and the temporal unpredictability of several disturbances common to Andean agroecosystems combine to favor potato types that possess the traits of ecological versatility.

My interpretation of the role of farm landscape change and its relation to the ecogeographical character of crop diversity is at odds with certain assumptions of conventional wisdom. One influential assumption is that the diverse crops, and especially the chief subunits known as landraces, function as ecological specialists. It is assumed that the landrace types—also referred to as landraces, native cultivars, and primitive, folk, and traditional crop varieties (PVs, FVs, TCVs)—are highly adapted to the specialized growing niches known as microenvironments (Brücher 1989; Fowler and Mooney 1990; Harlan 1975, 1992). The assumption of microscale specialization among landraces is based on the belief that the disturbances affecting such growing environments are either regular or insignificant with respect to the growth and yield of landrace types. Considering the regularity or insignificance of disturbances, the reasoning goes, landraces have evolved as specialists adapted to the microscale niches of agroecosystems.

This line of thinking hinges on broad assumptions about the nature of ecological disturbances and their role in farm landscapes. Disturbances affecting these landscapes are commonly referred to as hazards. Agricultural hazards include climatic events such as frost and drought, as well as biotic disturbances induced by pests and diseases. Such disturbances are a common problem in the marginal or suboptimal field sites of many diversity-growing farmers of the poorer countries. Yet we still know little about the disturbance regimes that distinguish these farm landscapes, and how they relate to the ecogeography and human ecology of the diverse crop landraces. The need to study these relations is supported by convincing evidence showing how the nature of biodiversity in many nonfarm landscapes is seen as closely related to disturbances (Pickett and White 1985; Vale 1982; Veblen 1992; White 1979).

Given the compelling reasons for protecting the diverse suites of both major and minor crops, our ideas about the ecogeography and human ecology of landraces and their relations to agroecosystem change are rife with implications for sound resource management and conservation (Altieri and Merrick 1987; Brush 1989; Cleveland et al. 1995; Frankel and Hawkes 1975; Harlan 1975; NAS 1972; NRC 1993; Oldfield and Alcorn 1987; Plucknett et al. 1983; Querol 1993; Rengifo 1988). If, for example, the design of new conservation and development strategies assume the predominance of fine-grain specialization and microenvironmental distributions, the ensuing programs would need to be framed around the collecting or continued cropping of landraces across a complete range of microenvironments in diverse agroecosystems. If those assumptions are wrong as they happen to be in the case of diverse potato farming in the Andes, then conservation strategies must incorporate other basic designs.

Diverse Potatoes and the Andean Landscape

Findings on potato agriculture among Indian peasants in the Andes of Peru are pioneering new perspectives on the relation of the ecogeographical character of diverse crops to the nature of agroecosystem change. Potatoes are a local staple and rank today as the world's premier vegetable and its third-most important food crop. Andean potatoes are also one of the world's most biologically diverse and well-studied food plants with many classic investigations by scientists and scholars of the Andean countries (Blanco Galdos 1981; Huamán 1986; Ochoa 1990). Small-scale farmers that include millions of Quechua and Aymara peasants continue to cultivate seven potato species and more than 6000 landraces in the central Andes Mountains of Peru, Bolivia, and Ecuador (table 10.1). Their diverse potato crops afford a crucial resource for themselves and for world agriculture.

Table 10.1 Potato species (*Solanum*) cultivated in the Andes Mountains.

Species	Ploidy	Distribution	Species abbreviation
S. stenotomum Juz. et Buk.	2n = 2x = 24	Central Bolivia–Central Peru	*S.s.*
S. phureja Juz. et Buk.	2n = 2x = 24	Central Bolivia–Venezuela (Eastern Andes)	*S.p.*
S. × *chaucha* Juz. et Buk.	2n = 3x = 36	Central Bolivia–Central Peru	*S.*×*ch.*
S. × *juzepczukii* Buk.	2n = 3x = 36	Central Bolivia–Central Peru	*S.*×*j.*
S. tuberosum subsp. tuberosum L.	2n = 4x = 48	Southern Chile/Southern Argentina–Venezuela	*S.t.t.*
S. tuberosum subsp. andigena Hawkes	2n = 4x = 48	Northwestern Argentina–Venezuela	*S.t.a.*
S. × *curtilobum* Juz. et Buk.	2n = 5x = 60	Central Bolivia–Central Peru	*S.*×*cu.*
S. ajanhuiri Juz. et Buk.	2n = 2x = 24	Northern Bolivia–Southern Peru (not cultivated in Paucartambo)	*S.a.*

Sources: Hawkes 1978, 1990; Hawkes and Hjerting 1989; Ochoa 1990; Ugent 1970.

The spatial patterning of the mountainous farm landscapes of the Andes is frequently portrayed as a mosaic. Regularity of the patterned mosaic has been stressed in several studies that have sought to bridge the analysis of mountain ecology, agriculture, and the biophysical variation of Andean environments *per se* (Sarmiento 1986; Thomas and Winterhalder 1976; Tosi 1960; Winterhalder and Thomas 1978; Zamora Jimeno 1991). The view of the Andean landscape as a patterned mosaic is built on the assumption that its defining environmental features vary with a strong degree of regularity across space, especially elevation. The mosaic model has been influenced by prominent ecogeographical ideas (Zimmerer n.d.a.). Decades earlier a few of the main modern-day precursors of the mosaic model were promoted by Troll and those using the Holdridge system of plant and climate classification, especially Joseph Tosi who undertook a definitive study of Peru (Tosi 1960; Troll 1968). Tosi's influential studies assumed both the regularity of environmental variation in the form of standardized elevations or "ecological floors" and the extreme predictability of vegetation communities based on the idea of climaxes that had been deeded by Frederic Clements.

Applied to farming and land use in the Andes, the influential view of the environmental mosaic is taken to infer the regular spatial variation of elevation-related parameters such as soils, crops and vegetation, and climate, especially the lapse of temperature with elevational change (lapse rate). From the human ecology perspective, certain studies of Andean farming go a major step further in arguing for the formative role of small-area mosaics or microenvironments that are thought to be controlled by the highly localized constellations of elevation-related effects. The studies staking claim to the formative role of microenvironmental mosaics assert that the small elevation-regulated niches are a primary basis for the adaptive crop selection strategies of the small farmers of the Andes (Brush 1976; Brush and Guillet 1985; Vargas 1949; Webster 1973).

Disturbance and its role in the Andean agroecosystems has tended to slip from sight when the focus of study is shifted to the microscale. Although microscale accounts may mention the impact of common disturbances such as frost, the disturbances tend to be regarded as events that are regular, or at the least predictable, in terms of spatial occurrence. However, most farm hazards are fairly predictable only across areas that are larger than the typical microenvironment. Such larger-size areas can serve much better as reliable predictors of where hazards may occur. For example, Troll's well-known cross-sectional depiction of the "climatoecological gradation of the high Andes of southern Peru and northern Bolivia" defines general areas of large enough size (traverses of at least several hundred meters) that much disturbance ecology is encompassed by these general units (Troll 1968). In fact it was the spatial and elevation-related limits of frost at "regional frost boundaries" that led to Troll's depiction of the large-size areas.

By contrast, the predictability of disturbances like frost is less certain in the small-size areas of farm microenvironments. Analysis of climate records and soils variation shows the occurrence of most microscale farm hazards to be irregular or unpredictable in terms of exact timing (Johnson 1976; Knapp 1991; Thomas and Winterhalder 1976; Winterhalder 1994; Winterhalder and Thomas 1978; Zimmerer and Langstroth 1994). Similarly, the records also show that the spatial distributions of disturbance events often do not match up with microenvironments. To be sure, specific processes like cold air drainage and geomorphic features such as slope slips are partly controlled by microscale mechanisms. Yet freeze events and earth flows conform only slightly to these factors. Microscale patterning should not therefore be granted explanatory significance *a priori* for the human ecology or ecogeography of Andean farming. In fact, the lessons of disturbance ecology caution against the assumption of microenvironmental forcing or control in such cases since disturbances commonly interact and thus their influence cannot be considered in isolation from other events.

The assumption of the fine-scale regular patterning of disturbances does, however, persist strongly in underpinning a prevalent concept of the diverse Andean potatoes as ecological specialists. Adopting this logic, one review of Andean agriculture summarizes that "Adaptation to microenvironments is managed . . . through . . . a wide variety of crops" and that "selection matches agronomic qualities to microenvironmental conditions" (Brush and Guillet 1985: 24). An appraisal by farm critic Wendell Berry avows that Andean potatoes "are delicately fitted into their appropriate ecological niches" (Berry 1977: 177). It is assumed that the landrace types function as ecological specialists. And that assumption hinges on the belief that agroecological disturbance is either unimportant overall or that it is accounted for at the microscale. Like the landraces of other diverse crops, the immensely diverse potatoes are being viewed *a priori* as ecological specialists where adaptive traits are fit to the regular variation of farm microenvironments.

Yet a series of recent studies are demonstrating that the diverse Andean potatoes show a moderate to high degree of ecological versatility that plays an important role in the livelihoods of small farmers (Brush 1992; Brush et al. 1995; Quiros et al. 1990; Zimmerer 1991c, 1992, 1996, 1998; Zimmerer and Douches 1991). These studies suggest a rethinking of the role of disturbances in the Andean farm landscape that can be extended to resource management and conservation prospects. The following sections illustrate this by presenting the new evidence at two geographical scales: the farm landscape of the region and that of the individual field. In each case the explanatory approach is a fusion of human and political ecology with biogeography. Each case assesses the attributes of the farm landscape and changes there as the outcomes of human activities and biophysical processes, with a specific focus on farm techniques and agroecological disturbances.

The Farm Landscape of the Paucartambo Andes

Most of my studies on agrodiversity have been conducted on the diverse Andean potatoes cultivated in the area of Cuzco, Peru, and in particular its eastern flank: the 500-square-kilometer region of the Paucartambo Andes (map 10.1). The Paucartambo segment of the tropical Andes straddles the core area of potato diversity that stretches through southern Peru into northern Bolivia. Due to its concentration of potato biodiversity, the region of the Paucartambo Andes holds a global renown. Since expeditions in the 1920s that were overseen by Nicholai Vavilov, the renowned leader of Soviet plant science and crop geography, a number of students and collectors of crop genetic resources have plied the region for its prizes of diverse potatoes and other crops (Cook 1925; Hawkes 1941, 1944; Ochoa 1975, 1990; Rhoades 1982; Vargas 1949, 1954).

At present more than 20,000 Quechua-speaking people who belong to scores of peasant communities and peasant groups farm the Paucartambo

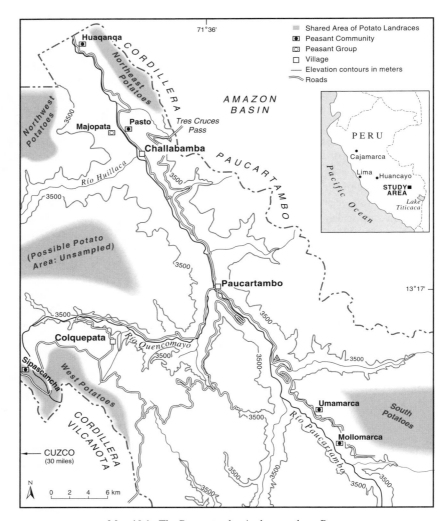

Map 10.1. The Paucartambo Andes, southern Peru.

region (a few communities appear in map 10.1). The Paucartambo farmers seed their potatoes together with a roster of 25–30 other food plants that are domesticated. About two-thirds of the farm families raise diverse potatoes as a subsistence crop—that is, one for self-consumption. Farmers of the region also pursue an ever-widening stream of commerce by planting high-yield potato varieties and malting barley and by migrating for off-farm work (Zimmerer 1991b, 1996).

Farmland of the Paucartambo Andes traverses a dissected terrain of varied plant habitats that grade among three chief vegetation types: thorn scrub, shrubland, and grassland. The general location and the genera of some fairly common species of each type are as follows: the subtropical thorn scrub (*Acacia, Trichocereus*) occupies the region's inner canyons near an elevation of 3000 meters; the shrubland or shrub savanna (*Baccharis, Senna*) is centered on midslopes around 3600 meters; and a cold grassland with small pockets of trees is found above 3800 meters (*Calamagrostis, Polylepis, Stipa*). The wide range of vegetation types and the notable role of human modification in the Paucartambo landscape bears a general similarity to other farm regions of the Peruvian Andes such as Cuzco, Cajamarca, Huancayo, and Ayacucho (map 10.1). Paucartambo's average rainfall is 633 millimeters and closely matches that of Cuzco. An eight-month rainy season lasts from October through May. Frequently the rains and unstable mountain slopes generate landslides, slumps, and other slope failures. Many farm fields show the evidence of fairly recent geomorphic disturbance.

Frost strikes only as low as 3900 meters or so during the rainy season; it descends to roughly 3300 meters during the dry season months of June through September. Water budget analyses indicate that a soil moisture deficit extends for three months or less at the higher growing sites, while in the inner canyons this deficit typically persists for more than six months. Farm soils also range widely across the landscape of the Paucartambo Andes. In general the soils contain a greater fraction of sand rather than clay at higher elevations. Still the Paucartambo soils, like the climate, vary only broadly as a function of elevation (Zimmerer and Langstroth 1994). Other major determinants of vegetation, soils, and climate include the influences of regional and local topography and historical factors.

Diverse potatoes as a group are planted widely in the Paucartambo Andes. Six cultivated species and 79 landraces range over elevations between roughly 3700 and 4100 meters in the three main areas of the region (map 10.1)—the northern Paucartambo Valley (the Challabamba area), the Quencomayo watershed (the Colquepata area), and the southern Paucartambo Valley (the Mollomarca area). Determining the degree of versatility of the diverse potato types requires taking separate accounts of species, landraces, and alleles, the major taxonomic and genetic building blocks of the Andean potatoes. The assessment also requires the consideration of regional-scale habitats. The regional scale in this case refers to the full area of the Paucartambo Andes and its elevational and areal habitats.

Species, landraces, and alleles of the diverse Andean potatoes are found capable of a moderate to high degree of versatility in the use of regional-scale habitats with respect to elevation and area (table 10.2). The evidence

Table 10.2. Summary of the biogeography of domesticated potatoes (*Solanum* spp.) in the farm landscape of the Paucartambo region (species abbreviations as shown in table 10.1).

Taxonomic unit	Elevational	Areal
Species	*Biogeography:* 3 spp. moderate-highly versatile and broadly distributed 3700–4050 m; (*S.s., S.×ch., S.t.*); 2 spp. broadly distributed 3900–4100 m (*S.×j.; S.×cu.*) 1 sp. distributed 2900–3300 m (*S.p.*).	*Biogeography:* 5 spp. moderate-highly versatile and broadly distributed across major areas of the region (*S.s., S.×ch., S.×j. S.t., S.×cu.;* see also map. 6.1); 1 sp. limited to cloud forest of the northern Paucartambo Valley (*S.p.;* see also "Farm Management" below).
	Study Methods: Field sampling and factorial field experiments.	*Study Methods:* Field sampling.
	Farm Management: Individual species within group not distinguished since farmers' management focuses on landrace units (see below).	*Farm Management:* Individual species within group not distinguished since farmers' management focuses on landrace units (see below). *S.p.,* which is fast-ripening and nondormant, is restricted to the area of year-round humidity.
Landrace (variety)	*Biogeography:* 78 landraces broadly distributed between 3700 and 4050 m. 1 landrace narrowly distributed between 3900 and 4050 m.	*Biogeography:* 17 landraces found across the region (cosmopolitan). 22 landraces found in about one half the region. 40 landraces restricted to 15–100 km² areas ("cultivar regions").
	Study Methods: Field sampling and field experiments.	*Study Methods:* Field sampling and field experiments.
	Farm Management: Farmers rotate seed among diverse field sites in order to lessen disease, to fallow fields, and as part of reducing the risk of crop loss.	*Farm Management:* Farmers create each "landrace area" via seed exchange networks and the supply of seed by high-elevation growers.
Alleles	*Biogeography:* Alleles at multiple loci broadly distributed.	*Biogeography:* Alleles at multiple loci broadly distributed; 99% in each subregion.
	Study Methods: Field sampling and isozyme analysis of widespread landraces.	*Study Methods:* Field sampling and isozyme analysis of widespread landraces.

Sources: Zimmerer 1991a, 1991c, 1992, 1996, 1998; on allele biogeography see Brush et al. 1995 (4 loci) and Zimmerer and Douches 1991 (10 loci).

of such ecological versatility is based on a variety of methods and analyses involving field sampling, field experiments, and laboratory isozyme analysis (Brush et al. 1995; Knapp 1991; Quiros et al. 1990; Zimmerer 1991a, 1991c, 1992, 1996, 1998; Zimmerer and Douches 1991). Research techniques include the random sampling, mapping, and identification of 200–225 potato plants per field shortly before harvest. Field experiments employ a reciprocal transplant or factorial design to test landrace adaptation. Laboratory isozyme analysis incorporates the assaying of allele type and diversity at various genetic loci.

The major findings of ecological versatility in the domesticated potatoes (table 10.2) of the Paucartambo region can be summarized as follows: (1) six of seven domesticated species are distributed among the main areas of the region and four species span an ample range of elevations (3700–4050 meters); (2) five potato species tested in factorial field experiments using field sites between 2800 meters and 4050 meters showed broad-based adaptation to elevation-related environments; (3) 78 out of 79 landraces are distributed widely between 3700 and 4050 meters; (4) nearly all conspecific landraces included in the factorial field experiments were not significantly different in terms of elevation-related adaptation; and (5) most alleles are distributed broadly with as much as 99% of allelic diversity in certain landraces being contained within each area (northern Paucartambo Valley, Colquepata watershed, and southern Paucartambo Valley).

The moderate-to-high degree of ecological versatility in the diverse Andean potatoes is the outcome of various forces. Chief among these are a couple of anthropogenic influences and natural disturbances. Farmers' rotations of planting sites supply one source of versatility. The cultivators rotate their potato plantings among fields from year to year to lessen the load of soil-born viruses and nematodes on seed tubers, and to restore the fertility of field soils. Their field rotation techniques result in potatoes being sown in sites that differ by 100 to 300 meters in elevation and occasionally by up to 500 meters. Given the extent of the elevational traverse, the versatility of the potatoes is crucial (Zimmerer 1991c, 1998). In addition to enabling field rotation, the moderate-to-high degree of versatility permits the scattering of a family's fields that helps reduce the risk of crop loss due to hazards like frost and drought. The crop rotation and field scattering in Paucartambo are common elsewhere in the Andes and other regions of peasant and indigenous farming (Goland 1993; Klee 1980; Mayer 1979; Netting 1993; Richards 1985; Wilken 1987).

Other husbandry techniques, particularly the exchange of seed tubers by farmers residing in different areas, also favor versatility. Their patterns of seed exchange strongly shapes the distribution patterns of different landraces. Widely exchanged landraces, known as cosmopolitans, account for one-third of the total landraces. These types range across the three main areas of the Paucartambo Andes (map 10.1). Some of the remaining landraces are grown in two of the region's three areas. Still other landraces cluster within just one of the 15 to 100-square-kilometer areas (referred to on the map as a "shared area of potato landraces") where local exchange networks are made up of valley slope farmers and nearby high-elevation growers that tend to be potato specialists (Zimmerer 1991c). The latter are the most dependable suppliers of seed that is less disease-ridden. Areal distribution patterns clustered on the high-elevation uplands of Paucartambo differ from other regions where potato seed exchange is centered on villages and valleys, and thus the nature of spatial patterning may vary regionally (Brush et al. 1981; Brush et al. 1995).

Each of the primary distribution patterns relies on the ecological versatility of potato types. Even the landrace types with localized areas depend on moderate to high versatility because they are planted across sizeable ranges in elevation within the single area. A similar pattern is found in genotypes. A pair of studies of the ecogeography of genetic variation shows that localized distributions are common in genotypes. Also the studies concur that the genotypes showing localized distributions did not display evidence of ecological specialization. Interestingly it is the estimate of genetic variation itself that differs between the two studies. The first investigation identified 30 genotypes in 139 tubers tested at ten loci (Zimmerer and Douches 1991); the second found 30 genotypes in a sample of 610 specimens tested at four loci (Brush et al. 1995).

Characteristic disturbances of Andean farm landscapes also exert pressure in favor of the ecological versatility of diverse potatoes. Several biotic disturbances, perceived by the farmers as major hazards to cropping, are found to be prevalent across broad swaths of elevation and area. Major pests, diseases, and field weeds of the potato crop recur between 2800 meters and 4050 meters with notably common occurrences taking place between 2800 meters and 3800 meters (table 10.3). For example, Andean weevils, the paramount pest plague of the potato, as well as the fungal late blight and the bacterial blackleg, the most serious disease problems, flare up frequently at fields throughout a broad elevational traverse. The major biotic disturbances are similarly widespread among the principal areas of the Paucartambo Andes (table 10.3).

Climatic disturbances similarly spread across wide traverses of the elevational and areal dimensions of the Paucartambo landscape (table 10.3). Drought, frost, hail, and the waterlogging of soils commonly beset a broad range of farm fields. Although the impact of certain farm hazards worsen noticeably in specific areas (for example, drought in exposed sites, frost in topographic depressions, and hail at high elevations), they commonly

Table 10.3. The range of principal hazards in potato agriculture of the Andes Mountains.

Elevation (m)	Common field pests	Common diseases	Common weeds (English Spanish or Quechua, scientific)	Main climate hazards	Main soil hazards
2800	Andean weevil (*Premnotrypes* spp.) Flea beetle (*Epitrex* spp.) Soil worms (family Noctuidae)	Late blight (*Phytophtera infestans*) Blackleg, soft rot (*Erwinia* spp.) Powdery scab (*Spongospora subterranea*) Potato viruses Potato cyst nemtodes, potato root eelworms (*Globodera* spp.) False root-knot nematode (*Nacobbus aberrans*)	Bind weed, *wilka* (*Convulvus* sp.) *Malvus, Bromus*	Drought	Waterlogging Soil-borne diseases
3300	Andean weevil Flea beetle Soil worms	Late blight Blackleg, soft rot Powdery scab Potato viruses Potato cyst nemtodes, potato root eelworms False root-knot nematode	Bind weed, *wilka* (*Convulvus* sp.)	Drought	Waterlogging Soil-borne diseases
3550	Andean weevil Flea beetle Soil worms	Late blight Blackleg, soft rot Powdery scab Potato viruses Potato cyst nemtodes, potato root eelworms False root-knot nematode	Kikuyu (*Pennisetum clandestinum*) wild mustard, *nabu* (*Brassica* sp.)	Drought	Waterlogging Soil-borne diseases

(continued)

Table 10.3. *Continued*

Elevation (m)	Common field pests	Common diseases	Common weeds (English Spanish or Quechua, scientific)	Main climate hazards	Main soil hazards
3800	Andean weevil—minor Flea beetle—minor levels Soil worms	Late blight—minor Blackleg, soft rot—minor Powdery scab Potato viruses Potato cyst nemtodes, potato root eelworms—reduced False root-knot nematode	None serious	Drought, frost, hail	Waterlogging Soil-borne diseases
4050	Soil worms—minor levels	Powdery scab Potato viruses—reduced Potato cyst nemtodes, potato root eelworms—minor False root-knot nematode—reduced	None serious	Frost, hail, drought—reduced	Waterlogging
Areal range in the Paucartambo Andes	Common pests range widely among all areas in the region	Common diseases range widely among all areas; fungal and bacterial diseases worsen in the humid eastern areas.	Common weeds range widely among all areas in the region	Widespread; drought, frost, and hail worsen in the semiarid western area	Widespread; waterlogging and soil-borne diseases worsen in the humid eastern areas

Sources: De Lindo et al. 1978; Ewell et al. 1990; Hooker 1981; IBTA-PROINPA 1994; Jatala 1994; O'Brien and Rich 1976.

affect various other areas as well. Overall, the distribution of such agricultural disturbances is patchy and largely unpredictable (Goland 1993). In short, the common forms of agroecosystem disturbance—biotic factors, climate, and soils—tend to range widely with a sizeable degree of uncertainty, rather than in a fashion that is narrow and regular. Such spatial patterning thus precludes the establishment of stable microenvironments that might otherwise spur the selection of potato types that could become ecologically specialized.

In summary, the findings of recent research in the Paucartambo Andes concur that the species, landraces, and alleles of diverse Andean potatoes demonstrate a moderate to high degree of ecological versatility and broad ecogeographical ranges at the regional scale. Interestingly, a survey of 25 uncultivated plant species in the Andes—that is, wild native plants—showed evidence of similarly broad ranges and thus presumed versatility (table 10.4). My findings differ sharply from the popular premise that native crops are specially adapted to particular microenvironments. The findings reported above also identify a number of conditions that select for the traits of ecological versatility and broad ecogeographical range in diverse potatoes, especially the wide occurrence of natural disturbances common in potato agroecosystems and the techniques and strategies of farmers. Combined results invite a rethinking of the ecological traits of crop diversity at other spatial scales, such as the individual field.

Ecological Versatility and Natural Disturbance at the Field Scale

The individual field is a second scale where the diverse Andean potatoes are claimed to show specialized, fine-grain distributions. Fields of the Paucartambo farmers vary in size from 120 to 4400 square meters, the mean area measuring about 1500 square meters. Potatoes are sown by hand in rows of elevated beds: as few as 7 and as many as 127 rows may be found in a field with the individual beds measuring about one-half meter in width. Each field contains a mixture of species, a pattern typical of the potato agriculture of small farmers in the central Andes (table 10.5) (Brush et al. 1981; Brush et al. 1995; Jackson et al. 1980; Quiros et al. 1990). The mixed species of a single field represent, in turn, a large assortment of diverse landraces, the average being 21 landraces per 225 plants (Zimmerer 1991a).

The within-field distributions of potato species and landraces demonstrate a substantial degree of versatility. Ecological versatility of these taxa at the field scale can be seen with respect to the typical sources of habitat variation. Mapped fields, typified by that of Juan Santos in Sipascancha (figure 10.1), were found to harbor at least one microenvironment that

Table 10.4. Range of plant species in the Andean highlands of Ecuador, Peru, and Bolivia.

Plant species	Number of accessions sampled	Lower and upper elevation limits (m) in sample	Elevational range (m)
Acmella oppositifolia	3	1700–2900	1200
Ageratina azangoroensis	5	2973–4080	1107
Ageratina pichichensis	3	2700–3500	800
Ageratina sternbergiana	3	3000–3900	900
Ageratum conyzoides	5	1000–3400	2400
Ambrosia arborescens	5	2200–3500	1300
Aristeguietia gayana	3	3600–4100	500
Austroeupatorium inulaefolium	3	1300–2200	900
Baccharis caespitosa	4	3600–4727	1127
Baccharis dracunculifolia	7	2249–3400	1151
Baccharis genistelloides	8	2500–3640	1140
Baccharis latifolia	17	2000–3600	1600
Baccharis prunifolia	7	2000–4500	2500
Barnadesia arborea	5	1600–3300	1700
Barnadesia horrida	7	3000–3600	1600
Barnadesia polycantha	6	2440–3600	1160
Bidens pilosa	6	1600–2700	1100
Grindelia boliviana	5	3500–3822	322
Hypochoeris meyeniana	7	3500–4693	1193
Matricaria matricoides	3	1080–3100	2020
Senecio rudbeckiaefolius	7	2570–4295	1725
Stevia lucia	3	3000–3900	900
Otholobium mexicanum	5	2200–3960	1760
Otholobium pubescens	7	2025–3800	1775
Sparticum junceum	6	2100–2870	770

Note: These plants were studied in the field in Paucartambo Province, Peru, and in collections at the Herbarium of the University of Wisconsin–Madison. The herbarium survey was made as comprehensive as possible within two representative plant families: Asteraceae and Fabaceae. In order to obtain comparative data the samples from the Herbarium collection were based on those species represented by at least three accessions. Sampling was restricted to five species per genus, thereby lessening the predominance of certain genera such as *Baccharis*. Because of the limited number of accessions this survey probably underestimates the full elevational range where these species actually occur.

exerts a typical constraint or hazard in potato farming. Yet the within-field distributions of species and landrace types, illustrated by figure 10.1, clearly do not match the locations of microenvironmental variation. Instead their distributions typically crosscut the major sources of variation as shown in figure 10.1. Potato species and landraces must therefore possess a certain degree of ecological versatility as a means of adapting to the within-field variability of environments (table 10.5).

Alleles likewise show versatility with respect to the within-field differ-

Sipascancha Field, Tubers Sampled by Row

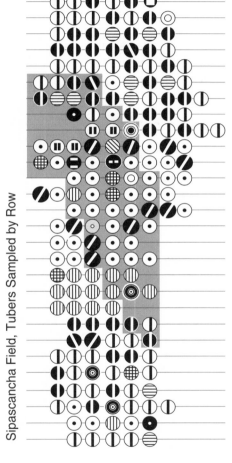

Keys

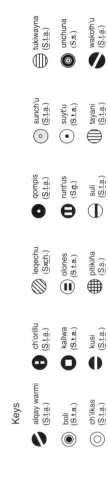

alqay warmi (S.t.a.)	ch'orillu (S.t.a.)	leqechu (Sxch.)	qompis (S.t.a.)	sunch'u (S.t.a.)	tukiwayna (S.t.a.)
boli (S.t.a.)	kallwa (S.t.a.)	olones (S.t.a.)	runt'us (S.g.)	suyt'u (S.t.a.)	unchuna (S.s.)
ch'ilkas (S.t.a.)	kusi (S.t.a.)	pitikiña (S.s.)	suli (S.t.a.)	tayani (S.t.a.)	wakoth'u (S.t.a.)

Fig. 10.1. Landraces and species in the potato field of Juan Santos, in Sipascancha, Paucartambo (n = 187 plants; species abbreviations as shown in table 10.1).

277

ences of habitat. The nature of allelic diversity within fields and within individual landraces merits close attention since even the tubers of single potato landraces within one field may harbor much allelic diversity (Quiros et al. 1990; Zimmerer and Douches 1991). In one analysis each landrace was found to contain two or more genotypes; one of these, *wako-th'u,* furnished 12 distinct genotypes (Zimmerer and Douches 1991). Genetic analysis thus proves that potato tubers belonging to the same landrace cannot be presumed clonal or identical in genetic terms even though they appear indistinguishable. Furthermore, the allelic diversity within single landraces may account for up to 75% of total diversity. Allelic diversity and the presence of individual alleles do not, however, match certain environments within fields. Like species and landraces, they range across varied field habitats and are thus ecologically versatile.

The ecological versatility of potatoes within fields is strongly reinforced by farm techniques and natural disturbances. The selection, storage, and planting of seeds rely upon, and in effect ensure, a certain degree of versatility. Most Paucartambo farmers, like their counterparts in other Andean regions, maintain the majority of their potato landraces in a mixed fashion throughout the processes of seed selection, storage, and sowing. The farmers select their landraces *en masse* as grouped mixtures while also expressing an overall aesthetic of diversity. In the selection process they devote no more than a modicum of attention to individual landraces (Zimmerer 1991a, 1996). A demonstration of that surprising feature is that the Paucartambo farmers seed only small numbers of those landraces most preferred for consumption, a seeming paradox common also among farmers in Puno, Peru (Jackson et al. 1980). In most cases only a few landraces, typically the common ones such as *kusi* and *suyt'u,* are handled as single types and planted separately within the field in block-style arrangements (a hint of this block-style patterning can be seen in the field shown in figure 10.1).

Overall, the farmers' rationales for choosing and sowing the diverse potatoes do not entail specialized ecological or genetic *desiderata.* Rather, their seed knowledge and techniques are aimed at reducing risk in a general way by mixing landraces. The farmers' seeding techniques thus exert a pressure in favor of those potato types that offer ecological versatility, for all planting material must prove viable and produce adequately across a range of within-field habitats. The landrace diversity of such plots tends to reduce the risk of crop failure in a general fashion rather than by recourse to ecological specialists.

Soils management by the Paucartambo farmers favors still further the versatility of landraces and diminishes the incentive for highly specialized types. Cultivators in Paucartambo and other Andean regions skillfully

Table 10.5. Summary of the biogeography of domesticated potatoes in fields (species abbreviations as shown in table 10.1).

Taxonomic unit	Individual field (mean area = 1800 m²)
Species	*Biogeography:* All species mixed and distributed broadly in field. *S.t.a., S.s.,* and *S.×ch.* in fields of what the farmers call "boiling potatoes" and *S.×cu.* and *S.×j.* in fields of "freeze-drying potatoes." Species are mostly intermixed within field and are not matched to certain microenvironments (frost, disease, drought, soil fertility).
	Study Methods: Field sampling, mapping of potato types (100–125 plants per field), measurement of elevation, and the mapping of major topographic and soils features (see figure 10.1).
	Farm Management: Farmers plant mixtures of species as part of a general strategy to reduce the risk of crop loss but they do not locate particular species or landrace types in certain field microenvironments.
	References: Brush et al. 1981; Jackson et al. 1980; Zimmerer 1991a, 1998.
Landraces	*Biogeography:* All landraces distributed broadly in field and most are thoroughly mixed. Both intermixed and single-type plantings of landraces are not matched to any particular microenvironment (frost, disease, drought, soil fertility).
	Study Methods: Field sampling, mapping of potato types (100–125 plants per field), measurement of elevation, and the mapping of major topographic and soils features (see map 10.1).
	Farm Management: Farmers plant multiple landraces as a general strategy to reduce risk but they do not locate particular species or landrace types in certain field microenvironments.
	References: Brush et al. 1981; Jackson et al. 1980; Zimmerer 1991a, 1998.

improve the quality of soils (fertility, moisture, texture) using techniques that include furrowing, mounding, pulverizing, fertilizing, and draining (Knapp 1991; Zimmerer 1996). The Andean farmer thus intensively manages soils as do many small farmers in other Latin America regions and elsewhere (Klee 1980; Netting 1993; Richards 1985; Wilken 1987). Their widespread management of soil fertility, moisture, and texture increases the uniformity of parcels, a trend that suits the versatile character of the potatoes. Such techniques soften the preexisting and sometimes sharp contrasts within field environments that might otherwise select for ecological specialists.

Natural disturbances add to the list of factors promoting ecological versatility at the field scale. Farmers in the Paucartambo Andes find that the abrupt changes of field climate and soils tend to occur either broadly or unpredictably at the scale of their individual parcels. Hazards such as frost and drought often impact whole field areas or strike unpredictably at areas within the parcel. The absence of a predictable conformity between natural disturbances and within-field habitat features acts to weaken the impetus for pinpoint-style planting and thus favors the reliance on potato types that are ecologically versatile.

Ecological Versatility and Disturbance: Implications for *In Situ* Conservation-with-Development

A new view of the ecogeography of diverse crops in the small-scale farming of developing countries and its relation to the character of natural disturbances is offered by the case of the cultivated Andean potatoes. Findings suggest that the spatial and temporal parameters of ecological disturbances play an important role in agroecosystem functioning, leading to the evolution of versatile rather than restricted adaptation in diverse crop types. The results advise us that it is unwarranted to assume that diverse crops like the Andean potatoes are ecological specialists *ipso facto.*

This critique of the faulty assumption serves to highlight the importance of determining the degree and kind of both landscape change and ecological adaptation. It does *not* imply that the versatility of diverse potatoes and the prevalence of disturbance in Andean agroecosystems somehow diminishes the usefulness of adaptation and ecosystem as central concepts. Rather, the critique indicates more comparative studies are needed. Indeed a suggestive comparison is found in the fields of highly varied Andean maize (Bird 1970; Brandolini 1970; Gade 1975; Grobman et al. 1961; Manglesdorf 1974; Zimmerer 1991a, 1996). The maize crop of the low- and middle-elevation Andes, which rivals its renowned diversity in Mexico, sports several landrace types that are relatively specialized with respect to local climate and soils. Agroecosystems of the maize types are relatively stable, moreover, since farmers expend much effort on irrigation, frost protection, and field tasks in general.

The history of ideas that are evident in previous studies of diverse crops and agroecosystems should also be mentioned. For at least the past two decades the landraces of the crops were deemed to be ecotypes within species (Zimmerer 1998). Yet, with respect to the role of natural disturbances, the analogy was seriously misconstrued. The landmark studies of ecotypes from the early 1900s were based on comparisons of the plant populations of widely-varied habitats, such as disparate woodlands, fields, and dunes in northern Europe (Turesson 1922) and across sharply contrasting elevation-related habitats in the California Sierra Nevada (Clausen et al. 1940). There the role of natural disturbance pales compared to the extent of permanent differences. By contrast, the assumptions applied to crop landraces tend to invoke the scale of small niches or microenvironments in which the role of disturbance cannot be easily subsumed.

Overall, this study's findings on ecological versatility and natural disturbances in farming deliver important implications for the possibility of combining the conservation of diverse crops with forms of sustainable development. First, ecological versatility adds to the possible viability of diverse crops being renewed *in situ* as continuing components of farming in developing countries. Recall that many peasant and indigenous farmers prefer the diverse crops as agronomic, culinary, and cultural resources (Altieri and Merrick 1987; Brush 1989; Cleveland et al. 1994; Oldfield and Alcorn 1987; Zimmerer 1991a, 1992, 1996). A key to *in-situ* conservation thus lies in the capacity of peasant and indigenous farmers to couple the cultivation of their diverse crops with the pursuit of commerce. Versatile capabilities of the diverse crops can enable conservation alongside the changes that are associated with the adoption of high-yielding crop varieties and other marketing activities.

Limits must be carefully assessed, however, in describing the potential for conservation-with-development that is conferred by versatility of agrodiversity. The versatility does not infer that the continued growing of diverse crops can be reconciled with all the forms of economic and social change that are implied in development. Certain sorts of development may involve the use of land or the deployment of other resource endowments such as farmers' labor time that do not permit the continued cropping needed for *in-situ* conservation (Zimmerer 1991b, 1992, 1996). In fact the crucial role of the availability of farm endowments indicates that the feasibility of such conservation be established not only on a crop-by-crop basis but also with respect to the particularities of regional economies and societies. The full participation of farmers themselves will aid in weighing the specific potential of *in-situ* conservation programs.

A final implication stems from the fundamental shaping of crop biogeography at human hands. The range of versatile crops like the Andean potatoes is molded as much by dispersal (seed exchange and field rotation, for example) as it is by the constraints stemming from ecological adaptation. As a result, conservation plans being designed for the diverse Andean potatoes must be closely guided by existing biogeographical patterns. The conservation plans should not adopt the credo of microenvironmental control. More generally, the assessment of the fundamental features of crop ecogeography needs to be seen as a prerequisite for successful conservation-with-development (Altieri and Merrick 1987; Brush et al. 1981; Brush 1989; Cleveland et al. 1994; Gade 1975; Johannessen 1970; Oldfield and Alcorn 1987; Zimmerer 1998). Standard knowledge in such conservation and development efforts needs to include an awareness that the spatial patterning of biodiversity and agroecosystem disturbances is likely to be complex at various spatial scales, including regional landscapes and smaller areas such as fields.

Acknowledgments

Components of this project were funded by the National Science Foundation (1992–1993); the Land Tenure Center of the University of Wisconsin–Madison (1993–1994); and the Graduate School of the University of Wisconsin–Madison (1994–1995). I gratefully acknowledge the cooperation of Hugh H. Iltis, herbarium director, and the assistance of Joseph McCann.

References

Alcorn, J. B. 1994. Noble Savage or Noble State? Northern Myths and Southern Realities in Biodiversity Conservation. *Etnoecología* 11(3): 7–19.

Altieri, M. A., and L. C. Merrick. 1987. *In situ* Conservation of Crop Genetic Resources through Maintenance of Traditional Farming Systems. *Economic Botany* 41(1): 86–96.

Berry, W. 1977. *The Unsettling of America: Culture and Agriculture.* New York: Avon Books.

Bird, R. M. 1970. *Maize and its Cultural and Natural Environment in the Sierra of Huánuco, Peru.* Ph. D. dissertation, University of California–Berkeley.

Blaikie, P. M. 1994. *Political Ecology in the 1990s: An Evolving View of Nature and Society.* CASID Distinguished Speaker Series No. 13. East Lansing: Center of Advanced Study of International Development, Michigan State University.

Blanco Galdos, O. 1981. Recursos genéticos y tecnología de los Andes altos. In *Agricultura de ladera en América Tropical,* ed. A. R. Novoa B. and J. L. Posner, pp. 297–303. Turrialba, Costa Rica: Centro Agronómico Tropical y Internacional.

Brandolini, A. 1970. Maize. In *Genetic Resources in Plants—Their Exploration and Conservation,* eds. O. H. Frankel and E. Bennet, pp. 273–309. Philadelphia: F. A. Davis Company.

Brookfield, H., and C. Padoch. 1994. Appreciating Agrodiversity: A Look at the Dynamism of Indigenous Farming Practices. *Environment* 36: 7–11, 37–45.

Brücher, H. 1989. *Useful Plants of Neotropical Origin.* Berlin: Springer-Verlag.

Brush, S. B. 1976. Man's Use of an Andean Ecosystem. *Human Ecology* 4(2): 147–166.

Brush, S. B. 1986. Genetic Diversity and Conservation in Traditional Farming Systems. *Journal of Ethnobiology* 6(1): 151–167.

Brush, S. B. 1989. Rethinking Crop Genetic Resource Conservation. *Conservation Biology* 3(1): 19–29.

Brush, S. B. 1992. Ethnoecology, Biodiversity, and Modernization in Andean Potato Agriculture. *Journal of Ethnobiology* 12(2): 161–185.

Brush, S. B., H. J. Carney, and Z. Huamán. 1981. Dynamics of Andean Potato Agriculture. *Economic Botany* 35(1): 70–88.

Brush, S. B., and D. W. Guillet. 1985. Small-Scale Agro-pastoral Production in the Central Andes. *Mountain Research and Development* 5(1): 19–30.

Brush, S. B., R. Kessili, R. Ortega, P. Cisneros, K. S. Zimmerer, and C. Quiros. 1995. Potato Diversity in the Andean Center of Crop Domestication. *Conservation Biology* 9(5): 1189–1198.

Brush, S. B., and D. Stabinsky, eds. 1996. *Valuing Local Knowledge.* Covelo: Island Press.

Clausen, J., D. Keck, and W. M. Heisey. 1940. *Experimental Studies on the Nature of Species.* Washington, D.C.: Carnegie Institution of Washington.

Cleveland, D. A., D. Soleri, and S. F. Smith. 1994. Do Folk Crop Varieties Have a Role in Sustainable Agriculture? *BioScience* 44 (11): 740–751.

Cook, O. F. 1925. Peru as a Center of Domestication, Part I. *Journal of Heredity* 16(2): 33–46.

De Lindo, L., and A. Kellman. 1978. *Erwinia* Spp. Pathogenic to Potatoes in Peru. *American Potato Journal* 55(7): 371–400.

Ewell, P. T., H. Fano, K. V. Raman, J. Alcázar, M. Palacios, and J. Carhuamaca. 1990. *Farmer Management of Potato Insect Pests in Peru.* Lima: International Potato Center.

Fowler, C., and P. Mooney. 1990. *Shattering: Food, Politics, and the Loss of Genetic Diversity.* Tucson: University of Arizona Press.

Frankel, O. H., and J. G. Hawkes, eds. 1975. *Crop Genetic Resources for Today and Tomorrow.* Cambridge: Cambridge University Press.

Gade, D. W. 1975. *Plants, Man, and the Land in the Vilcanota Valley of Peru.* The Hague: Dr W. Junk.

Goland, C. 1993. Field Scattering as Agricultural Risk Management: A Case Study from Cuyo Cuyo, Department of Puno, Peru. *Mountain Research and Development* 13(4): 317–338.

Gómez Molina, E., and A. V. Little. 1981. Geoecology of the Andes: The Natural Science Basis for Resource Planning. *Mountain Research and Development* 1(2): 115–144.

Grobman, A., W. Salhuana, and R. Sevilla. 1961. *Races of Maize in Peru: Their Origins, Evolution, and Classification.* Washington, D. C.: National Academy of Sciences.

Harlan, J. R. 1975. Our Vanishing Genetic Resources. *Science* 188: 618–621.

Harlan, J. R. 1992. *Crops and Man.* 2d edition. Madison: American Society of Agronomy and Crop Science Society of America.

Hawkes, J. G. 1941. *Potato Collecting Expeditions in Mexico and South America.* Cambridge: Imperial Bureau of Plant Breeding and Genetics.

Hawkes, J. G. 1944. *Potato Collecting Expeditions in Mexico and South America II. Systematic Classification of the Collections.* Cambridge: Imperial Bureau of Plant Breeding and Genetics.

Hawkes, J. G. 1978. Biosystematics of the Potato. In *The Potato Crop: The Scientific Basis for Improvement,* ed. P. M. Harris, pp. 15–69. London: Chapman and Hall.

Hawkes, J. G. 1990. *The Potato: Evolution, Biodiversity, and Genetic Resources.* Washington, D. C.: Smithsonian Institution Press.

Hawkes, J. G., and J. P. Hjerting. 1989. *The Potatoes of Bolivia: Their Breeding Value and Evolutionary Relationships.* Oxford: Clarendon Press.

Hooker, W. J. 1981. *Compendium of Potato Diseases.* St. Paul: American Phytopathological Society.

Horton, D. E. 1987. *Potatoes: Production, Marketing, and Programs for Developing Countries.* Boulder: Westview Press.

Huamán, Z. 1986. Conservación de recursos genéticos de papa en el CIP. *Circular* (Centro Internacional de la Papa) 14(2): 1–7.

IBTA–PROINPA. 1994. *Informe anual compendio.* Cochabamba, Bolivia: IBTA (Instituto Boliviano de Technología Agropecuario) and PROINPA (Programa de Investigación de la Papa).

Jackson, M. T., J. G. Hawkes, and P. R. Rowe. 1980. An Ethnobotanical Field Study of Primitive Potato Varieties in Peru. *Euphytica* 29(1): 107–113.

Jatala, P. 1994. Biology and Management of Nematode Parasites of Potato in Developing Countries. In *Advances in Potato Pest Biology and Management,* eds. G. W. Zehnder, M. L. Powelson, R. K. Jansson, and K. V. Raman, pp. 214–233. St. Paul: The American Phytopathological Society.

Johannessen, C. L. 1970. The Dispersal of Musa in Central America: The Domestication Process in Action. *Geographical Review* 60(4): 689–699.

Johnson, A. M. 1976. The Climate of Peru, Bolivia, and Ecuador. In *Climates of Central and South America,* ed. W. Schwerdtfeger, pp. 147–218. Amsterdam: Elsevier.

Klee, G. A. 1980. *World Systems of Traditional Resource Management.* New York: John Wiley and Sons.

Knapp, G. 1991. *Andean Ecology: Adaptive Dynamics in Ecuador.* Boulder: Westview Press.

Manglesdorf, P. C. 1974. *Corn: Its Origin, Evolution, and Improvement.* Cambridge: Harvard University Press.

Mayer, E. 1979. *Land Use in the Andes: Ecology and Agriculture in the Mantaro Valley of Peru with Special Reference to Potatoes.* Lima: Centro Internacional de la Papa.

NAS (National Academy of Sciences). 1972. *Genetic Vulnerability of Major Crops.* Washington, D. C.: National Academy of Sciences.

NRC (National Research Council). 1993. *Managing Global Genetic Resources: Agricultural Crop Issues and Policies.* Washington, D. C.: National Academy Press.

Netting, R. M. 1993. *Smallholders, Householders: Farm Families and the Ecology of Intensive Sustainable Agriculture.* Palo Alto: Stanford University Press.

O'Brien, M. J., and A. E. Rich. 1976. *Potato Diseases.* Washington, D. C.: United States Department of Agriculture.

Ochoa, C. 1975. Potato Collecting Expeditions in Chile, Bolivia and Peru, and the Genetic Erosion of Indigenous Cultivars. In *Crop Genetic Resources for Today and Tomorrow,* eds. O. H. Frankel and J. G. Hawkes, pp. 167–173. Cambridge: Cambridge University Press.

Ochoa, C. 1990. *The Potatoes of South America: Bolivia,* trans. Donald Ugent. Cambridge: Cambridge University Press.

Oldfield, M. L., and J. B. Alcorn. 1987. Conservation of Traditional Agroecosystems. *BioScience* 37(3): 199–208.

Pickett, S. T. A., and Peter S. White, eds. 1985. *The Ecology of Natural Disturbance and Patch Dynamics.* New York: Academic Press.

Plucknett, D. L., N. J. H. Smith, J. T. Williams, and N. Murthi Anishetty. 1983. Crop Germplasm Conservation and Developing Countries. *Science* 220: 163–169.

Querol, D. 1993. *Genetic Resources: A Practical Guide to Their Conservation.* London: Zed Books.

Quiros, C. F., S. B. Brush, D. S. Douches, G. Hewstes, and K. S. Zimmerer. 1990. Biochemical and Folk Assessment of Variability of Andean Cultivated Potatoes. *Economic Botany* 44(2): 254–266.

Rengifo, G. 1988. *Recursos fitogenéticos andinos.* Cajamarca: Proyecto Piloto de Ecosistemas Andinos.

Rhoades, R. E. 1982. The Incredible Potato. *National Geographic Magazine* (May): 668–694.

Richards, P. 1985. *Indigenous Agricultural Revolution: Ecology and Food Production in West Africa.* London: Unwin Hyman.

Sarmiento, G. 1986. Ecological Features of Climate in High Tropical Mountains. In*High Altitude Tropical Biogeography,* eds. F. Vuilleumier and M. Monasterio, pp. 11–45. New York: Oxford University Press.

Thomas, R. B., and B. P. Winterhalder. 1976. Physical and Biotic Environment of Southern Highland Peru. In *Man in the Andes,* eds. P. T. Baker and M. A. Little, pp. 21–59. Stroudsberg, Pennsylvania: Dowden, Hutchinson, and Ross.

Tosi, J. A. 1960. *Zonas de vida natural en el Perú: Memoria explicativa sobre el mapa ecológico del Perú.* Lima: Instituto Interamericano de Ciencias Agricolas de la OEA, Zona Andina.

Troll, C. 1968. The Cordilleras of the Tropical Americas: Aspects of Climatic, Phytogeographical and Agrarian Ecology. In *Geo-ecology of the Mountainous Regions of the Tropical Americas,* Proceedings of the UNESCO (Mexico) Symposium, 1966, pp. 15–56. Bonn: Ferd Dummlers Verlag.

Turesson, G. 1922. The Species and the Variety as Ecological Units. *Hereditas* 3: 100–113.

Ugent, D. 1970. The Potato. *Science* 170: 1161–1166.

Vale, T. R. 1982. *Plants and People: Vegetation Change in North America.* Washington, D. C.: Association of American Geographers.

Vargas, C. 1949. *Las papas sudperuanas. Parte I.* Cuzco: Universidad Nacional de San Antonio de Abad del Cusco.

Vargas, C. 1954. *Las papas sudperuanas. Parte II.* Cuzco: Universidad Nacional de San Antonio de Abad del Cusco.

Veblen, T. T. 1992. Regeneration Dynamics. In *Plant Succession in Theory and Prediction,* eds. D. C. Glenn-Lewin, R. K. Peet, and T. T. Veblen, pp. 152–187. New York: Chapman and Hall.

Webster, S. 1973. Native Pastoralism in the South Andes. *Ethnology* 12(2): 115–134.

White, P. S. 1979. Pattern, Process, and Natural Disturbance in Vegetation. *The Botanical Review* 45: 229–299.

Wilken, G. C. 1987. *Good Farmers: Traditional Agricultural Resource Management in Mexico and Central America.* Los Angeles and Berkeley: University of California Press.

Winterhalder, B. P. 1994. Rainfall Predictability and Water Management in the Central Andes of Southern Peru. In *Irrigation at High Altitudes,* eds. David Guillet and William P. Mitchell. Washington, D.C.: The American Anthropological Association.

Winterhalder, B. P., and R. B. Thomas. 1978. *Geoecology of Southern Highland Peru: A Human Adaptation Perspective* Occasional Paper 27. Boulder: University of Colorado, Institute of Arctic and Alpine Research.

Zamora Jimeno, C. 1991. El recurso suelo del Perú. *Boletín de la Sociedad Geográfica de Lima* 104/105: 93–122.

Zimmerer, K. S. 1991a. Managing Diversity in Potato and Maize Fields of the Peruvian Andes. *Journal of Ethnobiology* 11(1): 23–49.

Zimmerer, K. S. 1991b. Labor Shortages and Crop Diversity in the Southern Peruvian Sierra. *The Geographical Review* 82(4): 414–432.

Zimmerer, K. S. 1991c. The Regional Biogeography of Native Potato Cultivars in Highland Peru. *Journal of Biogeography* 18: 165–178.

Zimmerer, K. S. 1992. The Loss and Maintenance of Native Crops in Mountain Agriculture. *GeoJournal* 27(1): 61–72.

Zimmerer, K. S. 1996. *Changing Fortunes: Biodiversity and Peasant Livelihood in the Peruvian Andes.* Los Angeles and Berkeley: University of California Press.

Zimmerer, K. S. 1998. The Ecogeography of Major Crops: Versatility and Conservation. *BioScience* 48(6): 445–454.

Zimmerer, K. S. N.d.a. Overlapping Patchworks of Mountain Agriculture in Peru and Bolivia: Toward a Regional-Global Landscape Model. *Human Ecology.* In press.

Zimmerer, K. S., and R. P. Langstrogh. 1994. Physical Geography of Tropical Latin America. *Singapore Journal of Tropical Geography* 14(2): 157–172.

Zimmerer, K. S., and D. S. Douches. 1991. Geographical Approaches to Crop Conservation: The Patterning of Genetic Diversity in Andean Potatoes. *Economic Botany* 45(2): 176–189.

11

The Ecology, Use, and Conservation of Temperate and Subalpine Forest Landscapes of West Central Nepal

John J. Metz

Introduction: The Dilemma of Nepal's Forests

Deforestation in the Nepal Himalayas is a serious problem, although not, as many think, because it is causing catastrophic erosion and flooding (Bruijnzeel and Bremmer 1989; Ives and Messerli 1989; Metz 1991). Rather, forest destruction is important because it threatens endangered plants and animals with extinction (Yonzon and Hunter 1991) and because the resulting scarcity of forest products undermines the livelihood of subsistence farmers. In this paper I explore the contributions that biogeographical research can make to improving the management of the remaining upland forests, most of which are above 2600 meters.

Although small areas of upland forests continue to be converted to agriculture and other uses, the main threat to forests in the hills and mountains of Nepal comes from indigenous farmers whose subsistence activities interfere with tree regeneration, thereby transforming forests into depauperate woodlands and shrublands (Nield 1985; Stainton 1972). This is quite different from other parts of the world experiencing rampant deforestation (Repetto and Gillis 1988), including Nepal's southern lowlands, because here outside loggers, migrants, and ranchers are not entering forests and converting them to other uses. Nevertheless, outsiders and the feudal elite who ruled Nepal until 1950 and were kept in power by British imperialism, fostered deforestation by: (1) extracting from mountain farmers half of their crop plus corvee labor as taxes, (2) granting tax holidays to agricultural land converted from forest, and (3) creating the current highly-stratified social structure that blocks solutions to today's problems (Mahat et al. 1987; Metz 1991).

Throughout upland Nepal, forest ecosystems are integral parts of the landscapes people use for subsistence (Fox 1984; Metz 1990; Wyatt-Smith 1983). In addition to fuelwood, building materials, medicines, and wild foods, forests supply fodder to feed livestock through the long dry season between monsoons. Because over 75% of Nepal's people rely on food they grow themselves for survival, and since livestock manure is the main source of soil fertility, the energy and nutrients that flow from forest to field via livestock are essential to subsistence systems.

In most of Nepal a boundary at about 2600 meters marks a transition from a lower area that is densely populated, intensively used, and thinly forested, to an upper region with few people and extensive temperate and subalpine forests. In an overly simple but useful division the human communities can be divided into two types: a majority that lives below 1800 meters and does not access the forests above 2600 meters, and a minority that lives close to the forests and controls them (Metz 1989).

The landscapes inhabited by lower-elevation groups are composed of agroecosystems that cover 10% to 25% of the land, set within a matrix of degraded forests. Human activities have converted most forest ecosystems to shrublands of varying densities, but patches of remnant forests remain in protected areas on 10% to 30% of the land. In these lower areas people are responding to forest product shortages by planting trees on private land and developing management systems in their remnant forests. Several studies have found large increases over the last few decades of private fodder, fuel, and timber trees (Carter and Gilmour 1989; Gilmour and Nurse 1991; Robinson 1990). Other research has found that the users of many of the remnant forest patches have implemented some type of management, much of it begun within the last twenty-five years (Campbell and Bhattarai 1984; Fisher 1989; Gilmour and Fisher 1991; Messerschmidt 1986, 1987). However, these studies also show that "user groups" are able to limit current forest uses like grazing, or the types and timing of products extracted, much more easily than they can design and implement new silvicultural systems that manage for defined goals by removing products and dividing them among users (Gilmour and Fisher 1991:51).

The landscapes surrounding communities that control the upper forests have rain-fed widely scattered agricultural patches in a matrix of shrublands and degraded forests. Above 2600 meters agricultural fields disap-

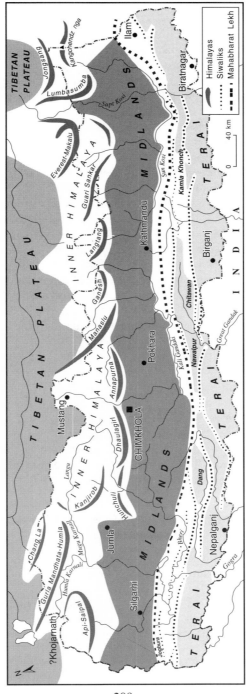

Map 11.1. West Central Nepal.

pear and the density and stature of the wild vegetation increase sufficiently to create a landscape of forest ecosystems. The control of upper-elevation communities over the relatively large areas of forest above 2600 meters allows them to rely heavily on communal forests rather than private trees. This easy availability of forest products discourages changes that may lead to more sustainable management.

In these upper forests livestock management practices are causing most forest degradation (Metz 1994). As the herbaceous and woody plants disappear in the postmonsoon dry season, herders cut leaves and small branches from evergreen broadleaf forest trees to supplement their animals' ration—this gradually reduces productivity and can kill the tree. In addition, most upper communities take their animals into the forests during the monsoon and, for some groups, during other times as well (Alirol 1979; Metz 1994; Schmidt-Vogt 1990). Within the forest, the livestock browse on whatever plants they can find and thereby hinder the regeneration of palatable species (Alirol 1979; Metz 1990, 1994).

In addition to these two general types of environmental use, a third system involves larger-scale herding for the production of dairy products and offspring. This system moves through the forest during the spring, into alpine pastures in summer, and then back into the forest and village environs in autumn and winter (Alirol 1979; Brower 1991; Schmidt-Vogt 1990; Stevens 1993). A few groups, like the Sherpa, live at high elevations and farm up to 3400 meters (Brower 1991; Stevens 1993). Although these herders are a small percentage of Nepal's total population, their fuelwood cutting and pasture burning activities depress the upper timberline.

In sum, human activities clearly are affecting forest regeneration, but scholars need to determine precisely how human uses are interacting with "natural" forest dynamics in order to design management systems that maintain forest cover. To that end, I now will review the ecological functioning of temperate and subalpine forest landscapes.

Ecological Functioning of the Forest Landscapes of Dhaulagiri Himal Region

Only since 1950 has Nepal's government allowed scholars to explore and describe Nepal's vegetation, although foresters have studied vegetation in the Indian Himalayas for over a century (Champion 1936; Omaston 1922). Early work in Nepal adapted the subjectively determined classification systems developed in the Indian Himalayas (Champion and Seth 1968; Schweinfurth 1957; Stainton 1972) and East Asia (Kawakita 1956; Numata 1983).

The system adopted by most (Dobremez 1976; Schweinfurth 1957;

Stainton 1972) is to divide Nepal or the entire Himalaya into east-to-west regions and to describe the vegetation "types" of each region, differentiated by elevation and sometimes by local moisture gradient. This strategy has proved to be an effective way to simplify the complexity and describe the broad patterns of Himalayan vegetation. Nevertheless, further progress will require quantitative studies of numerous stands at critical points along the range to verify and specify the forest types and their dynamics.

Several research programs and individual scholars have begun to describe the species composition, forest structure, and ecosystem functioning (such as productivity and nutrient cycling) of these regional forest ecosystems. Most comprehensive is the work of Kumaon University's botany department under J. S. Singh and S. P. Singh (Singh and Singh 1987, 1992; Singh et al. 1994). They and their students have studied forests of all elevations in the Indian Himalayas just west of Nepal. J. F. Dobremez and students have produced detailed studies of the composition, structure, and dynamics of the forests and the farming systems of people of Salme', 50 kilometers northwest of Kathmandu (Dobremez 1986; Richard 1980; Sueron 1985; Umans 1993; Wiart 1983). Japanese scholars from Chiba University and other institutions (Nakasuga 1972; Numata 1975; Yoda 1967, 1968) have studied forests in eastern Nepal and Bhutan (Ohsawa 1987, 1991). The Makalu-Barun conservation project is currently investigating the biology and human use of the area just east of Mt. Everest (Carpenter and Zomer 1996; Oliver and Sherpa 1993). Nepalese scholars from the Department of Medicinal Plants have contributed significantly to the efforts of Dobremez, the Japanese, and the Makalu-Barun project.

Nevertheless, detailed studies of disturbance patterns and forest regeneration are few. This is due, I believe, to an overly static view of forests, which results from assuming that the discrete forest types scholars name have higher reality status than the "intergrades," and that these forest types usually persist in some "climax" equilibrium from which they are seldom disturbed. The paper of Ohsawa et al. (1986) that seeks to explain why "western Himalayan" forest types are in eastern Nepal illustrates the first assumption. The second is evidenced by the Kumaon group (Singh and Singh 1987: 82, 1992) and the Japanese (Ohsawa 1983: 93) labelling forests "climax" without having carefully investigated their disturbance regimes or species dynamics. Singh and Singh (1987, 1991) are aware of the literature on disturbance and are beginning to examine forest dynamics more systematically (S. P. Singh, personal communication, 1996).

Studies in ecosystems around the world during the last 20 years have found disturbances to be frequent and decisive events (Botkin 1990; Denslow 1987; Glenn-Lewin et al. 1992; Pickett and White 1985). In some ecosystems, catastrophes like fires (Franklin and Hemstrom 1981; Heinselman 1973) or landslides (Veblen and Ashton 1978) occur within the lifespans of the dominant organisms and prevent tolerant species from replacing pioneers. In other systems, fires and/or windstorms occur every several generations (Henry and Swan 1974; Oliver and Stephens 1977), and depending on the disturbance and its severity lead to dramatically different dominant species in the regenerating forest (Henry and Swan 1974; Stewart 1986).

The importance of this "resetting of the ecological clock" by major disturbances has led several researchers to describe vegetation change as a predictable series of phases or stages through which a stand passes following disturbance (Borman and Likens 1979; Oliver 1981). The first phase, "stand initiation," follows a major disturbance as newly germinated plants, surviving seedlings, and saplings grow to form a new canopy. In the second "stem exclusion" or "thinning" phase, competition between individuals of the first cohort reduces the number of stems and casts such deep shade that even tolerant species are suppressed. The even-aged canopy trees of the first cohort often die within a short period and create the third "understory reinitiation" phase: Light, nutrients, and water become available to the suppressed understory, which grows to produce a new canopy. Only the final "steady-state" or "old-growth" phase approximates the traditional "climax" concept: As canopy individuals die, suppressed seedlings and saplings of tolerant trees grow rapidly into the gaps and compete to replace the dying second-cohort trees. Spatially, the cumulative process produces patches of varying sizes, each with groups of approximately even-aged individuals. Temporally, it causes pulses of regeneration and extended periods of slow growth with little recruitment of new individuals.

Scholars must apply the techniques developed to study these processes to Himalayan vegetation. Hitherto, only a few scholars (Brower and Dennis, this volume; Oliver and Sherpa 1990; Schmidt-Vogt 1990; Umans 1993) have begun to do so, and they have worked exclusively in subalpine fir-rhododendron forests. I am currently involved in similar work in West Central Nepal, in temperate and subalpine forest landscapes of the Chimkhola region, on the southeast flank of Dhaulagiri Himal. The first phase seeks to relate forest species composition and structure to environmental gradients and disturbance, to identify objectively forest "types," and to begin to explore via size-class analysis stand dynamics and regeneration patterns (Metz, n.d.). A second phase plans to explore disturbance regimes and regeneration patterns more precisely and to suggest silvicultural systems that local forest users can employ to promote forest regeneration. The remainder of this section will explore the implications of the first phase of my project.

The study site receives 85% of its 2700 millimeters of annual precipitation between 1 June and 15 October, based on two years of data from the 1750-meter village center. Midlatitude cyclones bring occasional precipitation during the winter and spring, and this is supplemented by afternoon convectional storms that produce increasing precipitation and, occasionally, crop-destroying hail in April and May. Average monthly maximum and minimum temperatures are 28°C in June and 6°C in January, respectively; absolute maximum and minimum temperatures are 32°C and 1°C, respectively. With the 8.17-kilometers-tall Dhaulagiri Himal 35 kilometers north of the village, topography is extremely steep. The bedrock is divided by the Main Central Thrust (MCT) that surfaces at about 3400 meters. Rocks below the MCT are weakly metamorphosed schists, phyllites, and slates; the bedrock above the MCT is gneiss and quartzite. Virtually all the soils are thin, rocky entisols and inceptisols, with inceptisols more common below the MCT (Greenberg 1986).

Field methods subjectively identified 77 stands of homogeneous, tall forest for sampling. Each one-tenth-hectare stand had eight circular quadrats, located in a stratified, systematic, unaligned manner (Berry and Baker 1968). Within the quadrats all trees and saplings were identified to species, measured for circumference, and evaluated for human impact on a 0 to 5 scale (0 = no impact, 5 = dead). We counted tall and short seedlings of woody plants in eight ten-square-meter and one-square-meter circular quadrats. We also measured stand elevation, aspect, slope angle, slope position, topographic configuration, forest floor cover, forest floor and A horizon thickness, and depth to bedrock. Finally, we collected samples of A soil horizon for chemical and mechanical analysis.

I used ordination, cluster analysis, and Twinspan (Gauch 1982) to identify seven forest types from the 77 stands (tables 11.1, 11.2). The types identified were similar to those found by other workers (Dobremez 1976; Ohsawa 1983; Stainton 1972): Predominantly broadleaf evergreen temperate forests are supplanted by *Abies-Rhododendron* forests at about 3000 meters (table 11.1). In three of the four broadleaf-evergreen types, however, my measure of "importance," the average of relative density and relative basal area identified subcanopy trees as the most important. For example, on the mesic sites between 2000 and 2600 meters *Symplocos ramocissima,* a tolerant, prolific subcanopy tree that is not used for fuel or fodder, is more important than the *Quercus oxyodon, Persea dutheii,* and *Dodecadenia grandifolia* canopy trees. On the mesic and xeric sites between 2500 and 2900 meters *Rhododendron arboreum* is most important, although *Q. semecarpifolia* and *Tsuga dumosa* form the canopy on xeric south- and east-facing sites; *Acer campbellii, Ilex dyprena,* and *Alnus nepa-*

Table 11.1. Forest landscape types of the Chimkhola region.

Temperate Forest Landscapes

1. *Quercus lanata–Lyonia ovalifolia* woodland. Only small remnants of a type that was widespread on southerly slopes in zone of cultivation (1500–2400 m) continue to exist; the community of Chimkhola preserves these in a very steep, south-facing bowl above the village to inhibit mass wasting. Trees are of low stature (average height is 8.1 m), relatively low density (1023 saplings and trees/ha), and low diversity (the top two species are 89% of tree importance). Only the ericad *Lyonia ovalifolia* is reproducing, making up 91% of sapling importance.

2. Mesic *Symplocos–Quercus–Lauraceous* forest. Low-elevation (2300–2600 m), highly diverse (tree-sapling Shannon Information Index [H'] = 1.708), tall forests (canopy trees up to 50 m) located on northwest-, north-, and northeast-facing slopes and valley bottoms. Subcanopy tree, *Symplocos ramocissima,* is the most important species, but *Quercus oxyodon, Persea dutheii, Dodecadenia grandifolia,* and *Ilex dipyrena* form canopy. The reproduction of all canopy trees except *Persea dutheii,* is inadequate, so shade-tolerant *S. ramocissima,* which is unusable for fodder or fuel, dominates reproduction (table 11.2).

3. Xeric *Rhododendron arboreum–Quercus semecarpifolia* forest. Mid-elevation (2500–2900 m), diverse (H' = 1.565), dry forests with a canopy of *Quercus semecarpifolia, Dodecadenia grandifolia, Lyonia ovalifolia,* and at some sites *Tsuga dumosa.* The most important species, *R. arboreum,* remains a subcanopy tree. *Q. semecarpifolia,* which can reach a height of 50 m and dbh of 1.5 m, is heavily lopped for fodder, as are the other oaks. Reproduction of oak and hemlock is inadequate (table 11.2).

4. Mesic *Rhododendron arboreum–Acer* forest. Mid-elevation (2500–2900 m), diverse (H' = 1.560), tall forests on west-, northwest-, north-, and northeast-facing slopes. *R. arboreum* is most important and again remains in the subcanopy. The canopy is composed of many species, but most important are *Acer campbellii, Ilex dipyrena, Alnus nepalensis, A. caesium,* and *Aesculus indica.* Reproduction of these forests is dominated by shrubs (*Viburnum erubescens* contributes 67% of sapling importance) and subcanopy trees (table 11.2).

Subalpine Forest Landscapes

5. Mesic *Abies spectabilis–Rhododendron barbatum* forests. Upper-elevation (2950–3300 m), tall forests of moderate diversity (H' = 1.191) with an overstory of *Abies spectabilis, Tsuga dumosa,* and *Acer campbellii* and an understory of *R. barbatum.* These forests are on west-, northwest-, north-, and northeast-facing slopes. The *Abies* and *Tsuga* trees can reach a height of 50 m and dbh of 2 m. *Rhododendron barbatum* and *Viburnum* dominate reproduction (table 11.2).

6. Xeric *Abies spectabilis–Rhododendron arboreum* forests. These forests are located on slightly higher and drier sites than the other *Abies* forests. *R. barbatum* is much less important than *R. arboreum,* and *T. dumosa* disappears on these east-, southeast-, south-, and southwest-facing slopes. These are the least diverse forests of my sample (H' = 0.914). Reproduction is dominated by *R. barbatum* and *R. arboreum.*

7. Mesic *Betula utilis–Rhododendron campanulatum* forests. This type is located on northerly slopes from 3350 to 3600 m. *Betula utilis* forms a canopy at a height of 10 m with *Sorbus ursina* as a subcanopy tree and *R. campanulatum* forming a shrub layer. *Betula* is not reproducing. At 3500–3600 m the birch disappears and a *Rhododendron campanulatum* shrubland extends to 3700 m. A *Juniper* woodland probably covered the south-facing slopes between 3400 and 3700 m, but fires and cutting by herders have replaced it with pasture.

Table 11.2. Importance percentages of major species in forest types at Chimkhola (tree layer).

Type:	Quercus lanata	Symplocos	Rhab-1	Rhab-2	Abies-1	Abies-2	Betula
Number stands:	N = 3	N = 18	N = 20	N = 10	N = 6	N = 14	N = 5
Tree layer (stems ≥10 cm dbh)							
SPP-1	Quln 56	Syrc 31	Rhab 34	Rhab 22	Abes 45	Abes 41	Btut 54
SPP-2	Lyov 33	Quox 16	Qusm 20	Accb 15	Rhbr 28	Rhab 38	Rhcp 24
SPP-3	Rhab 7	Prdu 13	Tsdm 7	Ildy 14	Tsdm 13	Rhbr 16	Sbur 19
SPP-4	Myes 2	Lnpl 7	Syrc 7	Vber 10	Accb 6	Btut 4	Vbnv 2
SPP-5	Crmp 1	Ddgf 4	Ddgf 5	Alnp 5	Btut 3	Acac 4	Abes 1
SPP-6	Sypn 1	Ildy 3	Lyov 3	Acce 4	Rhab 2	Vbnv 4	Lyov 2
Others	N = 1	N = 29 26	N = 44 26	N = 23 30	N = 6 3	N = 5 3	N = 1
Sapling layer (stems > 2.5 cm dbh, <10 cm dbh)							
SPP-1	Lyov 81	Syrc	Rhab 52	Vber 67	Rhbr 64	Rhbr 38	Rhcp 50
SPP-2	Rhab 8	Lnpl	Lnpl 15	Lnpl 4	Vber 11	Rhab 18	Sbur 22
SPP-3	Quln 7	Vber	Syrc 14	Syrc 4	Abes 7	Vbnv 11	Vbnv 21
SPP-4	Myes 2	Nlum	Vber 3	Ossv 3	Brer 4	Abes 8	Lyov 3
SPP-5	Rhwl 2	Ddgf	Qusm 2	Pier 2	Rhab 2	Vber 5	Btut 3
SPP-6	Sypn < 1	Quox	Vber 2	Ddgf 2	Lyov 2	Sbur 5	Vber 1
Others	N = 0	N = 40	N = 66 29	N = 35 18	N = 16 10	N = 15 15	N = 1 15

Abes	Abies spectabilis	
Acce	Acer caesium	
Crmp	Coriaria napalensis	
Lnpl	Lindera pulcherrima	
Ossv	Osmanthus suavis	
Quox	Quercus oxyodon	
Rhab	Rhododendron arboreum	
Rhwl	Rhus wallichii	
Syrc	Symplocos ramocissima	
Vber	Viburnum erubescens	
Acac	Acer acuminatum	Accb Acer campbellii
Alnp	Alnus nepalensis	Btut Betula utilis
Ddgf	Dodecadenia grandiflora	Ildy Ilex dipyrena
Lyov	Lyonia ovalifolia	Myes Myrica esculenta
Pier	Pieris formosa	Prdu Persea duthiei
Quln	Quercus lanata	Qusm Quercus semecarpifolia
Rhbr	Rhododendron barbatum	Rhcp Rhododendron campanulatum
Sbur	Sorbus ursina	Sypn Symplocos paniculatum
Vbnv	Viburnum nervosum	Tsdm Tsuga dumosa

lensis overtop *R. arboreum* on mesic north- and west-facing slopes. In general, the canopy trees do not seem to be reproducing adequately to maintain their populations.

My data allow a general description of forest regeneration and species size-class structures, but are insufficiently detailed to permit me to identify disturbance regimes and stand history. Size-class distributions can provide general predictions of future trends, but cannot explicate stand history because size and age correlations are often imprecise (Stewart 1986; Veblen 1992). Rather, we need to know the ages and spatial dispersion patterns of the trees to unravel disturbance and regeneration patterns (Stewart 1986). Such work is possible in the subalpine *Abies* and *Betula* forests, but the absence of reliable annual rings in many of the most important species of the broadleaf-evergreen forests hinders similar work in those communities. Nevertheless, the composition and structural data do show that canopy tree regeneration is poor to nonexistent in all the forest landscapes (table 11.2). Livestock browsing is usually assumed to be preventing regeneration of all but the most unpalatable species, and it undoubtedly has an impact. Yet I suspect that in all these forests significant regeneration of many if not most canopy species only occurs in openings in the canopy larger than .03 hectares, and research in subalpine forests confirms this suspicion in those ecosystems (Schmidt-Vogt 1990; Umans 1993). Umans (1993) examined age structures and the spatial dispersion patterns of *Abies spectabilis*–dominated forests in Central Nepal and found that within mature fir-rhododendron forests *Abies spectabilis* does not regenerate in gaps smaller than 300 square meters, and that its seedlings seldom survive in forest floor litter. Umans, whose results are supported by Schmidt-Vogt (1990) and Oliver and Sherpa (1993), attributes the poor regeneration to the shade intolerance of *A. spectabilis,* characteristics of the forest floor litter, and/or livestock browsing, although fir is not a preferred species. *Rhododendron arboreum* appears to live longer than fir and, without major disturbance, replaces it. Human uses of these forests (cutting *Abies* timber and grazing/browsing of livestock) further discourage fir relative to rhododendron. Larger scale catastrophic disturbances, like fires or mass wasting, occur at unspecified intervals and seem essential to the initiation of large-scale fir reproduction (Schmidt-Vogt 1990; Umans 1993). This interpretation varies somewhat from that of Taylor and Qin working in similar subalpine forests of southwestern China (this volume; 1996, 1988a, 1988b) and of Brower and Dennis (this volume). The stands of Brower and Dennis appear to be in the initiation and stem exclusion stages of development; hence, the canopy has recently or not yet closed, so regeneration behaviors should differ from the mature stands of my study. Taylor and Qin (1988a, 1988b) argue that canopy gap processes in Chinese subalpine forests are sufficient to maintain *Abies faxoniana* and *Betula utilis* populations without major disturbance; the long life of *A. faxoniana* allows regeneration in gaps from canopy tree death during the times of bamboo dieback, at about 50-year intervals. Taylor et al. (1996) used more thorough methods than have yet been applied to Himalayan forests. Himalayan scholars must use those methods to decide whether *A. spectabilis* can maintain dominance without major disturbance or whether *A. spectabilis* and *A. faxoniana* simply reproduce differently.

Although I do not have conclusive evidence to support it, I believe that the broadleaf evergreen forests also require major disturbances for oak and many other canopy species to regenerate. The scarcity of oak seedlings and saplings is at least in part due to livestock browsing, but I suspect that it is also due to limited natural regeneration even without livestock pressure. I conclude this based on the general lack of sites where significant regeneration is occurring, on the presence of significant numbers of saplings of *Ilex dipyrena,* the species most favored by livestock where oak is not regenerating (Metz, n.d.), and on the lack of oak and other canopy saplings and small trees in similar ecosystems of East Central Nepal from which livestock are excluded (Metz 1997). The conditions under which the oaks and other major species reproduce are extremely important questions requiring investigation by both ecologists and foresters.

Approaches To Learning How to Improve Forest Regeneration and Management

The evidence suggests that canopy species are not reproducing in most existing temperate and subalpine forests of Central Nepal. Because of the dense human populations using these forests and the politically unacceptable coercion that would be necessary to exclude them even from national parks (Yonzon and Hunter 1991), they must be included in the planning and implementation of any management plan that hopes to be successful. A few community forestry projects at lower elevations have successfully engaged forest users (Gilmour and Fisher 1991), but little effort has been made to convince upper-slope forest users of the need to manage their forests, although this will be a major thrust of the next phase of the Nepal Australia Community Forestry Project (Steven Hunt, personal communication, 1997). That is precisely the task those desiring improved forest cover in upper forests must undertake. To do so, forest scientists and forest-using villagers must solve the three related problems discussed below, a daunting challenge.

Research on Promoting Forest Regeneration

An essential first step is learning how to induce the regeneration of the desired species. Forest scientists need to locate relatively undisturbed stands of the major forest types and analyze regeneration with sophisticated methods. This would include spatial dispersions and ages (where possible) of the various-sized individuals of the major species (Stewart 1986; Taylor and Qin 1988a, 1988b, 1996), and the behaviors of species in forest gaps, especially over extended time periods (Brokaw 1985; Runkle 1985).

In addition, foresters and forest users should immediately begin to develop silvicultural systems that promote the regeneration of the species needed to fill local needs and to maintain forest ecosystems, because even after regeneration patterns are known, the information must be translated into silvicultural treatments. Focusing directly on ways to foster regeneration can also help resolve theoretical questions about forest dynamics.

The silvicultural research might consist of a series of experimental plots in each of the various forest types. Researchers would cut openings of various sizes in the forest, exclude livestock, and plant seedlings of desired species. The research should include replications to determine the affect of different treatments such as modifying the forest floor litter (do desired species need mineral soil or special seedbeds to reproduce?), changing the shape and orientation of the openings (north-south orientations maximize the area exposed to solar radiation in nonequatorial latitudes) and varying the placement seedlings within the opening (do the species need full or partial sunlight?).

Since the silvicultural systems are being designed for forest-user group use, researchers must include forest users in the research process. This "action research" consists of scholars working together with local people to design and run a series of trials (Raintree 1987; Rhoades and Booth 1982). The research process should include two parallel sets of experiments. One set would be in forests securely under forestry department control to eliminate unplanned cutting. Even at these "secure" research sites foresters need to collaborate with experienced forest users to insure that silvicultural recommendations are realistic for the village setting.

The second set of experimental plots should be in forests that remain under village control. The main challenge here will be in securing local cooperation and collaboration, but this is the same challenge that foresters will face later as they seek to introduce the improved silvicultural systems elsewhere in Nepal. Hence, these experimental plots will provide foresters with essential knowledge not only about how forest species behave, but also about how to secure forest-user participation.

Research on How to Secure Local Participation in Forest Management

Because maintaining intact forest ecosystems, which are essential habitat for endangered plants and animals, is impossible without local forest-user cooperation, researchers and foresters must first demonstrate that silvicultural management can provide the products villagers need, and that the forest users can easily learn and use the new silvicultural systems. After finding a community willing to try to improve forest management, the researchers must explain the goals and rationale of the experiments and enlist the community's active participation in the design and implementation of the trials. This will require that real, rather than token, authority be given to local research partners. Indeed, forest users will have important knowledge about their needs and useful suggestions about what actions are likely to work (Gilmour and Fisher 1991). It may be that no such system can be devised. Ironically, that researchers must admit that they don't have the answers and need the aid of forest users to design viable management systems may help secure local cooperation.

The process of securing local cooperation is a complicated procedure. As the Nepal Australia Community Forestry Project has shown in low-elevation communities (Gilmour and Fisher 1991), simply calling a meeting of the users to negotiate cooperation seldom works. Rather, the researcher must get to know the community, identify the various "interest groups" within the forest-user group, and figure out how to mitigate the hardships that these subgroups will endure when the silvicultural treatments are enacted.

Research into Motivating Department of Forest Personnel

The practices of "community" or "social" forestry I am advocating are antithetical to the traditional training and goals of foresters. In Nepal, as in forest ministries throughout the world (Cernea 1992; Miranda et al. 1992; Poffenberger 1990b), foresters are trained to produce commercial timber, and they view local forest users as "encroachers" who despoil national assets. Furthermore, foresters usually come from urban and advantaged backgrounds and are unfamiliar with rural life and minority cultures. Because their education justifies their high status, only the most self-confident and dedicated are likely to admit they don't have the answers to help forest users create new forest management systems. Finally, many forestry personnel supplement their low salaries with bribes and expropriations of ministry funds; genuine implementation of community forestry will end these resources.

These factors combine to produce foresters with exactly the wrong kinds of attitudes and skills for successful community forestry. In recent years Nepal's Department of Forests has profoundly modified its educational program to develop the skills community foresters will need, but the highly stratified social structure and inadequate salaries continue to discourage change. Fisher (1990) has diplomatically labelled these problems "institutional incompatibility"—the values and structures of the Department of Forests and the village are at variance. Similar problems exist in virtually all developing nations (Miranda et al. 1992; Poffenberger 1990b). Clearly, the Forest Ministry must develop a new reward system that provides financial and career rewards on the basis of successful collaboration with forest-user groups. This will require institutional changes that fly in the face of general social relations of power. It is both a great challenge and a great opportunity.

Conclusion

In Nepal, as in much of the developing world (Cernea 1992; Panatoyou and Ashton 1992), people who use forests for their subsistence must be included in the design and implementation of management systems that seek to sustain forests and biodiversity. Guaranteeing forest users long-term access to forests and their produce may secure their cooperation, but does not assure that they and government foresters will be able to create and implement systems of sustainable management.

Research I have done in the west-central region of Nepal has identified seven intergrading forest ecosystems. In these forest landscapes virtually all the canopy species are reproducing poorly. I argue that, contrary to conventional assumptions, the cessation of cutting and livestock browsing may not cause these species to reproduce in sufficient numbers to maintain their populations because they may need catastrophic disturbances, like fires or landslides, to initiate significant regeneration. Other research in fir-rhododendron forests supports this position, but we know too little about the broadleaf evergreen temperate forests to decide the matter. The information generated by forest scientists examining the disturbance regimes and dynamics of little-disturbed forests, and by foresters and forest users doing silvicultural trials, will help clarify what combinations of cutting, burning, and other treatments will induce forest regeneration and maintain forest cover.

Overly simplistic assumptions about the dynamics of Himalayan forests (for example, most canopy trees are "climax" species and will regenerate if livestock browsing and tree-cutting ends) can be translated into unrealistic prescriptions for forest management. Similarly, in other developing countries the lack of sophisticated knowledge about forest dynamics, and the need to develop multiple-use management plans to produce timber and nontimber forest products (Panayotou and Ashton 1992), are likely to simplify ecosystems to the point that many mature forests cannot survive. Numerous social forestry projects (Gilmour and Fisher 1990; Hafner and Apichatvullop 1990; Poffenberg 1990) and schemes to manage "production forests" (Palmer and Synnott 1992) seem to exemplify this problem (Botkin and Talbot 1992). And even in extractive reserves we know little about how indigenous shifting cultivators and collectors of nontimber forest products will affect long-term forest dynamics.

Even after foresters develop silvicultural systems that allow degraded forests to recover, they face the challenge of convincing forest users to adopt them. Securing agreement from all forest users to the inevitable restrictions on current use will require sophisticated skills of social analysis and communication.

In Nepal, as in all developing countries (Cernea 1992; Miranda et al. 1992; Poffenberger 1990b; Repetto and Gillis 1988), the training and outlook of Department of Forest personnel must change profoundly before they can effectively promote community forestry. The main impediment to this process is the difficulty of establishing a reward system within the forest bureaucracies that bases income and advancement on the forest officer's success in helping forest users establish improved management. This will require profound changes in institutional values and behavior.

Great uncertainty surrounds the entire project of maintaining forests and biodiversity in developing countries. What are the dynamics and disturbance patterns of the forest ecosystems? Is it possible to design sustainable forest-use systems that maintain forest habitat? Can local communities work together to adopt equitable and sustainable management systems? Will the government and forest ministry bureaucrats give real power to manage forests to lowly farmers? Can the forest ministries motivate their forest officers to do community forestry? Can foresters reach and train the thousands of forest-user groups that need to improve their management? The answers to all these questions are far from clear. The next decades will show whether the Nepalese people—and the people throughout the world facing similar challenges—can find the solutions to these problems and maintain the forest ecosystems on which so many species rely for survival.

References

Alirol, P. 1979. *Transhumant Animal Husbandry Systems in the Kalinchowk Region (Central Nepal): A Comprehensive Study of Animal Husbandry on the Southern Slopes of the Himalayas.* Bern: Swiss Association for Technical Assistance.

Berry, B. J. L., and A. M. Baker. 1968. Geographic Sampling. In *Spatial Analysis: A Reader in Statistical Geography,* ed. B. J. L. Berry and F. Marble, pp. 91–100. Englewood Cliffs, NJ: Prentice-Hall.

Bishop, Naomi. 1989. From Zomo to Yak: Change in a Sherpa Village. *Human Ecology* 17(2): 177–204.

Bormann, F. H., and G. E. Likens. 1979. *Pattern and Process in a Forested Ecosystem.* New York: Springer-Verlag.

Botkin, Daniel B. 1990. *Discordant Harmonies: A New Ecology for the Twenty-first Century.* New York: Oxford University Press.

Botkin, D. B., and L. M. Talbott. 1992. Biological Diversity and Forests. In *Managing the World's Forests,* ed. N. P. Sharma, pp. 47–74. Dubuque: Kendall-Hunt.

Brokaw, N. V. L. 1985. Treefalls, Regrowth, and Community Structure in Tropical Forests. In *The Ecology of Natural Disturbance and Patch Dynamics,* eds. S. T. A. Pickett and P. S. White, pp. 53–69. Orlando: Academic Press.

Brower, Barbara. 1991. *Sherpa of Khumbu: People, Livestock, and Landscape.* Oxford: Oxford University Press.

Bruijnzeel, L. A., and Bremmer, C. N. 1989. *Highland-Lowland Interactions in the Ganges-Brahmaputra River Basin: A Review of the Published Literature.* ICIMOD Occasional Paper No. 11. Kathmandu: International Center for Integrated Mountain Development.

Campbell, J. G., and Tara N. Bhattarai. 1984. *People and Forests in Hill Nepal: Preliminary Presentation of Findings of Community Forestry Household and Ward Leader Survey.* Project Paper 10. Nepal: HMG/UNDP/FAO.

Carpenter, C., and R. Zomer. 1996. Forest Ecology of the Makalu-Barun National Park and Conservation Area, Nepal. *Mountain Research and Development* 16(2): 135–148.

Carter, A. S., and D. A. Gilmour. 1989. Increase in Tree Cover on Private Farm Land in Nepal. *Mountain Research and Development* 9(4): 381–91.

Cernea, Michael. 1992. A Sociological Framework: Policy, Environment, and Social Actors for Tree Planting. In *Managing the World's Forests,* ed. N. P. Sharma, pp. 301–336. Dubuque: Kendall-Hunt.

Champion, H. G. 1936. A Preliminary Survey of Forest Types of India and Burma. *Indian Forest Record (N. S.) Botany* 1: 1–36.

Champion, H. G., and S. K. Seth. 1968. *A Revised Survey of the Forest Types of India.* Delhi: Manager of Publications.

Denslow, J. S. 1987. Tropical Rainforest Gaps and Tree Species Diversity. *Annual Review of Ecology and Systematics* 18: 431–51.

Dobremez, J. F. 1976. *Le Nepal: Ecologie et Biogeographie.* Paris: Centre National de la Recherche Scientifique.

Dobremez, J. F. 1986. *Les Collines du Nepal Central. Ecosystemes Structures Sociales et Systems Agraires.* Paris: Institut National de la Recherche Agronomic.

Fisher, Robert J. 1989. *Indigenous Systems of Common Property Forest Management in Nepal.* Working Paper No. 18. Honolulu: East-West Center, Environment and Policy Institute.

Fisher, Robert J. 1990. Institutional Incompatibility in Community Forestry: The Case of Nepal. In *Community Organizations and Government Bureaucracies in*

Social Forestry, eds. J. M. Fox and R. J. Fisher. Working Paper No. 22. Honolulu: East-West Center, Environment and Policy Institute.

Fox, J. M. 1984. Firewood Consumption in a Nepali Village. *Environmental Management* 8: 243–250.

Fox, J. M. 1993. Forest Resources in a Nepali Village 1980–1990: The Positive Influence of Population Growth. *Mountain Research and Development* 13(1): 89–98.

Franklin, J. F. and M. Hemstrong 1981. Aspects of Succession in Coniferous Forests of the Pacific Northwest. In *Forest Succession: Concepts & Application,* eds. D. C. West, H. H. Slugert, & D. B. Botkin. Berlin & New York: Springer-Verlag.

Gauch, Huge G. 1982. *Multivariate Analysis in Community Ecology.* Cambridge: Cambridge University Press.

Gilmour, D. A., and R. J. Fisher. 1991. *Villagers, Forests, and Foresters: The Philosophy, Process and Practice of Community Forestry in Nepal.* Kathmandu: Sahayogi Press.

Gilmour, D. A., and M. C. Nurse. 1991. Farmer Initiatives in Increasing Tree Cover in Central Nepal. *Mountain Research and Development* 11(4): 329–337.

Glenn-Lewin, D. C., R. K. Peet, and T. T. Veblen. 1992. *Plant Succession Theory and Prediction.* London: Chapman and Hall.

Greenberg, Wendy. 1986. *Soil Survey of Kali Gandaki Area Myagdi and Mustang Districts, Nepal.* Kathmandu: Resources Conservation and Utilization Project, United States Agency for International Development.

Hafner, J. A., and Y. Apichatvullop. 1990. Migrant Farmers and Shrinking Forests of Northeast Thailand. In *Keepers of the Forest, Land Management Alternatives in Southeast Asia,* ed. Mark Poffenberger. West Hartford: Kumarian Press.

Heinselman, M. L. 1973. Fire in the Virgin Forests of the Boundary Waters Canoe Area, Minnesota. *Quarternary Research* 3: 329–382.

Henry, J. D., and J. M. Swan. 1974. Reconstructing Forest History from Live and Dead Plant Material—An Approach to the Study of Forest Succession in Southwest New Hampshire. *Ecology* 55: 772–783.

Ives, Jack D., and Bruno Messerli. 1989. *The Himalayan Dilemma: Reconciling Development and Conservation.* London: Routledge.

Kawakita, Jiro. 1957. Vegetation. In *Scientific Results of the Japanese Expeditions to Nepal Himalaya, 1952–53, Land and Crops of Nepal Himalaya,* ed. Hiro Kihara, pp. 1–66. Kyoto: Fauna and Flora Research Society, Kyoto University.

Mahat, T. B. S., D. M. Griffin, and K. R. Shepherd. 1987. Human Impact on Some Forests of the Middle Hills of Nepal. Forest is the Subsistence Economy of Sindhu Palchok and Kabhre Palanchok. *Mountain Research and Development* 7: 53–70.

Messerschmidt, Donald. 1986. People and Resources in Nepal: Customary Resource Management Systems of the Upper Kali Gandaki. In *Common Property Resource Management,* Proceedings of the International Conference, pp. 455–480. Washington, D. C.: National Academy Press.

Messerschmidt, Donald. 1987. Conservation and Society in Nepal: Traditional Forest Management and Innovative Development. In *Lands at Risk in the Third*

World: Local Level Perspectives, eds. P. D. Little and M. M. Horowitz, pp. 373–397. Boulder: Westview Press.

Metz, J. J. 1989. A Framework for Classifying Subsistence Production Types of Nepal. *Human Ecology* 17(2): 147–176.

Metz, J. J. 1990. Forest Product Use in Upland Nepal. *Geographical Review* 80(3): 279–287.

Metz, J. J. 1991. A Reassessment of the Causes and Severity of Nepal's Environmental Crisis. *World Development* 19(7): 805–820.

Metz, J. J. 1994. Forest Product Use at an Upper Elevation Village in Nepal. *Environmental Management* 18(3): 371–390.

Metz, J. J. 1997. Vegetation Dynamics of Several Little Disturbed Temperate Forests of East Central Nepal. *Mountain Research and Development* 17(4): 333–351.

Metz, J. J. N.d. *Composition, Structure, and Dynamics of Temperate and Subalpine Forests of West Central Nepal.* Unpublished manuscript.

Miranda, M. L., O. M. Corrales, M. Regan, and W. Ascher. 1992. Forestry Institutions. In *Managing the World's Forests,* ed. N. P. Sharma, pp. 269–300. Dubuque: Kendall-Hunt.

Nakasuga, Tsuneo. 1972. An Investigation on Forests in Nepal. *Research Bulletin of the College Experiment Forests* 29(2): 155–174.

Nield, R. S. 1985. *Fuelwood and Fodder—Problems and Policy.* Kathmandu: Water and Energy Commission.

Numata, Makoto. 1975. *Mountaineering of Mt. Makalu II and Scientific Studies in Eastern Nepal, 1971.* Chiba, Japan: Laboratory of Ecology, Chiba University.

Numata, Makoto, ed. 1983. *Structure and Dynamics of Vegetation in Eastern Nepal,* Chiba, Japan: Laboratory of Ecology of Chiba University.

Ohsawa, Masahiko. 1983. Distribution, Structure and Regeneration of Forest Communities in Eastern Nepal. In *Structure and Dynamics of Vegetation in Eastern Nepal,* ed. M. Numata, pp. 89–120. Chiba, Japan: Laboratory of Ecology of Chiba University.

Ohsawa, Masahiko. 1987. *Life Zone Ecology of the Bhutan Himalaya.* Chiba, Japan: Chiba University Press.

Ohsawa, Masahiko. 1991. *Life Zone Ecology of the Bhutan Himalaya II.* Chiba, Japan: Chiba University Press.

Ohsawa, M., P. R. Shakya, and M. Numata. 1986. Distribution and Succession of West Himalayan Forest Types in the Eastern Part of the Nepal Himalaya. *Mountain Research and Development* 6(2): 143–157.

Oliver, C. D. 1981. Forest Development in North America Following Major Disturbances. *Forest Ecology and Management* 3: 153–68.

Oliver, C. D., and L. N. Sherpa. 1990. *The Effects of Browsing and Other Disturbances on the Forest and Shrub Vegetation of the Hongu, Inkhu, and Dudh Koshi Valleys.* Working Paper 9. Kathmandu: Makalu-Barun Conservation Project.

Oliver, C. D., and E. P. Stephens. 1977. Reconstruction of a Mixed Species Forest in central New England. *Ecology* 58: 562–72.

Omaston, A. E. 1922. Notes on the Forest Communities of Garhwal Himalayas. *Journal of Ecology* 10: 129–167.

Palmer, J., and T. J. Synnott. 1992. The Management of Natural Forests. In *Man-*

aging the World's Forests, ed. N. P. Sharma, pp. 337–374. Dubuque: Kendall/Hunt.

Panayotou, Theodore, and Peter Ashton. 1992. *Not by Timber Alone. Economics and Ecology for Sustaining Tropical Forests.* Covelo: Island Press.

Pickett, S. T. A., and P. S. White, eds. 1985. *The Ecology of Natural Disturbance and Patch Dynamics.* Orlando: Academic Press.

Poffenberger, M. 1990a. *Joint Management for Forest Lands: Experiences from South Asia.* New Delhi: Ford Foundation.

Poffenberger, M. 1990b. Facilitating Change in Forestry Bureaucracies. In *Keepers of the Forest: Land Management Alternatives in Southeast Asia,* ed. Mark Poffenberger. West Hartford: Kumarian Press.

Repetto, R., and M. Gillis, eds. 1988. *Public Policy and the Misuse of Forest Resources.* New York: Cambridge University Press.

Raintree, John B. 1987. The State of the Art of Agroforestry Diagnosis and Design. *Agroforestry Systems* 5: 219–250.

Rhoades, Robert, and Robert H. Booth. 1982. Farmer-Back-To-Farmer: A Model for Generating Acceptable Agricultural Technology. *Agricultural Administration* 11: 127–137.

Richard, D. 1980. *Variations de la Structure, de l'Architecture et de la Biomasse des Forets du Centre Nepal.* Ph. D. dissertation, Grenoble University.

Robinson, P. J. 1990. Some Results of the Dolakha Private Tree Survey. In *Proceedings of the Third Working Group Meeting on Fodder Trees, Forest Fodder, and Leaf Litter* (Occasional Paper 2/90), pp. 40–48. Kathmandu: FRIC.

Runkle, J. R. 1985. Disturbance Regimes in Temperate Forests. In *The Ecology of Natural Disturbance and Patch Dynamics,* eds. S. T. A. Pickett and P. S. White, pp. 17–33. Orlando: Academic Press.

Schmidt-Vogt, Dietrich. 1990. *High Altitude Forests in the Jugal Himal (Eastern Central Nepal): Forest Types and Human Impact.* Stuttgart: Franz Steiner Verlag.

Schweinfurth, U. 1957. Die Horizontale und Vertikale Verbreitung der Vegetation im Himalaya. Bonner Geographische Abhandlungen, Heft 20. Bonn: F Dumummlers Verlag.

Singh, S. P., B. S. Adhikari, and D. B. Zobel. 1994. Biomass, Productivity, Leaf Longevity, and Forest Structure in the Central Himalaya. *Ecological Monographs* 64(4): 401–421.

Singh, J. S., and S. P. Singh. 1987. Forest Vegetation of the Himalaya. *The Botanical Review* 53(1): 80–192.

Singh, J. S., and S. P. Singh. 1992. *Forests of Himalaya, Structure, Functioning and Impact of Man.* Nainital, India: Gyanodaya Prakashan.

Stainton, J. D. A. 1972. *Forests of Nepal.* New York: Hafner Publishing Company.

Stevens, Stanley. 1993. *Claiming the High Ground: Sherpas, Subsistence, and Environmental Change in the Highest Himalaya.* Berkeley: University of California Press.

Stewart, G. H. 1986. Population Dynamics of a Montane Conifer Forest, Western Cascade Range. *Ecology* 67: 534–544.

Sueron, C. 1985. *Regeneration, Structure, et Production d'une Foret "Naturelle": la*

Sapiniere a Abies spectabilis (D.Don) Mirb du Nepal Central. Ph. D. dissertation, University of Grenoble.

Taylor, A. H., and Qin Zisheng. 1988a. Regeneration Patterns in Old-Growth *Abies-Betula* Forests in Wolong Natural Reserve, Sichuan, China. *Journal of Ecology* 76: 1204–1218.

Taylor, A. H., and Qin Zisheng. 1988b. Tree Replacement Patterns in Subalpine *Abies-Betula* Forests in Wolong Natural Reserve, China. *Vegetatio* 78: 141–149.

Taylor, A. H., Qin Zisheng, and Liu Jie. 1996. The Structure and Dynamics of Subalpine Forests in Wang Lang Natural Reserve, China. *Vegetatio* 124: 25–38.

Umans, Laurent. 1993. The Unsustainable Flow of Himalayan Fir Timber. *Mountain Research and Development* 13(1): 73–88.

Veblen, T. T. 1992. Regeneration Dynamics. In *Plant Succession Theory and Prediction,* eds. D. C. Glenn-Lewin, R. K. Peet, and T. T. Veblen, pp. 153–187. London: Chapman and Hall.

Veblen, T. T., and D. H. Ashton. 1978. Catastrophic Influences on the Vegetation of the Valdivian Andes, Chile. *Vegetatio* 36: 149–167.

Wiart, Jacques. 1983. *Ecosysteme Villageois Traditionnel en Himalaya Nepalais: la production forestiere suffit-elle aux besoins de la population?* Ph. D. dissertation, Universite Scientifique et Medicale de Grenoble.

Wiart, Jacques, and J. F. Dobremez. 1986. Les prelevements do produits forestiers. In *Les Collines du Nepal Central Ecosystemes Structures Sociales et Systemes Agraires. Tome II: Millieux et Activites dans un Village Nepalais,* ed. J. F. Dobremez, pp. 37–65. Paris: Institut National de la Recherche Agronomique.

Wyatt-Smith, J. W. 1983. *The Agricultural System in the Hills of Nepal: Ratio of Agricultural to Forest and the Problem of Animal Fodder.* Occasional Paper No. 1. Kathmandu: Agricultural Projects Research Organization and Science Centre.

Yoda, Kyoji. 1967. A Preliminary Survey of the Forest Vegetation of Eastern Nepal II: General Description, Structure, and Floristic Composition of the Sample Plots Chosen from Different Vegetation Zones. *Journal of the College of Arts and Sciences, Chiba University* 5(1): 99–139.

Yoda, Kyoji. 1968. A Preliminary Survey of the Forest Vegetation of Eastern Nepal III: Plant Biomass in the Sample Plots Chosen from Different Vegetation Zones. *Journal of the College of Arts and Sciences, Chiba University* 5(2): 277–302.

Yonzon, Pralad, and Malcolm Hunter. 1991. Conservation of Red Panda (*Ailurus fulgens*). *Biological Conservation* 57: 1–11.

12

Ethnobotanical Knowledge and Environmental Risk
Foragers and Farmers in Northern Borneo

Robert A. Voeks

Rain Forest Knowledge and Environmental Risk

Moist tropical forests, long assumed to be the biological consequence of benign and stable environmental forces, are increasingly perceived as landscapes of disturbance—patchwork patterns of plants and animals molded by episodes of drought, cold, fire, and wind (Colinvaux 1993; Goldammer and Seibert 1990; Nicholls 1993; O'Brien et al. 1992). Within this view, biogeographical patterns are seen to exist *because of,* rather than *despite,* natural disturbance. This also applies to the role of forest-dwelling people. No longer seen as the exterminators of our precious biodiversity, indigenous societies are viewed more and more as complementary or even necessary components of landscape-level diversity patterns (Balée 1994: 135–138; Clay 1988). Disturbance, whether natural or cultural, has displaced stability as the overarching paradigm of tropical forest biogeography.

Along the continuum from cultivated to pristine, tropical forested landscapes constitute storehouses of material and other cultural resources (Denevan et al. 1985; Smith et al. 1992; Wilson 1988). Native foods, fibers, medicinals, and timber species, long fundamental to the subsistence economies of indigenous hunter-gatherers and small-scale cultivators, are increasingly viewed as components in the design and management of sustainable tropical agroecosystems, as well as commercial, export-based enterprises (Alcorn 1989; Gliessman 1990; Voeks 1996a). As the value of tropical genetic resources becomes increasingly evident, the combined forces of destructive forest exploitation and erosion of ethnobotanical knowledge threaten their anticipated contribution to appropriate development, as well as the continued sustenance of rural tropical people (Smith and Schultes 1990; Voeks 1996b).

Ethnobotanical inventories drawn from traditional societies, remarkable both for their breadth and diversity, exhibit quantitative similarities as well as differences. The sizes of folk plant lexicons possessed by traditional societies, for example, are reasonably consistent, even among groups that are widely separated geographically. Lévi Strauss (1966) first noted that ethnobotanical lists compiled for nonliterate societies ranged from roughly 300 to 600 named taxa, with 2000 names projected as an upper limit to the powers of individual memory. Berlin (1992) reported a similar mean of 520 named plant taxa among traditional cultivating peoples with a range from 238 to 956.

Significant differences in ethnobotanical knowledge occur, however, when the mode of subsistence of the various groups is considered. In a comparative survey of plant-name lists, Brown (1985) discovered that groups pursuing a hunting and gathering subsistence strategy possessed on average only 179 named plant classes. For small-scale agriculturalists, however, this figure climbed dramatically to about 890, or roughly five times as many as their foraging counterparts. Berlin (1992) came to similar conclusions regarding subsistence strategy and number of folk taxa, although he underscored several methodological problems with Brown's assessment and took issue with his "knowledge based on necessity" conclusions.

The notion that hunter-gathers possess less ethnobotanical knowledge than cultivators seems counterintuitive, given that foragers would appear to depend more intimately on nature's wild resources than would cultivating societies. Clearly, some of this disparity is a product of utility and opportunity. Whereas protean landscapes of disturbance are associated with sustained, small-scale cultivation regimes, nearly pristine and hence more homogenous habitats result from the activities of foraging people. Cultivation creates a patchwork of successional stages, from fresh swidden patches to long fallow forests, thus producing novel and often highly productive opportunities for useful plant collection (Grenand 1992; Toledo et al. 1992; Voeks 1995). Ethnobotanical lists likewise are amplified by the acquisition of domesticated crop species and their varieties, a flora with which foragers should have relatively limited familiarity. The myriad weedy plant invaders associated with cultivation—species with little or no salience to wild food gatherers—are likely to receive the attention necessary among cultivators to warrant names as well as ecological knowledge (Ellen in press).

The sedentary nature of small-scale cultivators may also be a factor in ethnobotanical acquisition. With their greater population densities, cultivating societies may be forced by heightened risk of viral, bacterial, and parasitic disease to become better acquainted with the medicinal properties of the native flora than their hunting and gathering counterparts are (Black 1975; Dunn 1968). In principle, this feature could result in a greater understanding of medicinal species and a concomitant larger plant pharmacopoeia (Brown 1985; Kohn 1992; Voeks 1996b). Finally, the adoption of a settled as opposed to a nomadic existence is associated with a significantly expanded material culture—such as permanent dwellings, bulky furnishings, musical instruments, and toys. These would naturally require a detailed understanding of the properties of lumber-yielding species, again a flora for which nomadic foragers would have limited utility.

Brown (1985) introduced another perhaps more problematic factor—the element of environmental risk and famine. He reasoned that the principal benefits associated with a sedentary agrarian existence (increased food supplies and the resultant enhanced carrying capacity) also carried the liability of heightened risks of famine (see also Dunn 1968; Hunn and French 1984). Unlike hunter-gatherers who depend on wild resources that are reasonably well-adapted to occasional environmental fluctuations, cultivators by choice draw on food resources that are at the mercy of periodic crop failure from droughts, floods, insect outbreaks, and other natural calamities. In this view, periodic hunger is seen as a driving force behind the acquisition of knowledge about temporary famine foods, and thus an expanded ethnobotanical inventory. Groups that pursue a foraging form of subsistence are seldom if ever required to exploit such marginal food resources, because of their low population densities and reliance on a relatively reliable subsistence base. Thus, at least in terms of knowledge of marginal foods, environmental fluctuation rather than stasis underpins the process of ethnobotanical acquaintance and acquisition.

Brown's conclusions can be criticized on several methodological and theoretical grounds. For example, establishing a clear demarcation between what constitutes a hunter-gatherer existence and a shifting cultivator existence is problematic, given that the former frequently engage in incipient horticultural enterprise, that the latter are often skilled hunters and gatherers, and that switching from one mode of subsistence to another as opportunities appear is known to have occurred (Layton et al. 1991). At the same time, Ellen (in press) has questioned the use of named plant taxa as an index of ethnobotanical knowledge, arguing that populations that depend more on hunting and gathering may not lexically encode the breadth of their systematic knowledge, a situation that simple ethnobotanical inventories would fail to reveal. Most importantly, many of the small-scale cultivators in Brown's inventory inhabit species-rich, moist, tropical regions where the opportunity to exploit different taxa are legion; most of the hunter-gatherer records are for groups occupying less speciose, more temperate zone habitats (Bulmer 1985). Clearly, the validity of comparing groups from such widely different biomes as tundra and tropical forest is open to question. Indeed, in one of the few inventories of hunter-gatherer-named plant categories from a moist tropical landscape, Headland (1985) compiled an ethnobotanical lexicon of 603 plant terms, well within the range of that reported for cultivating groups. More recently, Ichikawa (1992) identified over 500 species being used by the Mbuti hunter-gatherers in the Republic of the Congo, a figure very near the mean reported for small-scale cultivators.

This paper examines the role of environmental risk and uncertainty in the acquisition and retention of ethnobotanical knowledge. I focus on the quantitative knowledge of wild esculent plants. In order to diminish some of the methodological difficulties associated with geographically separated groups, I worked with sympatric hunter-gatherers and cultivators: people with the opportunity to exploit essentially the same range of organisms and habitats. I also avoided the question of whether complete ethnobotanical lists were produced for either group by employing a plot-based sampling method. Because they are repeatable in other habitats and among other human groups, plot-based sampling represents a useful and relatively unbiased methodology for ethnobotanical study (Phillips et al. 1994; Prance et al. 1987; Voeks 1996b).

Two local informants were used to census a one-hectare plot for named plants and useful species (map 12.1). These were Udi, a 44-year-old male Penan, and Kilat, a 64-year-old male Dusun. Among their respective peers, both men are thought to be quite knowledgeable about local flora. Although no longer nomadic, Udi still spends considerable time hunting and foraging for food and other forest products. Kilat, after fifty years as a hill rice cultivator, has recently retired from traditional farming.

The plot, located in old-growth, mixed, dipterocarp forest, contains a total of 1086 tagged and identified trees. It is floristically dominated in order of importance by Dipterocarpaceae, Euphorbiaceae, Ebenaceae, and Anacardiaceae. With 303 tree species or morpho-species, this site represents one of the most species-rich forest plots enumerated to date (Gentry 1988). The soils are ultisols developed on soft shale substrate. Mean

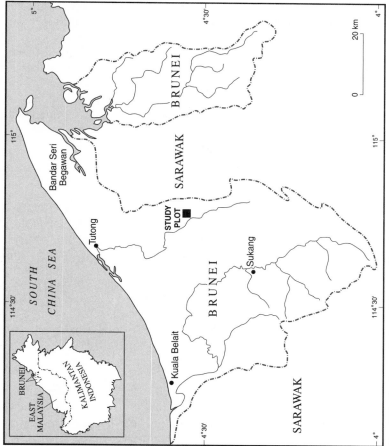

Map 12.1. Study area in Brunei. Penan village is located in Sukang. Dusun settlements occur along the major rivers in the Belait and Tutong districts.

311

annual precipitation is roughly 3000 millimeters, with no pronounced dry season (Sirinanda 1990).

Old-growth tropical rain forest represents the preferred foraging habitat of the Penan hunter-gatherers surveyed in this study. By the Penan's own accounts, they did not (until recently) cut trees larger in diameter than their forearms, or hunt and collect in the secondary forests created by their swidden neighbors. This habitat preference contrasts markedly with the Dusun and other Bornean cultivators, who utilize the full range of seral communities, from the highly disturbed to nearly pristine. In terms of the quantitative objectives of this study, this narrow habitat preference in effect gives the hunter-gatherer group the advantage. If the Penan's ethnobotanical knowledge is as substantial as that of the Dusun cultivators, then surely it will be elicited in a census of their principal foraging habitat. The question of differential knowledge of cultivars, weeds, medicinals, and timber species is considered elsewhere.

Culture and Landscape

The regional focus of this study is a village of roughly fifty Penan named Sukang on the west bank of the Belait River in Brunei Darussalam (map 12.1). Once charting a nomadic pathway through what is now southwestern Brunei and the Baram River drainage of Sarawak, this group of Penan was convinced to settle in Sukang by local Dusuns with whom they had maintained a long trading relationship. Since 1962 they have occupied a permanent longhouse. They now carry out limited hill rice cultivation, and many have recently become employed in government-sponsored forestry and road-building projects. Although the principal subsistence strategy of the Penan at Sukang is in transition, they nevertheless spend considerable time foraging for game and plant products in surrounding old-growth forested landscapes.

The Dusun of Brunei represent one of the oldest settled agricultural groups in the region. Although they traditionally lived in village longhouses occupied by up to twenty families, this practice was largely abandoned in favor of single-family dwellings after World War II. Their subsistence economy, at least until recently, depended on community-based hill rice cultivation, supplemented by fishing, snare-trapping of forest game, and collection of extractive products. Like other indigenous Bruneian groups, the Dusun are increasingly drawn to wage-earning jobs in government and the private sector, and are in the process of abandoning traditional agricultural activities (Antaran 1993; Ellen and Bernstein 1994).

Although the Penan and Dusun occupy a floristic landscape characterized by lowland peat forests, riparian forests, heath forests, and mixed

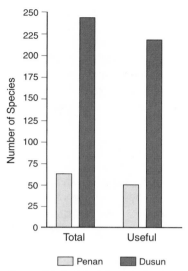

Fig. 12.1. Total named and/or useful plant taxa in a one-hectare plot. Plants identified by Penan or Dusun possess folk names and/or recognized material value.

dipterocarp forests, the latter community represents their main area of exploitation. Mixed dipterocarp forest, which is characterized by tall, broadleaf evergreen trees often exceeding 40 meters in height, is dominated by the speciose Dipterocarpaceae family. The understory is rich in palms, including a host of rattan climbers (for example, *Calamus* and *Daemonorops*) and shrubby licuala (*Licuala*). The forest floor is covered with tree seedlings and herbaceous ginger (Zingiberaceae). Rich in economically valuable dipterocarp species, especially *meranti* (*Shorea*) and *kapur* (*Dryobalanops*), these forests have traditionally represented a focus of both logging and shifting cultivation activities in Borneo.

Census Results

The Penan informant identified a total of 63 species distributed among 44 folk names in the plot (figure 12.1). Forty-six of these are adult trees, representing roughly 15% of the total tree species identified. Fifty-one species were determined to be useful, 32 of which, or about 73%, are trees. Most of the remaining named taxa are understory palms and rattan. The range of perceived values include timber, rattan, edible fruit, sago, palm cabbage, fuelwood, medicine, poles for temporary huts, insecticide, vegetable oil, blowpipe wood, sandpaper for blowpipes, and leaf packets for carrying food. Significantly, eight of the tree species are used in the construction of permanent houses, an activity that has only been initiated since the Penan settled in Sukang. The informant indicated that these species, which consist mostly of canopy dipterocarps, were not used traditionally by the Penan, but had been learned about from settled groups. Whether these species had Penan names prior to this assimilation process is unclear. Eight of the remaining useful taxa are palms, both climbing rattans and shrubs, and are utilized as food sources or material for weaving mats and baskets. No herbs, lianas, or epiphytes were reported to be useful.

The Dusun informant recognized 241 taxa distributed among 82 folk names, of which 219 were considered useful (figure 12.1). Of these, 191 are adult tree species representing approximately 63% of the total tree species in the plot. About 84% of these tree taxa (161 species) were considered useful. It is important to note, however, that many species that lack material or magical utility, and thus were omitted from the results, are sufficiently culturally significant to enter into the Dusun plant lexicon. Trees in the Anacardiaceae family, for example, cause mild to severe skin eruptions when handled and are avoided. Several of the understory taxa are considered to be jinx plants—species that should never be used to construct animal snares or be near set traps. In addition, tree species that provide food for game animals, principally bearded pigs, barking deer, and sambar deer, are well known to Dusun hunters because they are good locations to place snares. Thus, as an indicator of species salience among the Dusun, strict material benefit constitutes a rather superficial criterion.

The Dusun informant identified three to four times more folk taxa in the plot than the Penan, as well as four to five times more materially useful species. Although preliminary, these results support Brown's contention that small-scale cultivators possess a larger inventory of plants than hunter-gatherers. This enhanced recognition of the floristic landscape applied to a wide variety of plant life, with the Dusun recognizing more species of trees, treelets, shrubs, rattan and nonrattan palms, climbers, lianas, and herbs.

The Extractive Landscape: Food Resources and Risk

Brown's prediction that cultivators maintain a more extensive inventory of famine food sources than foragers do appears to be supported by the plot census results as well (table 12.1). The Penan informant recognized 20 species that supply food of some sort, including fruit, vegetable oil, sago flour, and palm heart (figure 12.2). Twelve of these are forest trees; the remainder are understory plants. The Dusun informant recognized a

Table 12.1. Penan edible plant species in a one-hectare forest plot.

Penan name	Species	Use
abang	*Shorea* cf. *acuta* [#422] *Shorea amplexicaulis*	Oil extracted from fruit
belulang	*Ptychopyxis* sp. [#418]	Edible fruit
daun itot	*Licuala* sp. [#427]	Vegetable from pith of lower trunk; leaves for hats
daun upak	*Borassadendron borneense* [#426]	Pith from lower trunk for food
janan	*Calamus ornatus* [#432]	Edible fruit, rattan
karot	*Garcinia mangostana* [#413]	Edible fruit
keramo	*Dacryodes laxa* [#423]	Edible fruit
keruong	*Dacryodes expansa* [#415]	Edible fruit
manan mitus	*Santiria tomentosa*	Edible fruit
medong	*Artocarpus odoratissimus*	Edible fruit
mesilat	*Nephelium cuspidatum* [#419]	Edible fruit
nyivung	*Oncosperma horridum*	Edible palm heart
peripun	*Artocarpus melinoxylus*	Small type of tarap, edible fruit
perutang	*Pommetia pinnata* [#416]	Edible fruit
pidau	*Baccaurea angulata* [#414]	Edible fruit
sin savit belengon	*Daemonorops korthalsii* [#431]	Palm heart from apex
sin savit medok	*Daemonorops periacantha* [#429]	Fruit eaten uncooked; heart is cooked
tebangat	*Xerospermum noronhianum*	Edible fruit
uvat	*Gironniera hirta*	Type of edible sago

Notes: All plants listed have been vouchered. Plants whose names are marked with asterisks (*) were collected by the author and housed at the Brunei Forestry Herbarium. The author's voucher numbers are in brackets. Species without asterisks were vouchered when the plot was originally established by the Biology Department, Universiti Brunei Darussalam, and are also on file at the herbarium.

total of 29 esculent species, covering roughly the same range of uses (table 12.2). Twenty-five of these are trees, followed by only four understory edibles. The Penan informant thus recognized the food value of 45% fewer species than his cultivating counterpart did.

The significance of these results in terms of Brown's overarching hypothesis requires further consideration. First, it should be noted that the 20 Penan edible species represent 62% of the informant's total list of perceived useful species. The 29 Dusun food plants, on the other hand, represent only 13% of his useful taxa. Thus, although quantitatively small, the Penan food list constitutes a much greater share of total ethnobotanical knowledge than that of the Dusun. Second, of the 178 taxa in the plot recognized by the Dusun informant that were not noted by the Penan, only nine (about 5%) are edibles. Thus, only a tiny portion of the total

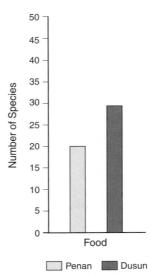

Fig. 12.2. Edible plant species in a one-hectare plot. Plants listed have at least one perceived food use among the Penan or Dusun.

difference in named plants can be attributed to this apparent disparity in marginal food knowledge. Clearly other use categories outweigh the importance of wild foods.

Closer scrutiny of the assumptions upon which the "famine hypothesis" are based reveals further problems. Is it necessarily true, for example, that hunter-gatherers are less subject to famine than small-scale cultivators? For the Penan and the Dusun, there is no direct evidence of serious famine, either in the past or recent times. Poor harvests have been recorded in nearby regions (Rousseau 1990: 131), but these do not appear to have been of devastating consequences. At any rate, the Dusun and other Bornean groups cultivate a range of rice varieties and other scarcity crops that are intended to mitigate the effects of food supply fluctuations (Christensen and Mertz 1993; Dove 1985; Lian 1987: 125–127; Strickland 1989). In an Iban community in Sarawak, Christensen (1994) identified 181 varieties of the nine principal cultigens, as well as 68 wild species that were cultivated in small numbers. Bornean farmers also prepare for environmental uncertainty by planting more rice than is necessary (Rousseau 1990: 131), and when food shortages occur in one family, others in the longhouse are expected to share their surplus (Sutlive 1988: 78).

Neither does the climatic record suggest a pattern of feast and famine. Although droughts lasting more than 30 days occur about once every two

Table 12.2. Dusun food species in a one-hectare forest plot.

Dusun name	Species	Use
abang	*Shorea amplexicaulis*	Timber for upper house; vegetable
	Shorea mecistopteryx	oil from fruits
adal	*Canarium* (2 species)	Snack food
bidang	*Licuala* sp. [#504]	Edible palm heart; leaf thatch
bubuk	*Dryptes kikir*	Edible fruit; spinning tops
	Dryptes macrostigma	
durian mentui	*Durio grandiflorus*	High-quality fruit
kamangis	*Nephelium uncinatum*	Edible fruit
kapayang	*Scaphium macropodum*	Edible fruit; medicinal
latid	*Dipterocarpus geniculatus*	Edible fruit
lilas	*Swintonia* (2 species)	Edible fruit; timber for upper
	Mesua (3 species)	house
mau gbuk	*Mallotus wrayi*	Edible fruit
	Dimocarpus sp. 1	
medang latung	*Artocarpus odoratissimus*	High-quality fruit
natu	*Palaquium* (2 species)	Edible oil from fruit; planks
	Payena obscura	
nguluon	*Dacryodes incurvata*	Edible fruit
nibung	*Oncosperma horridum*	Edible heart; wood for beams
ombokot	*Nephelium cuspidatum* sp. [#517]	Edible fruit; good luck
pogos	*Elateriospermum tapos*	Edible fruit
sagang	*Alpinia* cf. *glabra* [#509]	Edible fruit
sanggara	*Galearia fulva* det [#505]	Leaves edible; root medicinal
uwai lambat	*Daemonorops pericantha* [#452]	Edible heart; fronds for fish traps

years (Becker 1992), the northern part of Borneo inhabited by the Penan and Dusun is aseasonal even by tropical rain forest standards. Mean annual rainfall ranges from 2600 millimeters to over 5000 millimeters in Brunei Darussalam, with only a slight tendency for moisture deficits during February, March, and August (Sirinanda 1990). In addition, Brunei does not appear to suffer the devastating El Niño-related droughts and conflagrations that plague southern Borneo (Goldammer and Seibert 1990; Walsh 1996). Thus, the case for environmentally-induced famine in this immediate region appears rather weak.

Aside from environmental fluctuations, the principal disruption of food supplies in previous times may well have been head-hunting raids and intertribal wars. Integral to the individual prestige and religious beliefs of cultivating groups, head-hunting was carried out to promote human fertility, health, and abundant rice harvests (King 1993: 237). Although warfare for the purpose of territorial expansion, especially on the part of the feared Iban, may have interrupted food production in the past (Hose 1988: 130–147), raids as a rule were executed after sowing or harvest were completed (Rousseau 1990: 265).

Assuming for the moment that cultivators *did* in fact suffer from periodic crop failures as a result of environmental perturbation or other problems, does it necessarily follow, as Brown suggests, that such famine episodes would have expanded the ethnobotanical knowledge of cultivators relative to that of their foraging neighbors? Would cultivators be more driven than hunter-gatherers to experiment with and learn the value of marginal food sources? Bornean swidden cultivators, as noted above, go to considerable lengths to prepare for food scarcity by planting a wide array of cultigens, especially rice. Penan subsistence, on the other hand, is traditionally based on collection and preparation of sago starch from *nangah* (*Eugeissona utilis*), a native, clump-forming palm. As much a staple to the Penan as rice is to the Dusun, this species is carefully husbanded in order to achieve maximum long-term benefits. Older stems are harvested and processed, but the younger stems (*uvud*) are marked and protected for future exploitation (Brosius 1991; Langub 1989). Although this seemingly sustainable system of protective extraction, termed *molong,* is meant to insure a dependable yield of sago, it is no guarantee that resource scarcity never occurs. There are reports that overexploitation of *nangah* has occurred in recent times, although this may reflect the impact of recent population pressure (Lian 1993). In any event, dependence on a single starch-producing tree, wild or cultivated, carries the inherent risk of temporary shortages, leaving the Penan—like their cultivating neighbors—vulnerable to lean times.

Uncertainty of availability is the rule for the Penan's wild fruit and seed crops as well. Rather than a reliable source of subsistence, forested landscapes provide resource bases that are both patchy and unpredictable in time and space (Urban et al. 1987; Worster 1990). Southeast Asia's dipterocarp forests are renowned for their mast fruiting characteristics, with years of little or no flower and fruit production punctuated by the infrequent bumper crop (Ashton et al. 1988; Toy 1991). Mast fruiting necessarily places stress on a society whose subsistence is derived solely from collection. Fruit and seed shortage during off years is further exacerbated by the fluctuating population of bearded pigs, the Penan's preferred protein and fat source, which appears to wax and wane with the condition of the fruit crop. Thus, complete dependence on the extraction of wild species, even under favorable climatic conditions, is a chancy mode of subsistence in tropical moist forests, and seemingly as risky as that pursued by groups depending on cultivated crops.

The precarious nature of wild tropical forest food resources is underscored by the flurry of recent revisionist research on the topic. It is now an open question as to whether a purely hunting and gathering mode of subsistence is, or ever was, practiced in tropical forest regions. Forager subsistence in these "biological deserts", according to this view, is possible only if food can be acquired (usually through trade) from neighboring sedentary cultivators (Bailey et al. 1989; Headland 1987; Hoffman 1986). Although the validity of this hypothesis has been seriously challenged, perhaps even put to rest (Bellwood 1993; Brosius 1991; Colinvaux and Bush 1991), the overriding theme of tropical forest food scarcity seriously undermines the notion that foragers are somehow less subject to episodes of famine than forest farmers.

The Dusun and the Penan are skilled hunters who spend considerable time in the pursuit of forest and riverine game. Whereas the Penan are admired for their use of blowpipe and spear, the Dusun concentrate their efforts on carefully constructed snares and fish traps. In both cases, gradual acquaintance with minor food sources would seem to be a natural consequence of tracking and fishing. Individuals in both groups stop frequently to snack on marginal foods as they walk through the forest. Although open to interpretation, it is interesting that neither the Penan nor the Dusun recognize the myriad of supplemental food sources as famine resources, but rather as occasional wild snacks. Thus, however appealing a famine hypothesis is, the incremental acquisition of ethnobotanical knowledge as part of the ordinary comings and goings of everyday life seems a more plausible explanation for knowledge of marginal food sources.

Conclusions and Recommendations

Our perception of tropical forest biogeography and environmental history has shifted radically in recent decades. No longer the products of environmental stability, biogeographical configurations and species richness patterns are explained increasingly within a nonequilibrium paradigm. Stochastic processes such as windthrow, sunflecks, waif dispersal, and insect predation produce dispersion patterns that are naturally patchy and uneven in space and time (Bazzaz 1991; Hubbell and Foster 1992; Primack and Hall 1992). The perceived role of human-derived disturbance in tropical landscapes has witnessed an even sharper turn. Dispersion patterns once thought to be the outcome of natural processes are explained increasingly as vestiges of past human activities (Clark et al. 1995; Gomez-Pompa et al. 1987; Sauer 1967 [1958]; Voeks and Vinha 1988). No longer the pristine creations of natural ecological processes (Denevan 1992), tropical forests and their patterns of diversity are viewed, at least in the extreme, as little more than "cultural artifacts" (Bailey 1996: 323).

Human survival in the rain forest is posited on a profound understanding of its plant resources. Brown (1985) and others have suggested that the ethnobotanical knowledge possessed by societies following a cultivating existence is quantitatively more extensive than that of hunter-gatherers. The preliminary results of this plot-based survey strongly support Brown's primary hypothesis. Dusun hill rice cultivators recognize and have vernacular names for considerably more species than the neighboring Penan hunter-gatherers. They also perceive the utility of over four times the number of plant taxa as the Penan.

Brown's explanation for this disparity includes the role of environmental disturbance and resource fluctuation. Enhanced risk of famine among cultivating societies, according to this view, necessitates a broader understanding of edible plants than that possessed by hunter-gatherer groups. The data presented here support this prediction. The Dusun recognized a total of 29 species compared to 20 for the Penan. Nevertheless, the suggestion that the risk of famine is an underlying cause of this discrepancy is less than compelling on several fronts.

The Dusun and the Penan, indigenous groups that occupy the same floristic landscape, exploit their mutual habitats quite differently. Their partially dichotomous modes of subsistence—hill rice cultivator and pig-sago forager—allow for a complementary partition of forest resources. The sedentary Dusun direct their rain forest knowledge at species that are useful to a cultivating existence. Aside from the wild foods described in this study, their ethnobotanical lexicon includes an extensive array of medicinal plants, timber species used in longhouse construction, and plants employed for snares, fish traps, insecticides, musical instruments, household utensils, fuelwood, baskets, and the whole gamut of material and magic needs associated with a settled lifestyle. Much of this knowledge, as will be shown in subsequent work, is focused on landscapes of disturbance—pastures, fresh swidden plots, long fallows, and kitchen gardens.

The traditionally nomadic Penan, exclusive denizens of the primary forest, relied entirely (at least until recently) on extraction of forest products for food, medicine, shelter, and trade items. Their migratory way of life allowed for few material possessions, a situation clearly evident from their relatively limited ethnobotanical lexicon. Food, fiber, fuelwood, and rattan encompassed most of their resource needs. Added to this list were the rain forest exotica—camphor from primary forest trees, bezoar stones from the gallbladders of monkeys, and quality rattan—highly-sought-after extractive products for which the Penan were the principal collectors.

Called "professional primitives" by some, the Penan navigated forest paths similar to those of their cultivating neighbors. The array of resources they came to perceive, however, was narrower and more specialized—bearded pigs, sago palms, and a limited number of marginal food resources. Their breadth of ethnobotanical knowledge, as indicated by the number of named taxa and perceived useful species, is several times less than that of their cultivating neighbors. Nevertheless, this relatively shallow knowledge may have been compensated for by "substantive knowledge" of those few wild resources absolutely essential to a foraging mode of subsistence (Ellen in press). Successful subsistence for the Penan, I suggest, is contingent upon a profound understanding of a relatively limited range of resources.

Life is changing rapidly for the indigenous people of Borneo. In Brunei Darussalam, oil wealth has created a radically new set of material expectations among its native populations. The traditional modes of subsistence of both the Dusun and the Penan are, as a consequence, in a state of rapid transition. For the Dusun, access to formal education and government sector jobs has all but eliminated subsistence cropping patterns. Among the young, for whom rain forest wisdom is associated with the "primitive" lifestyles of their ancestors, ethnobotanical knowledge is increasingly perceived as a useless anachronism. The process of forgetting the forest is occurring rapidly and irrevocably.

As traditionally cultivating groups increasingly abandon agricultural subsistence, Brunei is in the unique position of witnessing gradual afforestation of its previously swiddened mixed dipterocarp forests. Combined with its progressive forest policy, which includes a strict logging quota and a prohibition on timber exports, Brunei harbors one the richest and least threatened moist tropical forests on Earth. In order to capitalize on this situation, Brunei has embarked on an ambitious project to attract ecotourists, including the establishment of the 50000-hectare Batu Apoi National Park. It is in the area of ecotourism that the Penan could realize an important role.

Small in numbers and lacking political connections, the Penan are not finding the same degree of opportunity in the "new Brunei" as other indigenous groups are. Long perceived as ignorant jungle people, the Penan have been reduced to accepting unskilled jobs in road construction and forestry. With limited understanding of the rudimentary laws of horticulture, their attempts at rice cultivation have been less than successful. The Penan have, however, retained much of their substantive knowledge of the forest, such as the feeding habits of the silver langur monkey, the reproductive cycle of the rhinoceros hornbill, and the nocturnal forays of army ants—forest wisdom that could find a creative outlet in Brunei's experiment in ecotourism. Nearby Sarawak's employment of Penan guides in

Gunung Mulu National Park, although no model for sustainable development, suggests the type of meaningful employment that could ease the Penan's transition into modern society. Although not a long-term answer to the "Penan problem," integration into Brunei's tourism trade would go a long way towards enhancing the Penan's image and self-esteem, as well as improving their material livelihoods. Most important from the perspective of the Penan, it would allow them to continue pursuing the pigs and palms of the primary forest.

Acknowledgments

I thank Roy Ellen and an anonymous reviewer for helpful comments on this manuscript. I thank the Biology Department, Universiti Brunei Darussalam, especially Peter Becker, for orientation and use of the enumerated plot. The field assistance provided by Samhan Nyawa and Peter Sercombe was greatly appreciated. Species determinations would not have been possible without the able assistance provided by the Brunei Forestry Herbarium staff, especially Idris M. Said and Joffre H. A. Ahmad. Kelly Donovan kindly produced the map and graphs. Field research was funded by a grant from the National Geographic Society.

References

Alcorn, J. B. 1989. Process as Resource: The Traditional Agricultural Ideology of Bora and Huastec Resource Management and its Implications for Research. In *Advances in Economic Botany,* eds. D. A. Posey and W. Balée, pp. 63–77. New York: New York Botanical Garden.

Antaran, B. 1993. The Brunei Dusun: An Ethnographic Study. M. A. dissertation, University of Hull.

Ashton, P. S., T. J. Givnish, and S. Appanah. 1988. Staggered Flowering in the Dipterocarpaceae: New Insights into Floral Induction and the Evolution of Mast Fruiting in the Aseasonal Tropics. *American Naturalist* 132: 44–66.

Bailey, R. C., G. Head, M. Jenike, B. Owen, R. Rechtman, and E. Zechenter. 1989. Hunting and Gathering in Tropical Rain Forest: Is It Possible? *American Anthropologist* 91: 59–82.

Bailey, R. C. 1996. Promoting Biodiversity and Empowering Local People in Central African Forests. In *Tropical Deforestation: The Human Dimension,* eds. L. E. Sponsel, T. N. Headland, and R. C. Bailey, pp. 316–341. New York: Columbia University Press.

Balée, W. 1994. *Footprints of the Forest: Ka'apor Ethnobotany–the Historical Ecology of Plant Utilization by an Amazonian People.* New York: Columbia University Press.

Bazzaz, F. A. 1991. Regeneration of Tropical Forests: Physiological Responses of Pioneer and Secondary Species. In *Rain Forest Regeneration and Management,* eds. A. Gomez-Pompa, T. C. Whitmore, and M. Hadley, pp. 91–118. Paris: Parthenon.

Becker, P. 1992. Seasonality of Rainfall and Drought in Brunei Darussalam. *Brunei Museum Journal* 7: 99–109.

Bellwood, P. 1993. Cultural and Biological Differentiation in Peninsular Malaysia: The Last 10,000 Years. *Asian Perspectives* 32: 37–60.

Berlin, B. 1992. *Ethnobiological Classification: Principles of Categorization of Plants and Animals in Traditional Societies.* Princeton: Princeton University Press.

Black, F. L. 1975. Infectious Diseases in Primitive Societies. *Science* 187: 515–518.

Brosius, J. P. 1991. Foraging in Tropical Rain Forests: The Case of the Penan of Sarawak, East Malaysia. *Human Ecology* 19: 123–150.

Brown, C. H. 1985. Mode of Subsistence and Folk Biological Taxonomy. *Current Anthropology* 26: 43–53.

Bulmer, R. 1985. Comments. *Current Anthropology* 26: 54–55.

Christensen, H. 1994. *The Potential of Wild Plants for Permanent Cultivation.* Paper presented at the Meeting of the Borneo Research Council, Pontianac, Kalimantan, Indonesia.

Christensen, H., and O. Mertz. 1993. The Risk Avoidance Strategy of Traditional Shifting Cultivation in Borneo. *Sarawak Museum Journal* 44: 1–14.

Clark, D. A., D. B. Clark, R. Sandoval, and M. Castro. 1995. Edaphic and Human Effects on Landscape-Scale Distributions of Tropical Rain Forest Palms. *Ecology* 76: 2581–2594.

Clay, J. W. 1988. *Indigenous Peoples and Tropical Forests.* Cambridge: Cultural Survival, Inc.

Colinvaux, P. A. 1993. Pleistocene Biogeography and Diversity in Tropical Forests of South America. In *Biological Relationships between Africa and South America,* ed. P. Goldblatt, pp. 473–499. New Haven: Yale University Press.

Colinvaux, P. A., and M. B. Bush. 1991. The Rain-Forest Ecosystem as a Resource for Hunting and Gathering. *American Anthropologist* 93: 153–160.

Denevan, W. M. 1992. The Pristine Myth: The Landscape of the Americas in 1492. *Annals, Association of American Geographers* 82: 369–385.

Denevan, W. M., J. M. Treacy, J. B. Alcorn, C. Padoch, J. Denslow, and S. F. Paiton. 1985. Indigenous Agroforestry in the Peruvian Amazon: Bora Indian Management of Swidden Fallows. In *Change in the Amazon Basin. Man's Impact on Forests and Rivers,* ed. J. Hemming, pp. 137–155. Manchester: Manchester University Press.

Dove, M. R. 1985. *Swidden Agriculture in Indonesia: The Subsistence Strategies of the Kalimantan Kantu'.* Berlin: Mouton.

Dunn, F. L. 1968. Epidemiological Factors: Health and Disease in Hunter-Gatherers. In *Man the Hunter,* eds. R. B. Lee and I. Devore, pp. 221–228. Chicago: Aldine.

Ellen, R. F., and J. Bernstein. 1994. Urbs in Rure: Cultural Transformations of the Rainforest in Modern Brunei. *Anthropology Today* 10: 16–19.

Ellen, R. F. In press. Modes of Subsistence and Ethnobiological Knowledge: Between Extraction and Cultivation in Southeast Asia. In *Folkbiology,* eds. D. Medin and S. Atran. Cambridge: Massachusetts Institute of Technology Press.

Gentry, A. H. 1988. Tree Species Richness of Upper Amazonian Forests. *Proceedings National Academy of Science* 85: 156–159.

Gliessman, S. R. 1990. Applied Ecology and Agroecology: Their Role in the Design of Agricultural Products for the Humid Tropics. In *Race to Save the Tropics: Ecology and Economics for a Sustainable Future,* ed. R. Goodland, pp. 33–47. Washington, D.C.: Island Press.

Goldammer, J. G., and B. Seibert. 1990. The Impact of Droughts and Forest Fires on Tropical Lowland Rain Forest of East Kalimantan. In *Fire in the Tropical Biota: Ecosystem Processes and Global Challenges,* ed. J. G. Goldammer, pp. 11–31. Berlin: Springer-Verlag.

Gomez-Pompa, A., J. Salvador Flores, and V. Sosa. 1987. The "Pet Kot": A Man-Made Forest of the Maya. *Interciencia* 12: 10–15.

Grenand, P. 1992. The Use and Cultural Significance of the Secondary Forest among the Wayapi Indians. In *Sustainable Harvest and Marketing of Rain Forest Products,* eds. M. Plotkin and L. Famolare, pp. 27–40. Washington, D.C.: Island Press.

Headland, T. N. 1985. Comments. *Current Anthropology* 26: 57–58.

Headland, T. N. 1987. The Wild Yam Question: How Well Could Independent Hunter-Gatherers Live in a Tropical Rain Forest Ecosystem? *Human Ecology* 15: 463–491.

Hoffman, C. 1986. *The Punan: Hunters and Gatherers of Borneo.* Ann Arbor: UMI Research Press.

Hose, C. 1988 [1926]. *Natural Man: A Record from Borneo.* Oxford: Oxford University Press.

Hubbell, S. P., and R. B. Foster. 1992. Short-term Dynamics of a Neotropical Forest: Why Ecological Research Matters to Tropical Conservation and Management. *Oikos* 63: 48–61.

Hunn, E., and D. French. 1984. Alternatives to Taxonomic Hierarchy: The Sahaptin Case. *Journal of Ethnobiology* 3: 73–92.

Ichikawa, M. 1992. Traditional Use of Tropical Rain Forest by the Mbuti Hunter-Gatherers in Central Africa. In *Topics in Primatology: Behavior, Ecology, Conservation,* volume 2, ed. N. Itoigawa, pp. 305–317. Tokyo: University of Tokyo Press.

King, V. T. 1993. *The Peoples of Borneo.* Oxford: Blackwell.

Kohn, E. O. 1992. Some Observations on the Use of Medicinal Plants from Primary and Secondary Growth by the Runa of Eastern Lowland Ecuador. *Journal of Ethnobiology* 12: 141–152.

Langub, J. 1989. Some Aspects of Life of the Penan. *The Sarawak Museum Journal* 40: 169–184.

Layton, R., R. Foley, and E. Williams. 1991. The Transition between Hunting and Gathering and the Specialized Husbandry of Resources. *Current Anthropology* 32: 255–274.

Lévi Strauss, C. 1966. *The Savage Mind.* London: Weidenfeld and Nicolson.

Lian, F. J. 1987. Farmers' Perceptions and Economic Change—The Case of Kenyah Farmers of the Fourth Division, Sarawak. Ph. D. dissertation, Australian National University.

Lian, F. J. 1993. On Threatened Peoples. In *South-East Asia's Environmental Future: The Search for Sustainability,* eds. H. Brookfield and Y. Byron, pp. 322–340. Kuala Lumpur: Oxford University Press.

Nicholls, N. 1993. ENSO, Drought and Flooding Rain in South-East Asia. In *South-East Asia's Environmental Future: The Search for Sustainability,* eds. H. Brookfield and Y. Byron, pp. 154–175. Kuala Lumpur: Oxford University Press.

O'Brien, S. T., B. P. Hayden, and H. H. Shugart. 1992. Global Climatic Change, Hurricanes, and a Tropical Forest. *Climatic Change* 22: 175–190.

Phillips, O., A. H. Gentry, C. Reynel, P. Wilkin, and C. Galvez-Durand. 1994. Quantitative Ethnobotany and Amazonian Conservation. *Conservation Biology* 8: 225–248.

Prance, G. T., W. Balée, B. M. Boom, and R. I. Carneiro. 1987. Quantitative Ethnobotany and the Case for Conservation in Amazonia. *Conservation Biology* 1: 296–310.

Primack, R. B., and P. Hall. 1992. Biodiversity and Forest Change in Malaysian Borneo. *BioScience* 42: 829–837.

Rousseau, J. 1990. *Central Borneo: Ethnic Identity and Social Life in a Stratified Society.* Oxford: Clarendon Press.

Sauer, C. O. 1967 [1958]. Man in the Ecology of Tropical America. In *Land and Life: A Selection from the Writings of Carl Ortwin Sauer,* ed. J. Leighly, pp. 182–193. Berkeley: University of California Press.

Sirinanda, K. U. 1990. Rainfall Seasonality in Negara Brunei Darussalam. *Asian Geographer* 9: 39–52.

Smith, N. J. H., and R. E. Schultes. 1990. Deforestation and Shrinking Crop Gene-Pools in Amazonia. *Environmental Conservation* 17: 227–234.

Smith, N. J. H., J. T. Williams, D. L. Plucknett, and J. P. Talbot. 1992. *Tropical Forests and Their Crops.* Ithaca: Cornell University Press.

Strickland, S. S. 1989. Kejaman Adaptive Strategies in the Context of Development Efforts. *Sarawak Museum Journal* 40: 251–270.

Sutlive, V. H. 1988. *The Iban of Sarawak: Chronicle of a Vanishing World.* Prospect Heights, IL: Waveland Press.

Toledo, V. M., A. I. Batis, R. Becerra, M. Esteban, and C. H. Ramos. 1992. Products from the Tropical Rain Forests of Mexico: An Ethnoecological Approach. In *Sustainable Harvest and Marketing of Rain Forest Products,* eds. M. Plotkin and L. Famolare, pp. 99–109. Washington, D. C.: Island Press.

Toy, R. J. 1991. Interspecific Flowering Patterns in the Dipterocarpaceae in West Malaysia: Implications for Predator Satiation. *Journal of Tropical Ecology* 7: 49–57.

Urban, D. L., R. V. O'Neill, and H. H. Shugard. 1987. Landscape Ecology. *BioScience* 37: 119–127.

Voeks, R. A., and S. G. Vinha. 1988. Fire Management of the Piassava Fiber Palm (*Attalea funifera*) in Eastern Brazil. *Yearbook of the Conference of Latin Americanist Geographers* 14: 7–13.

Voeks, R. A. 1995. Candomblé Ethnobotany: African Medicinal Plant Classification in Brazil. *Journal of Ethnobiology* 15: 257–280.

Voeks, R. A. 1996a. Extraction and Tropical Rainforest Conservation in Eastern Brazil. In *Tropical Rainforest Research—Current Issues,* eds. D. S. Edwards, W. E. Booth, and S. C. Choy, pp. 477–487. Dordrecht: Kluwer Academic Publishers.

Voeks, R. A. 1996b. Tropical Forest Healers and Habitat Preference. *Economic Botany* 50(3): 354–373.

Walsh, R. P. D. 1996. Drought Frequency Changes in Sabah and Adjacent Parts of Northern Borneo Since the Late Nineteenth Century and Possible Implications for Tropical Rain Forest Dynamics. *Journal of Tropical Ecology* 12: 385–407.

Wilson, E. O., ed. 1988. *Biodiversity.* Washington, D. C.: National Academy Press.

Worster, D. 1990. The Ecology of Order and Chaos. *Environmental History Review* 14: 1–18.

Conclusion
Biological Conservation in Developing Countries

Kenneth R. Young and Karl S. Zimmerer

Policies and programs meant to promote biological conservation must be based on knowledge about the dynamic nature of the resources in question. In this volume, we have been concerned with biogeographical landscapes, their components, the regions in which they are located, and their sustainable use and conservation. The conservation of biological diversity is not limited to plant and animal species. Also included are the genetic diversity found among both domesticated and wild species, the ecological diversity contained in the ecosystems, and the biotic communities in a wide spectrum of landscapes and regions, often inhabited by people.

Efforts to assemble this information are only beginning (UNEP 1995), so attempts must be based on incomplete data and despite great uncertainties. Already it is apparent that development strategies seeking to account for biological conservation require a good understanding of the rhythms and scales of natural processes, in addition to an appreciation of how these processes have been altered by human activities. These processes function differently in landscapes and regions (Saunier and Meganck 1995) compared to local or population levels (as examples, see Echelle and Echelle 1997; Song 1996; Turner et al. 1996). In addition, the issues of concern at a global scale or in terms of general worldwide issues (Leis and Viola 1995; Miller 1995; Tomich et al. 1995) are quite different than those discussed in this volume. Finally, many development projects are implemented at landscape and regional scales.

Conservation-with-development programs must consider the aspirations, needs, and full participation of the people involved, although seldom are assessment, program design, and support simple tasks (Caldecott 1996; Cusworth and Franks 1993). Also difficult is finding an effective balance with the social organization of projects at local, regional or national levels. For example, which formal and informal social groups, government entities, or coalitions of social actors should be making policy and decisions (Fowler 1991; Ghai and Vivian 1992; McNeely 1995; Princen and Finger 1994; Vivian 1994; Western and Wright 1994)? When outside agencies are involved, the appropriate mechanisms for funding and institutional linkages are also difficult to choose. It is no surprise that conservation-with-development programs often fail by being overtaken by rapidly changing economic and social conditions or by conflicts among decision-makers.

Given these social complexities, it is interesting and a bit ironic that assumptions about the nature of ecosystems and organisms are often uncritically accepted. Yet, current theoretical considerations about the dynamic nature of biogeographical landscapes can help inform both development workers and the inhabitants of developing countries. We believe that a better understanding of the multifaceted functioning of biogeographical landscapes and regions would be of paramount importance for supporting and designing programs appropriate for protected areas, such as nature reserves and national parks, and for the inhabited lands surrounding those protected areas. Moreover, it is also necessary for long-term use of managed or manipulated ecosystems, including those that are planted, grazed, or harvested.

In this concluding essay, we revisit the themes introduced at the start of this volume in order to draw together the unifying concepts that emerge from the twelve case studies and a careful consideration of recent scientific and scholarly thinking on the complex dynamics of nature's geography.

Biogeographical Landscapes and Regions

Biogeographical landscapes and regions are conceptual frameworks used by biogeographers and practitioners from cognate disciplines. Although a conceptual framework necessarily offers a selective reflection of the actual complexity of patterns and processes found in a particular site or area, it provides a useful device that organizes the types of research questions to be considered and clarifies the scale or scales of investigation.

Biogeographers coming from traditions found in academic geography departments will often explicitly include the complexities of human impacts in landscapes and regions (this volume; see also Mather and Sdasyuk 1991; Simmons 1989; Turner et al. 1990). They often embark on studies that deal with the entire physical environment, including soil-plant (chapter 6, Pérez; chapter 9, Turner), climatic (chapter 5, Horn), and geomorphic processes (Parker 1995; Swanson et al. 1988). This orientation is not exclusive, of course, to the discipline of geography, but its emphasis in geography differs from a common practice in ecology. At least as found in the teaching and research of many U.S. biology departments, ecological situations are typically selected for research that exclude or simplify the added complexities of human-caused and physical processes (Christensen 1989). Among others, Gómez-Pompa and Kaus (1992) criticized this approach as applied to biological conservation, referring to it as a "wilderness myth" whereby ecosystems free from human influence ("wilderness" or "virgin forests") are assumed (1) to exist in developing countries, and (2) to be best conserved by preventing human use or intervention.

The ubiquity of past and present human impact resonates through all twelve case studies. Other unifying concepts found in the case studies and amply discussed in recent scientific literature are (1) the importance of nonequilibrium conditions acting at different scales (Botkin 1990; Sprugel 1991; Zimmerer 1994), (2) the relevance of patch dynamics when including fragmentation and disturbance at the scale of landscapes and regions (Pickett et al. 1997; Pickett and White 1985; Veblen et al. 1992a, 1994; Wu and Loucks 1995), and (3) the usefulness of evaluating the uncertainty that arises as the above factors are considered and as management decisions are made in natural resource conservation (Carroll and Meffe 1994; Ives 1988; Lemons 1995; Thompson and Warburton 1988).

The case studies in this volume touch on all these issues collectively, but in ways that correspond to the type of natural region studied and to the degree of resource extraction practiced by humans. Since certain dynamics of environmental change are shared among situations, there are also some general solutions that can be offered. Similar physical environments and cultural practices often share nonequilibrial patterns and processes. Below, we review these topics in relation to forested environments, grasslands and shrublands, and managed or occupied ecosystems.

Forests

Forested landscapes are composed of patches made up of vegetation at different stages of recovery following disturbance. Forest patches can differ in terms of their tree species composition and structure such as their stature, layering, and size class distributions of the arboreal species. Patch size will be large in landscapes with catastrophic disturbances that affect large areas, and small when disturbances affect individual trees. Patch shape will vary with the type of disturbance. For example, riverine forests have linear patches that correspond to the shape of land areas affected by movements of rivers (chapter 1, Medley).

Forest disturbances can also release new resources. Establishment and growth of shade-intolerant tree species in some of China's forests are dependent on the dieback of understory bamboos, as was discovered by Taylor and Qin (chapter 2). Tree regeneration in some forests in Nepal apparently requires catastrophic opening of the canopy (chapter 11, Metz). Because of continual change at some scale in the landscape mosaic formed by forest patches, equilibrial conditions must be fleeting in occurrence. In fact, much recent research on forests has been focused on the more common nonequilibrial states (Lertzman 1995).

In addition to the patchwork structure of forested landscapes caused by natural disturbances, there is also a complicated additional mosaic associated with the spatial heterogeneity of physical factors that can underlie changes in the structure and composition of forests. Thus, soil characteristics are influenced by the nature of the bedrock and by the landscape position of the site; in turn, forests are much affected by heterogeneity in soil texture and chemistry. Climate is another physical factor that affects vegetation at several scales. Regional differences in vegetation are tied to changes in macroclimate and/or elevation. Microclimatic differences can result in differences in forest type between north- and south-facing slopes or in plant species composition between the understory and canopy. Long-lived trees in some cases provide living records of growth and establishment under environmental conditions that no longer exist (Villalba 1994). Fossils can be used to reconstruct past dynamics of forest in relation to other vegetation types (chapter 5, Horn).

Studies of forest dynamics grapple with these issues by characterizing the patch structure of forested landscapes and by examining change through time along important environmental gradients. This was done in this volume in reference to the riverine forests of eastern Kenya (chapter 1, Medley) and the moist montane forests of southwest China (chapter 2, Taylor and Qin), the central Andes (chapter 3, Young), and southern Ecuador (chapter 4, Echavarria).

Medley found that fluvial dynamics created most of the spatial heterogeneity in forest composition and structure. River movements and floods destroy some forests and create substrates for new forests as plants colonize the newly deposited sediments. People exploit this habitat mosaic by using some forest patches more than others. But people also add to nonequilibrium by changing composition, structure, and future tree regeneration through selective harvesting.

The forests described in Taylor and Qin's chapter are in part spatially structured by the changes in tree species composition and dominance that are associated with elevation. Also critical is the history of forest disturbance, because sites affected by past bamboo dieoffs develop into different forest types due to differential tree species establishment and growth. Humans again are important additional sources of heterogeneity—timbering and deforestation create new habitat conditions for the plant species within forested landscapes and disconnect once-continuous forests on a regional scale.

In his chapter, Young characterized the spatial heterogeneity of central Andean forests in relation to the natural fragmentation imposed by topography, and the additional fragmentation created by human-caused deforestation and conversion to scrub, agricultural, or pastoral land covers. Much of this conversion has undoubtedly occurred over centuries and the remaining native forests are the result of these changes, plus chronic additional disturbances as forests are utilized. In the new deforestation frontiers associated with roads and mechanization, forest loss is especially rapid. Regionally cohesive forests are converted into a series of isolated forested watersheds.

Echavarria reported on the nature of this rapid, new conversion in a montane landscape in southern Ecuador. He used remote sensing to illustrate and quantify some of the spatial patterns created by deforestation in just the past few decades. Once technical difficulties are resolved, it is likely that the techniques he has helped pioneer will be used routinely by resource managers to keep informed on the location and timing of forest conversion.

The case studies of Medley, Taylor and Qin, and Young all relate the patchy nature of forests to the wildlife found in the study landscapes. The availability of food resources and of particular habitat types (such as interior forest) affect the types, abundances, and distributions of animal species, including monkeys in East Africa, pandas in China, and birds and spiders in Andean forests. In turn, the native animals can add to or modify spatial heterogeneity. Introduced animal species can also cause additional changes in forest composition and structure (Veblen et al. 1992b).

When people use or modify forests, they are affecting the forests' composition, structure, and regeneration dynamics. Obviously deforestation adds large patches of nonforested land cover to the landscape and results in forest fragmentation: The remaining forests have less area and are more isolated from each other. These attributes can be illustrated by mapping and are measured by tools such as change detection (chapter 4, Echavarria). Other anthropogenic modifications may be less visible from a remote sensing perspective, but still result in changes in biotic composition. These modifications may include the effects of local extinctions (chapter 3, Young) or of the extraction of forest products (chapter 1, Medley; chapter 11, Metz; chapter 12, Voeks).

In conclusion, if forests are constantly changing at the scales of individual trees, forest stands, and landscapes with forest patches, then conservation programs in developing countries need to plan for this type of change. This is done by including surfaces within protected areas that are several times larger than the grain size of patches caused by natural disturbances (called "minimum dynamic area" by Pickett and Thompson [1978]).

Outside of protected areas, this dynamic perspective can be used to distinguish natural fluxes from those caused by humans. Then the choice can be made whether to modify or manipulate the system for a given goal or set of goals. Medley, Taylor, and Young recommend in this volume that forest restoration and the establishment of regional forest corridors be among the techniques considered. As a generalization, when forest use is traditional, observed changes may have taken place centuries earlier. However, when forests are cut or otherwise transformed catastrophically, loss of biological diversity is to be expected and programs should be developed to mitigate the negative consequences.

Grasslands and Shrublands

Biogeographical landscapes containing grasslands, shrublands, or other nonarboreal vegetation can also be conceptualized in terms of their respective patch dynamics. Natural disturbances that help to create patchiness in these landscapes include soil disturbances caused by hooved or burrowing mammals, grazing and browsing by herbivores, and fire. Patch size may be equal to that of a wallow made by a large herbivore, or it may cover hundreds of hectares of recently burned terrain.

Plant species vary in their ability to withstand these disturbances or grow back or recolonize afterwards (Bond and van Wilgen 1996). Grasses have a particular advantage in that the meristem tissue that initiates new growth is at or just below ground level and so often is protected from grazers or fire. Other plants participate by resprouting, germinating from dormant seeds, or seed dispersal to open sites. Again, nonequilibrium conditions are to be expected, especially at the scale of individual plants and of local assemblages of plants (Collins and Glenn 1991; Platt 1975; van der Valk 1992).

At landscape and regional scales, additional spatial heterogeneity in species composition and biomass originates with edaphic factors, such as soil moisture and nutrients, and with climatic factors that act to limit species distributions. In semiarid and arid environments, climatic fluctuations over several years or decades are notorious and add irregularity to the vegetation and landscape dynamics, as was described in this volume for the Near East (chapter 8, Blumler) and the Sahel (chapter 9, Turner).

High-elevation variants on these themes include additional environmental constraints that originate with cooler temperatures and other kinds of disturbances. Some of these are needle ice formation (Pérez 1987), frost and hail, and geomorphic disturbances such as landslides (chapter 10, Zimmerer). High mountains include nonforest vegetation types dominated by striking plant life forms such as the rosette plants and cryptogamic crusts described by Pérez (chapter 6). Some of the ecological complexity present at landscape scales in high-mountain environments was described in this volume for regions in Costa Rica (chapter 5, Horn), Venezuela (chapter 6, Pérez), and Nepal (chapter 7, Brower and Dennis).

Horn used paleoinformation to situate the tropical alpine landscape she studied in a temporal context. Fire is not only a human-caused disturbance that currently affects the appearance and floristic composition of the vegetation, but also a natural disturbance that has occurred episodically over the last 10000 years. This kind of information very much alters perceptions of what is or is not "natural" for a given landscape or region. Presumably the plant species present are adapted to this disturbance regime, and the landscape patterns result from both those disturbances and the limitations imposed by the physical environment.

The chapter by Pérez also situates his tropical alpine landscape temporally to the history of human use and settlement in highland Venezuela, much of which has taken place in the last two centuries. He found that elevation and edaphic variation explained much of the spatial heterogeneity present in vegetation. In addition, natural disturbances are present, although often overlooked, because they involve subtle geomorphic processes, such as soil movements caused by needle ice. The result is a series of unique and dynamic plant communities that are now experiencing new disturbance types, such as livestock and off-road vehicles.

Brower and Dennis reported their observations on the shrublands and associated vegetation types of the subalpine zone in a site in Nepal. They were able to document considerable spatial variations in vegetation composition and species regeneration that were only in part explicable in reference to the physical environment. Also important were the types and histories of human use, including grazing and harvesting. Given the length of time that people have interacted with landscapes here, it is clear that their influence must be considered part of the disturbance regime, and thus part of the resulting nonequilibrium conditions.

Often changes in disturbance regimes originate with shifts in local land-use systems. For example, a switch to use of higher elevations in Venezuela for grazing cattle and recreation is documented by Pérez (chapter 6) and has repercussions for both the location of timberline and the composition of highland plant communities. Furthermore, Turner (chapter 9) demonstrates that the complex interactions of climatic variability and grazing by domesticated livestock are potent sources of spatial heterogeneity. Presumably wildlife populations also respond to these dynamics.

By legal definition, the management goals of many protected areas in developing countries include the removal or reduction of deleterious effects due to human-caused disturbances or resource use. For example, this often requires modification of the levels of grazing and/or burning. Difficult issues revolve around the extent to which these impacts differ from natural disturbance regimes, and also to what extent it is practical, possible, or beneficial to change those impacts, as was discussed by Brower and Dennis (chapter 7), Horn (chapter 5), and Pérez (chapter 6). Yet the goals of the protected areas, whether overtly stated or not, at least provide general guidelines.

Very different are the economic development goals for landscapes that are not managed and protected for specific conservation reasons. Here it appears to be necessary to accommodate development strategies to the actual patchwork structure of the vegetation types, their constituent species and dynamics (Westoby et al. 1989), wildlife (Western 1989), and resource extraction (Galaty and Johnson 1990). Turner (chapter 9), for example, suggests that stocking levels for livestock be adjusted to the spatial and temporal heterogeneity found at landscape levels, even though most development planning for pastoral systems has been done at regional levels using aggregated data. The identification of important controlling factors at plant, community/ecosystem, landscape, and regional levels is critical to reducing risk and accommodating the inherent uncertainties.

Land Use and Settled Areas

The role of change processes and the impact of human modification are especially important in those biogeographical landscapes that are put to use by people for the purposes of livestock grazing, agriculture, and the extraction of forest products. Yet, what matters for conservation-with-development is not simply whether these livelihood activities lead to landscape change. Instead, the studies in this volume examine more difficult questions concerning the nature and degree of such changes and modifications. There is no doubt that the biogeographical nature of land use and the rural resource systems of settled areas are complex and frequently misunderstood.

This volume demonstrates that the modifications stemming from land use and settlement deliver lessons that are crucial for the prospects of conservation-with-development. Such lessons pertain to a variety of resource systems and world regions. This volume's studies of land use and settlement include the forest, savanna, and grassland landscapes of pastoralists and farmers in the Near East (chapter 8, Blumler), dryland thorny scrub utilized by herders in the African Sahel (chapter 9, Turner), mixed agricultural landscapes managed by agropastoralists in the tropical Andes (chapter 10, Zimmerer), utilized temperate and subalpine forest landscapes in the Himalaya of Nepal (chapter 11, Metz), and the lowland dipterocarp and heath forest landscapes of tropical Southeast Asia (chapter 12, Voeks).

The first lesson that emerges from these case studies is that new approaches in ecology and biogeography inspire a rethinking of the human place in environments that are subject to land use and settlement. The prominence of disturbances and other nonequilibrium processes of ecological change combine to create a new view of the geography of utilized nature, and how to make it environmentally sustainable. This view is far more complex and varied than could be acknowledged using the old ideas of nature as regular and deterministic in its variation. It is the specific characteristics of such change processes, moreover, that matter most critically for understanding the environmental and conservation dimensions of land use and settlement. Our case studies highlight how the nature of changes such as climate variation, soil development, and vegetation dynamics must be evaluated in terms of important parameters such as frequency of occurrence and spatial scale.

The second lesson is that human activities have imparted an environmental impact that is profound and pervasive. Yet, in thinking about biological conservation and sustainable development, this finding should not be confused with the mistaken idea of the human "creation" or "production" of all environmental functions. To be sure, the nature of land use and settled areas evidences the conspicuous imprint of human endeavors. Many ecological features are a direct artifact of the desire of people to make their homes in these environments and to earn their livelihoods from them. Yet, at the same time, it is clear that even in such areas there are key environmental processes that act independently of direct human influence. Those processes are shaped strongly by nonanthropogenic atmospheric dynamics, soil-organism relations, and plant and animal interactions.

Blumler's chapter on the landscapes of the Near East shows that vegetation succession and soil development are distinctly nonlinear and noncyclical in a variety of major environments. The complex processes that he demonstrates are at odds with the overly simplistic notions of linear change that are still widely applied to these environments. He also shows that much floristic diversity has been maintained over many millennia in the Near East, thus refuting the claim that human settlement is *ipso facto* inimical to biological richness. At the same time, his chapter also challenges the quite opposite yet still oversimplified charge that there has evolved an inherent compatibility between our species and the richness of others. This demonstration of long-term human impacts on biogeographical landscapes highlights the role of various historical contingencies in human-induced environmental change.

Turner's chapter on Sahel drylands shows how the biomass production of rangeland plants is limited by the sharp interannual variation of rainfall that is characteristic of the Sahel. This climatic irregularity combines with spatial differences in soils, including those induced by the soil nutrient-altering role of livestock, to produce a patchwork of range resources. The pastoralists of the inner Niger Delta region of Mali skillfully utilize the landscape mosaic for raising their cattle. This extensive local environmental knowledge has been reported in many other landscapes in developing countries (Warren 1995; Warren 1996; Woodgate 1994). Turner's treatment of the human-environmental setting, adds an important additional insight by showing that it is the occurrence of abrupt temporal change, spatial variation in the physical environment, and human interaction with those processes, that have combined to create the patchiness exploited by the land users.

The chapter by Zimmerer illustrates how the spatial patchiness and temporal change characteristic of mountain environments present a major environmental challenge for small-scale farmers, as well as for the management of their biologically diverse crops. Quechua Indian cultivators of the Paucartambo region of southern Peru rely on scores of diverse potato varieties to farm at upper elevations. Zimmerer evaluates the uncertain occurrence of natural hazards in the region's farm landscape. Complex interactions and a high degree of unpredictability are found in the distributions of climate factors, soils, agricultural pests, and geomorphologic disturbances such as landslides. The spatial patches of disturbances and their temporal variation create nonequilibrium conditions within Andean farm landscapes. In order to cope, Andean farmers have manipulated their diverse potatoes so that most landrace types display a sizeable degree of ecological versatility and are thus able to produce tolerable yields in a wide range of environments.

Metz's chapter on the forest landscapes of Nepal Himalaya discusses how the inhabitants of the Chimkhola region in West Central Nepal extract a wide array of necessary products. In some cases, it appears that they may be degrading these forests. Metz describes how the nonequilibrium dynamics of forest change must be taken into account by community forestry projects meant to counteract the degradation, but also support the livelihood needs of local people. However, if those projects mistakenly persist in advancing a static view of forest succession, they will not gain an adequate understanding of the change processes due to the interaction of extractive activities and natural disturbances. Effective forest management would thus be thwarted.

Voek's study demonstrates how the Penan and Dusun peoples of Borneo are successful in knowing their diverse forests intimately and using them widely, as has been shown in other tropical regions (Medley 1993; Phillips et al. 1994; Voeks 1996). Furthermore, he shows that the Dusun people, who are agriculturalists, are capable of identifying 241 tree species according to their own nomenclature, and that they utilize 219 of them. Still, the impact on these tropical environments is only partial as forest composition and structure are mostly influenced by processes other than those created by human actions. Plants, animals, and their interactions create and shape much of the forest and its dynamics.

These varied lessons and insights from our volume offer an important message for conservation-with-development. The complexity of changing biogeographical landscapes under human use urges the rejection of styles of conservation management derived from conventional (and often incorrect) assumptions: that environments are static and vary regularly, and that people, and rural residents in particular, should be excluded from conservation efforts. In particular, the case studies advise *against:* (1) the use of animal destocking programs as the sole or chief means of range management (chapter 9, Turner); (2) the dismissal of the *in-situ* conservation of agrodiversity as impractical due to the specialization of crop landraces (chapter 10, Zimmerer); and (3) the prohibition of human use as a necessary step in the conservation of forests and other landscapes (chapters by Blumler, Brower and Dennis, Medley, Taylor, Voeks, Young, and Zimmerer).

Our volume is in accord with a new trend that recognizes and assesses the nature of environmental complexities. We believe that many current challenges are as much environmental as social and political. It follows that conservation-with-development must promote land-use practices that are environmentally sound, yet also sufficiently flexible to accommodate a variety of changes, whether they be environmental or broadly social. Our studies furnish specific insights of this type with respect to forest use, livestock-raising, and farming. We are careful to point out, at the same time, that these practices and their flexibility do indeed have limits beyond which they cannot accommodate changes in either the environmental or social realms. Identifying such limits is also an integral part of conservation-with-development.

Conclusions

These themes help illuminate approaches to biological conservation in developing countries. In some circumstances, it may be possible to conserve large protected areas by delimiting the boundaries of the protected area and allowing natural processes free reign inside (Alverson et al. 1994). Development programs in this context are used to foster better relationships with people living within the region surrounding the protected area. This sort of reserve protection is not always successful and rarely if ever is it uncomplicated (Amend and Amend 1995; Ghimire 1994; Poole 1989; Wells et al. 1990; Zube and Busch 1990).

Most areas protected for biological conservation in developing countries, however, are not adequately safeguarded in this government-imposed manner. They are too small, too accessible, too impacted by human activity, or too contrary to the interests of local residents. Sustainable development programs in such settings would include working with and fully involving people living nearby the protected areas. These programs must also consider ecological restoration and other active intervention, such as the augmentation of populations of rare species, habitat modifications, and reintroduction programs for locally extinct species. Other approaches are also needed that provide linkages among landscapes within the region, such as through the design of regional conservation corridors.

An activist approach is also needed for considering biological conservation-with-development programs that do not include protected areas. The goal might be to look for ways that resource extraction can continue in a traditional or modernized manner that do not cause long-term degradation of the resource or resource zone (Bruenig 1996; Green 1989; Halladay and Gilmour 1995). New types of disturbances should be of special concern, as they may cause profound and long-lasting changes. This volume suggests doing this in relation to the dynamic mosaics of physical and biological processes that form biogeographical landscapes. This can be done holistically by including wildlife and biological diversity concerns within the planning done for improvements to agriculture, public health, or infrastructure (Young in press). The activism needed will include the professional level—that is, convincing project directors, technicians, or advisors to match their expectations to the spatial and temporal heterogeneity inherent in the physical and biological environments involved. Activist approaches will benefit from the knowledge of the traditional farmers or hunter-gatherers about these factors. Indeed, the viability of traditional knowledge is as important as that of biological diversity.

All the essays in this volume concern rural regions in developing countries. There is a great need in the future to also consider the conservation needs of the biogeographical landscapes of urban and urban-periphery localities (Medley et al. 1995). One of the most potent ways to do this is to search for projects that combine recreational uses with the conservation of biological resources. Islands, not discussed in this volume, offer additional challenges because of their unique biological diversity and their particular vulnerability to exotic species and socioeconomic change (Cuddihy and Stone 1990; Paulson 1994; Savidge 1987; Vitousek et al. 1987, 1995).

As the dynamics of biogeographical landscapes become better understood in general, it will be possible to search for other ways to use this kind of change creatively. At the least, all practices based on an assumption of equilibria need to be reconsidered. What is more, scientific approaches dependent on equilibrial and static frameworks need to be recast. This task is less burdensome than it might at first appear, because solutions spring from the spatial and scale insights coming from contemporary geography (Abler et al. 1992; National Research Council 1997).

As far as the issue of scientific uncertainty, Echavarria's essay offers an interesting contrast to the others because his remote sensing technique provides a quantitative means to document the uncertainty inherent in measuring deforestation. Other quantitative solutions to dealing with nature's geography and human-induced changes are rapidly being developed for analyses of biogeographical landscapes and their dynamics (Kent et al. 1997; Lambin 1997) and for applications to biological conservation (Lewis 1995). These solutions include, but are not limited to, the imaginative combination of multiple data layers in geographic information systems (GIS).

A possible approach for planning for economic development could be to evaluate likely future scenarios of landscapes and regions. In many cases, it will be possible to at least qualitatively indicate the level of certainty that exists for each scenario. This would allow participants in the development process to better evaluate the advantages and disadvantages of managing for desired outcomes by using techniques such as restoration practices and the creation of habitat corridors, or by harvesting resources in relation to the sustainability of landscape mosaics. Projects that are designed for regional goals might have to be assembled from a series of goals for the component landscapes, much like the multilevel approach proposed for watershed management by Dixon (1989). Attention to these spatial patterns, their changes through time, and the additional inherent heterogeneity within landscapes are fundamental for operationalizing this new approach.

References

Abler, R. F., M. G. Marcus, and J. M. Olson, eds. 1992. *Geography's Inner Worlds: Pervasive Themes in Contemporary American Geography.* New Brunswick, New Jersey: Rutgers University Press.

Alverson, W. S., W. Kuhlmann, and D. M. Waller. 1994. *Wild Forests: Conservation Biology and Public Policy.* Washington, D. C.: Island Press.

Amend, S., and T. Amend, eds. 1995. *National Parks without People? The South American Experience.* Quito, Ecuador: The World Conservation Union (IUCN).

Bond, W. J., and B. W. van Wilgen. 1996. *Fire and Plants.* London: Chapman and Hall.

Botkin, D. B. 1990. *Discordant Harmonies: A New Ecology for the Twenty-first Century.* New York: Oxford University Press.

Bruenig, E. F. 1996. *Conservation and Management of Tropical Rainforests: An Integrated Approach to Sustainability.* Wallingford: CAB International.

Caldecott, J. 1996. *Designing Conservation Projects.* Cambridge: Cambridge University Press.

Carroll, C. R., and G. K. Meffe. 1994. Meeting Conservation Goals in an Uncertain Future. In *Principles of Conservation Biology,* eds. G. K. Meffe and C. R. Carroll, pp. 531–557. Sunderland, Massachusetts: Sinauer.

Christensen, N. L. 1989. Landscape History and Ecological Change. *Journal of Forest History* 33: 116–125.

Collins, S. L., and S. M. Glenn. 1991. Importance of Spatial and Temporal Dynamics in Species Regional Abundance and Distribution. *Ecology* 72: 654–664.

Cuddihy, L. W., and C. P. Stone. 1990. *Alteration of Native Hawaiian Vegetation: Effects of Humans, Their Activities and Introductions.* Manoa: University of Hawaii Cooperative National Park Resources Studies Unit, University of Hawaii.

Cusworth, J. W., and T. R. Franks, eds. 1993. *Managing Projects in Developing Countries.* Harlow: Longman Scientific and Technical.

Dixon, J. A. 1989. Multilevel Resource Analysis and Management: The Case of Watersheds. In *Environmental Management and Economic Development,* eds. G. Schramm and J. J. Warford, pp. 185–200. Baltimore: Johns Hopkins University Press.

Echelle, A. A., and A. F. Echelle. 1997. Genetic Introgression of Endemic Taxa by Non-natives: A Case Study with Leon Springs Pupfish and Sheepshead Minnow. *Conservation Biology* 11: 153–161.

Fowler, A. 1991. The role of NGOs in Changing State-Society Relations: Perspectives from Eastern and Southern Africa. *Development Policy Review* 9: 53–84.

Galaty, J. G., and D. L. Johnson, eds. 1990. *The World of Pastoralism: Herding Systems in Comparative Perspective.* New York: Guilford Press.

Ghai, D., and J. Vivian, eds. 1992. *Grassroots Environmental Action: People's Participation in Sustainable Development.* London: Routledge.

Ghimire, K. B. 1994. Parks and People: Livelihood Issues in National Parks Management in Thailand and Madagascar. In *Development and Environment: Sustaining People and Nature,* ed. D. Ghai, pp. 195–229. Oxford: Blackwell/UNRISD.

Gómez-Pompa, A., and A. Kaus. 1992. Taming the Wilderness Myth. *BioScience* 42: 271–279.

Green, B. H. 1989. Conservation in Cultural Landscapes. In *Conservation for the Twenty-First Century,* eds. D. Western and M. C. Pearl, pp. 182–198. New York: Oxford University Press.

Halladay, P., and D. A. Gilmour, eds. 1995. *Conserving Biodiversity Outside Protected Areas.* Gland, Switzerland: The World Conservation Union (IUCN) Forest Conservation Programme.

Ives, J. D. 1988. Development in the Face of Uncertainty. In *Deforestation: Social Dynamics in Watersheds and Mountain Ecosystems,* eds. J. Ives and D. C. Pitt, pp. 54–74. London: Routledge.

Kent, M., W. J. Gill, R. E. Weaver, and R. P. Armitage. 1997. Landscape and Plant Community Boundaries in Biogeography. *Progress in Physical Geography* 21: 315–353.

Lambin, E. F. 1997. Modelling and Monitoring Land-Cover Change Processes in Tropical Regions. *Progress in Physical Geography* 21: 375–393.

Leis, H. R., and E. J. Viola. 1995. Towards a Sustainable Future: The Organizing Role of Ecologism in the North-South Relationship. In *Greening Environmental Policy: The Politics of a Sustainable Future,* eds. F. Fischer and M. Black, pp. 33–49. New York: St. Martin's Press.

Lemons, J. 1995. *Scientific Uncertainty and Environmental Problem Solving.* Cambridge: Blackwell Science.

Lertzman, K. P. 1995. Forest Dynamics, Differential Mortality and Variable Recruitment Probabilities. *Journal of Vegetation Science* 6: 191–204.

Lewis, D. M. 1995. Importance of GIS to Community-Based Management of Wildlife: Lessons from Zambia. *Ecological Applications* 5: 861–871.

Mather, J. R., and G. V. Sdasyuk, eds. 1991. *Global Change: Geographical Approaches.* Tucson: University of Arizona Press.

McNeely, J. A., ed. 1995. *Expanding Partnerships in Conservation.* Washington, D. C.: Island Press.

Medley, K. E. 1993. Extractive Forest Resources of the Tana River National Primate Reserve, Kenya. *Economic Botany* 47: 171–183.

Medley, K. E., M. F. McDonnell, and S. T. A. Pickett. 1995. Forest-Landscape Structure along an Urban-to-Rural Gradient. *Professional Geographer* 47: 159–168.

Miller, M. A. L. 1995. *The Third World in Global Environmental Politics.* Boulder: Lynne Rienner Publishers.

National Research Council. 1997. *Rediscovering Geography: New Relevance for Science and Society.* Washington, D.C.: National Academy Press.

Parker, K. C. 1995. Effects of Complex Geomorphic History on Soil and Vegetation Patterns on Arid Alluvial Fans. *Journal of Arid Environments* 30: 19–39.

Paulson, D. D. 1994. Understanding Tropical Deforestation: The Case of Western Samoa. *Environmental Conservation* 21: 326–332.

Pérez, F. L. 1987. Needle-Ice Activity and the Distribution of Stem-Rosette Species in a Venezuelan Paramo. *Arctic and Alpine Research* 19: 135–153.

Phillips, O., A. H. Gentry, C. Reynel, P. Wilkin, and C. Galvez-Durand. 1994. Quantitative Ethnobotany and Amazonian Conservation. *Conservation Biology* 8: 225–248.

Pickett, S. T. A., R. S. Ostfeld, M. Shachak, and G. E. Likens, eds. 1997. *The Ecological Basis of Conservation: Heterogeneity, Ecosystems, and Biodiversity.* New York: Chapman and Hall.

Pickett, S. T. A., and J. N. Thompson. 1978. Patch Dynamics and the Design of Nature Reserves. *Biological Conservation* 13: 27–37.

Pickett, S. T. A., and P. S. White, eds. 1985. *The Ecology of Natural Disturbance and Patch Dynamics.* New York: Academic Press.

Platt, W. J. 1975. The Colonization and Formation of Equilibrium Plant Species Associations on Badger Disturbances in a Tall Grass Prairie. *Ecological Monographs* 45: 285–305.

Poole, P. 1989. *Developing a Partnership of Indigenous Peoples, Conservationists and Land Use Planners in Latin America.* Washington, D. C.: The World Bank.

Princen, T., and M. Finger. 1994. *Environmental NGOs in World Politics: Linking the Local and the Global.* London: Routledge.

Saunier, R. E., and R. A. Meganck. 1995. *Conservation of Biodiversity and the New Regional Planning.* Washington, D. C.: Organization of American States and the IUCN.

Savidge, J. A. 1987. Extinction of an Island Avifauna by an Introduced Snake. *Ecology* 68: 660–668.

Simmons, I. G. 1989. *Changing the Face of the Earth.* Oxford: Blackwell.

Song, Y. -L. 1996. Population Viability Analysis for Two Isolated Populations of Haionan Eld's Deer. *Conservation Biology* 10: 1467–1472.

Sprugel, D. G. 1991. Disturbance, Equilibrium, and Environmental Variability: What Is "Natural" Vegetation In a Changing Environment. *Biological Conservation* 58: 1–18.

Swanson, F. J., T. K. Kratz, N. Caine, and R. G. Woodmansee. 1988. Landform Effects on Ecosystem Patterns and Processes. *BioScience* 38: 92–98.

Thompson, M., and M. Warburton. 1988. Uncertainty on a Himalayan Scale. In *Deforestation: Social Dynamics in Watersheds and Mountain Ecosystems,* eds. J. Ives and D. C. Pitt, pp. 1–53. London: Routledge.

Tomich, T. P., P. Kilby, and B. F. Johnston. 1995. *Transforming Agrarian Economies: Opportunities Seized, Opportunities Missed.* Ithaca: Cornell University Press.

Turner II, B. L., W. C. Clark, R. W. Kates, J. F. Richards, J. T. Mathews, and W. B.

Meyer, eds. 1990. *The Earth as Transformed by Human Action.* Cambridge: Cambridge University Press.

Turner, I. M., K. S. Chua, J. S. Y. Ong, B. C. Soong, and H. T. W. Tan. 1996. A Century of Plant Species Loss from an Isolated Fragment of Lowland Tropical Rain Forest. *Conservation Biology* 10: 1229–1244.

UNEP (United Nations Environmental Programme). 1995. *Global Biodiversity Assessment.* Cambridge: Cambridge University Press.

Van der Valk, A. 1992. Establishment, Colonization and Persistence. In *Plant Succession: Theory and Prediction,* eds. D. C. Glenn-Lewin, P. K. Peet, and T. T. Veblen, pp. 60–102. London: Chapman and Hall.

Veblen, T. T., K. S. Hadley, E. M. Neal, T. Kitzberger, M. Reid, and R. Villalba. 1994. Disturbance Regime and Disturbance Interactions in a Rocky Mountain Subalpine Forest. *Journal of Ecology* 82: 125–135.

Veblen, T. T., T. Kitzberger, and A. Lara. 1992a. Disturbance and Forest Dynamics along a Transect from the Andean Rain Forest to Patagonian Shrublands. *Journal of Vegetation Science* 3: 507–520.

Veblen, T. T., M. Mermoz, C. Martin, and T. Kitzberger. 1992b. Ecological Impacts of Introduced Animals in Nahuel Huapi National Park, Argentina. *Conservation Biology* 6: 71–83.

Villalba, R. 1994. Tree-Ring and Glacial Evidence for the Medieval Warm Epoch and the Little Ice Age in Southern South America. *Climatic Change* 26: 183–197.

Vitousek, P. M., L. L. Loope, and H. Adersen, eds. 1995. *Islands: Biological Diversity and Ecosystem Function.* Berlin: Springer.

Vitousek, P. M., L. R. Walker, L. D. Whiteaker, D. Mueller-Dombois, and P. A. Matson. 1987. Biological Invasion by *Myrica faya* Alters Ecosystem Development in Hawaii. *Science* 238: 802–804.

Vivian, J. 1994. NGOs and Sustainable Development in Zimbabwe: No Magic Bullets. In *Development and Environment: Sustaining People and Nature,* ed. D. Ghai, pp. 167–193. Oxford: Blackwell/UNRISD.

Voeks, R. A. 1996. Tropical Forest Healers and Habitat Preference. *Economic Botany* 50: 381–400.

Warren, A. 1995. Changing Understandings of African Pastoralism and the Nature of Environmental Paradigms. *Transactions of the Institute of British Geographers NS 20:* 193–203.

Warren, D. M. 1996. Indigenous Knowledge Systems for Sustainable Agriculture in Africa. In *Sustainable Development in Third World Countries: Applied and Theoretical Perspectives,* ed. V. U. James, pp. 15–24. Westport, Connecticut: Praeger.

Wells, M., K. Bradon, and L. Hannah. 1990. *People and Parks: Linking Protected Areas Management with Local Communities.* Washington, D. C.: World Bank.

Western, D. 1989. Conservation Without Parks: Wildlife in the Rural Landscape. In *Conservation for the Twenty-First Century,* eds. D. Western and M. C. Pearl, pp. 158–165. New York: Oxford University Press.

Western, D., and R. M. Wright, eds. 1994. *Natural Connections: Perspectives in Community-Based Conservation.* Washington, D. C.: Island Press.

Westoby, M., B. Walker, and I. Noy-Meir. 1989. Opportunistic Management for Rangelands not at Equilibrium. *Journal of Range Management* 42: 266–273.

Woodgate, G. 1994. Local Environmental Knowledge. Agricultural Development and Livelihood Sustainability in Mexico. In *Strategies for Sustainable Development: Local Agendas for the Southern Hemisphere,* eds. M. Redclift and C. Sage, pp. 133–170. Chichester: John Wiley and Sons.

Wu, J., and O. L. Loucks. 1995. From Balance of Nature to Hierarchical Patch Dynamics: A Paradigm Shift in Ecology. *Quarterly Review of Biology* 70: 439–466.

Young, K. R. In press. Wildlife Conservation in the Cultural Landscapes of the Central Andes. *Landscape and Urban Planning* 38.

Zimmerer, K. S. 1994. Human Geography and the "New Ecology": The Prospect and Promise of Integration. *Annals of the Association of American Geographers* 84: 108–125.

Zube, E. H., and M. Busch. 1990. Park-People Relationships: An International Review. *Landscape and Urban Planning* 19: 117–131.

Index